Communications in Computer and Information Science 479

Marco Kalz Yasemin Bayyurt
Marcus Specht (Eds.)

Mobile as Mainstream – Towards Future Challenges in Mobile Learning

13th World Conference
on Mobile and Contextual Learning, mLearn 2014
Istanbul, Turkey, November 3-5, 2014
Proceedings

Springer

Volume Editors

Marco Kalz
Marcus Specht
Open University of the Netherlands
Welten Institute
Research Centre for Learning, Teaching and Technology
P.O. Box 2960, 6401 DL Heerlen, The Netherlands
E-mail:{marco.kalz, marcus.specht}@ou.nl

Yasemin Bayyurt
Bogazici University
Faculty of Education
Department of Foreign Language Education
Bebek - Istanbul, 34342, Turkey
E-mail: bayyurty@boun.edu.tr

ISSN 1865-0929 e-ISSN 1865-0937
ISBN 978-3-319-13415-4 e-ISBN 978-3-319-13416-1
DOI 10.1007/978-3-319-13416-1
Springer Cham Heidelberg New York Dordrecht London

Library of Congress Control Number: 2014954450

Typesetting: Camera-ready by author, data conversion by Scientific Publishing Services, Chennai, India

Printed on acid-free paper

Springer is part of Springer Science+Business Media (www.springer.com)

Preface

The 13th World Conference on Mobile and Contextual Learning in 2014 was entitled "Mobile as a mainstream - towards future challenges in mobile learning". On the one hand, the title was meant to give reference to the fact that learning with multiple and mobile devices is becoming more and more a mainstream rather than a special interest or special situation. Innovative higher education institutions are already following the "mobile first" development strategy with regard to further expansion of the technological infrastructure for learning and teaching. On the other hand, the title also expresses that there is a need for the research community to redefine their field in the light of this mainstreaming trend. Mike Sharples and John Traxler contributed their vision for such a redefinition of the field during the keynotes. Besides traditional topics, mLearn 2014 also attracted a number of interesting cross-disciplinary papers in which mobile learning contributes to addressing the grand challenges of our time. There were 31 individual papers, four posters, three panels (including a plenary panel), and two plenaries presented at 42 sessions during the conference. The sessions were organized around the following broad themes:

- Theories and methodologies of mobile and seamless learning in context
- Theories and models for mobile and contextual learning
- Quality assurance of mobile learning apps
- Evaluation of mobile learning support
- Ethics and privacy in mobile and contextual learning
- New technologies and developments in mobile learning
- Wearable and sensor technologies
- Augmented reality for learning
- Mobile seamless learning support
- Toys and smart objects for learning
- Ubiquitous and ambient learning and technology
- Interoperability and standards for mobile learning
- Innovative approaches to learning of current and emerging mobile technologies
- Adaptive, virtual or collaborative environments for mobile learning
- Instructional design for mobile and seamless learning
- Orchestration of mobile learning
- Mobile learning in and across formal and informal settings
- Open and distance education with mobile devices
- Interaction design and usability for mobile learning
- Mobile learning across cultures and educational sectors
- Challenges for mobile learning in developing countries
- Social empowerment through mobile learning
- Using mobile technologies for development

- Mobile language learning
- Mobile learning strategies in schools, higher institutions, industry, and organizations

Geographically, Turkey is a bridge between Asia and Europe. Hence, the organization of mLearn 2014 in Istanbul was relevant and timely for the promotion of mobile learning research and bringing together scholars from Asia, Europe, the Americas, and other parts of the world. Since studies in mobile learning take place intensively in Europe, the Americas, and Asia, hosting this conference in Istanbul was ideal to enable people from these continents and other parts of the world to meet and exchange ideas about the use and application of mobile learning and mobile technologies in their local contexts.

We would like to take this opportunity to thank all the reviewers, contributing authors, local organizers, and sponsoring institutions for their support.

September 2014

Marco Kalz
Yasemin Bayyurt
Marcus Specht

Organization

mLearn2014 was organized by Bogazici University in cooperation with Kadir Has University and supported by the professional conference organization company PCO (`http://www.figur.net`).

Executive Committee

General Chair

Marcus Specht	Open University of the Netherlands, The Netherlands

Program Chairs

Marco Kalz	Open University of the Netherlands, The Netherlands
Yasemin Bayyurt	Bogazici University, Turkey

Industry Chair

Taner Arsan	Kadir Has University, Turkey

Workshop Chairs

Senem Yildiz	Bogazici University, Turkey
Zeki Bozkuş	Kadir Has University, Turkey

Doctoral Chair

Senem Yildiz	Bogazici University, Turkey

Local Organizing Chairs

Yasemin Bayyurt	Bogazici University, Turkey
Taner Arsan	Kadir Has University, Turkey

Scientific Committee

Mohamed Ally	Athabasca University, Canada
Inmaculada Arnedillo-Sanchez	Trinity College Dublin, Ireland
Taner Arsan	Kadir Has University, Turkey
Hasnain Baloch	International Medical University, Malaysia
Yasemin Bayyurt	Boğaziçi University, Turkey
Zeki Bozkuş	Kadir Has University, Turkey

Tom Brown	Unisa, South Africa
Helen Crompton	Old Dominion University, USA
Gulcan Ercetin	Boğaziçi University, Turkey
Hamdi Erkunt	Boğaziçi University, Turkey
Christian Glahn	Swiss Federal Institute of Technology, Switzerland
Davinia Hernández-Leo	Universitat Pompeu Fabra, Spain
Ulrich Hoppe	University of Duisburg-Essen, Germany
Gwo-Jen Hwang	National Taiwan University of Science and Technology, Taiwan
Marco Kalz	Open University of the Netherlands, The Netherlands
Feza Kerestecioğlu	Kadir Has University, Turkey
Agnes Kukulska-Hulme	The Open University, UK
Cheri Macleod	CNAQ, Canada
Marcelo Milrad	Linnaeus University, Sweden
Diler Öner	Boğaziçi University, Turkey
Agnieszka Palalas	George Brown College, Canada
David Parsons	Massey University, New Zealand
Christoph Pimmer	University of Applied Science and Arts, Switzerland
Rob Power	College of the North Atlantic, Qatar
Demetrios Sampson	CERTH, Greece
Mike Sharples	The Open University, UK
Marcus Specht	Open University of the Netherlands, The Netherlands
Brendan Tangney	Trinity College Dublin, Ireland
Mark van tHooft	Kent State University, USA
Mike Tissenbaum	University of Toronto, Canada
John Traxler	Wolverhampton University, UK
Inge de Waard	The Open University, UK
Jocelyn Wishart	University of Bristol, UK
Senem Yildiz	Boğaziçi University, Turkey

Additional Reviewers

A. Giemza
A. Suarez
J. Schneider
B. Tabuenca

Sponsoring Institutions

Qualcomm, Inc., San Diego, USA

Table of Contents

Technologies and Interaction

Tablets and eBook Readers

Learning and Teaching Inside and Outside the Classroom

Learning Design and Design Implications

Evaluation and Review Studies

Development and National Perspectives

Inquiry-Based Learning and Science Applications

Work-Based Learning

Theory

Language Learning

Learner Perspectives

Mobile Learning: An Ergonomic Alternative for Long-Awaited Educational Changes?

Thomas Petit and Gilberto Lacerda Santos

Faculty of Education, University of Brasilia, Brasilia, Brazil

Abstract. Mobile learning, until now, was often defined and presented as a learning that could occur anytime and anywhere. However, now we clearly need to go over this definition if we want to take pedagogical benefits from this new learning method. In this paper, we present results of documentary research which demonstrates that the novelty of mobile learning does not consist in a new educational theory, but more in the possibility of new educational practices. Thus, we propose a contextual and technology-oriented framework that takes account of the user's experiences and interactions in and with the environment of his networked mobility. The affordances of the handheld devices, and more specifically smartphones, seem to have an important role to play in this context, to turn appropriate places and moments of everyday life into learning opportunities, through which the mobile user become a learner.

1 Introduction

In this early twenty-first century, we are witnessing, more than ever, radical changes in our everyday life, due to the convergence of two new ages. On the one side, there is the digital age, characterised by the growth of the digital technologies, which have invaded our daily activities, from home to work. On the other side, we are also living in the mobile age, in which the mobility concerns knowledge as much as people [1].

Nowadays, smartphones are the only devices representing a convergent point between both our digital and our mobile life. Indeed, these handheld devices, unlike other mobile and/or digital technologies such as laptops, are not just carried from one point to another but can actually be used while moving around. Beyond being the convergence between the digital and the mobile age, the smartphone is a device that also integrates tools of many other devices: phone, digital camera, computer, recorder, GPS, and so on.

Accordingly, we consider that mobile learning (mLearning) theory and practice cannot be identical depending on whether we use a cellphone or a smartphone. The affordances of the latter are much more numerous, and offer a greater diversity of actions and activities than the former. In many countries the smartphone market is bigger than the "regular" cellphones', and according to the Unesco forecast it will exceed the sales of computers in a near future [2].

In the next sections of this paper, we present documentary research based on two sorts of source materials. On the one hand, considering the affordances

M. Kalz et al. (Eds.): mLearn 2014, CCIS 479, pp. 1–14, 2014.

and the democratisation of the smartphone, we did a literature review on mobile learning focused on scientific articles from the past few years, which corresponds approximatively to the appearance of the first smartphone. On the other hand, we used relevant books written by two great educational thinkers of the past century: John Dewey and Paulo Freire. Thus, we are attempting to bridge recent mobile learning literature with older educational ideals so as to identify convergences and innovations.

Finally, from this theoretical dialogue, we propose a contextual and technology-oriented framework for mobile learning. The main goal of this approach is to focus on new practices in order to answer a central question: how can a shared and specific mobile learning theory lead us to long-awaited educational changes?

2 Mobile Learning Literature Review

Though mLearning is a relatively new field in educational research, the scientific material has grown exponentially since 2010. This material comes from events, groups, associations, journals, revues, books and so on, dedicated to mLearning. This proves that it is becoming a considered area on the international education scene. Our aim, in this section, is to share some relevant informations that could help us better understand mLearning theory to propose our framework.

According to our findings, we propose a division in two categories: (i)visions and definitions, (ii)frameworks and applicability. Through both categories, we are attempting to contemplate two levels of consistency for the bases of mLearning as a proper educational theory: (i)conceptual and (ii)pedagogical. In this way, we propose to take a step backwards in the academic literature in order to present a future-oriented theory that will be in accordance with the development of new practices from now on.

2.1 Visions and Definitions: Towards a Conceptual Consistency

If there is one difficult thing to achieve, it is defining what mobile learning really is. Visions and definitions differ from one author to another and many scientific articles begin with the almost unavoidable question "What is mobile learning?". The definitions have also evolved as technology has progressed. The researches before 2006, being mostly technology-oriented, were too obsolete to be integrated in this documentary research. According to Endrizzi [3], mLearning researches are relatively recent and increased only over the past few years.

In a recent book chapter, Parsons [4] wrote about the "Top 5 Mobile Learning Myths and Misunderstandings", the first one being the famous "anytime anyplace". The author says that "[o]ne of the major affordances of a mobile device is that it can be brought to use in a specific context, a concept not acknowledged by 'anywhere, anyplace'" (p. 219). It seems that the anytime anyplace vision of mLearning is, above all, an unfortunate expression used to refer to learning out of school through digital and mobile devices, to express a dimension of freedom. Nevertheless, we cannot stay in this mere vision or definition anymore, because taking advantage of the smartphone affordances, as much as learning, involves being in a specific and appropriate context, in time and in place.

A first step to take, in order to define mLearning, would be to distinguish it from other learning theories [3]. Indeed, although mLearning has similarities or can integrate elements of eLearning, ubiquitous learning, lifelong learning, informal learning and so on, it became necessary to highlight what makes it new, different and, therefore, unique. In 2006, Patten et al.[5] were already in the vanguard of mLearning reflections to the extent that they pointed out the necessity of broadening "our own horizons and create new innovative learning opportunities which would not be possible without handheld devices" (p. 307). In this regard, Laurillard [6] rightly noted that "[i]n defining the pedagogies for mobile learning, it is important to be clear about what exactly m-learning contributes that is new and different from previous technologies of learning" (p. 154). Nowadays, in the "smartphone age", these recommendations make a lot of sense.

In 2007, Sharples et al. [1] defined a theory of learning in the mobile age as "the process of coming to know through conversations across multiple contexts amongst people and personal interactive technologies" (p. 4). This definition, by its general dimension, did not take into account the new affordances of the smartphone, which did not yet exist. However, it brought for mLearning the concept of interactivity. Furthermore, the adapted framework they used in the same research, as we will see later, is still a good reference.

In their definition, Martin and Ertzberger [7] described mLearning as: "[...] learning that occurs when learners have access to information anytime and anywhere via mobile technologies to perform authentic activities in the context of their learning. Here and now mobile learning gives students the opportunity to be in the context of their learning and have access to information that is related to what they are seeing and experiencing at the moment" (p. 77). Accessibility, authenticity and experience are three important elements. Moreover, the authors already considered the technology users as learners, which means a certain intentionality of learning in mLearning, with which we agree.

The language area is also helping to define mLearning. With mobile language learning, for instance, Kukulska-Hulme [8] and Al-Shehri [9] emphasised the participation of learners in the design of learning experiences. The first researcher insisted on the importance of appropriate time and space to help create spontaneous relevant support. Similarly, the second author highlighted the context awareness paradigm that allows an authentic contextualisation of learning. Surroundings represent unlimited situations of authenticity [9].

Finally, a reflection about the technology is central in some visions and definitions of mLearning. Kukulska-Hulme [10] highlighted the differences of methodologies and of opportunities according to whether a computer or a smartphone is being used. According to her, this makes mLearning theory unique, by the continuity, the spontaneity and the interaction allowed by the use of smartphone. Cochrane and Bateman [11] also emphasised the affordances of the smartphone and its influence for contextual learning and the creation of personalised content.

Thus, we consider in our research that the technology is central to the definition of mLearning, unlike Ally's and Prieto-Blázquez's affirmation that mLearning "is not about the technology, it is about the learner" [12]. We believe precisely that one can not be separated from the other. On this point, Pachler et al. [13], in their socio-cultural ecological model linking the mobile complex to learning from the perspective of appropriation, already gave importance to the "everyday use of personally owned, multifunctional mobile devices" (p. 11). In the mLearning theory, we should now assume that our learner is the owner and the user of a smartphone, familiar and connected with this handheld technology. It is from this connection that mLearning becomes possible. It is a learning theory about both technology and the learner, considered together.

Nevertheless, we do not fully agree when Parsons [4] says that learning occurring on mobile devices is necessarily mLearning. In an exploratory research, Petit and Lacerda Santos [14] demonstrated that the use of the smartphone with a traditional learning method without taking into consideration the affordances in specific environments, is not mLearning. In that case, reading a book in a park would be the same. In other words, the handheld device itself does not automatically imply mLearning. In this sense, researchers observed that few mobile applications are taking benefits from the mLearning potential [15,5]. It means also that the mLearning method is not yet quite put into practice.

2.2 Frameworks, Applicability and Pedagogical Consistency

Most of the time, analysing frameworks results more interesting and less philosophical than the definition's exploration. The researchers generally offer a concrete vision of what they mean by mLearning, and it is easier to identify contexts of applicability. For our documentary research, we identified three complementary frameworks that could help us to create our own.

The first relevant framework is the FRAME proposed by Koole [16]. In her FRAME, as we can see in Figure 1, she showed that mLearning was the final intersection of other ones between the device, the learner and the social aspects. This framework is very interesting to the extent that it is neither centered on the learner nor on the technology, but on both, and on a social aspect. Important concepts such as wireless networking, authentic situations and collaborative learning come from this contextual framework explanation.

The second framework, in Figure 2, were proposed by Sharples et al. in 2007 [1], adapted from Engeström's expansive activity model. In this adapted framework for analysing mobile learning, they proposed a superimposition of two layers that correspond to semiotic and technological frameworks. The layers superimposition considers the subject, the object, the control, the context, the communication and the mediating artifacts.

One of the most relevant elements of this framework is the dialectical relation that the authors found between the semiotic and the technological spaces when applied to an educational context. In the case of a lack of usability in the technological layer, for example, there are consequences at the semiotic level. Moreover, and it is a very interesting fact, the learner experience is not worthwhile

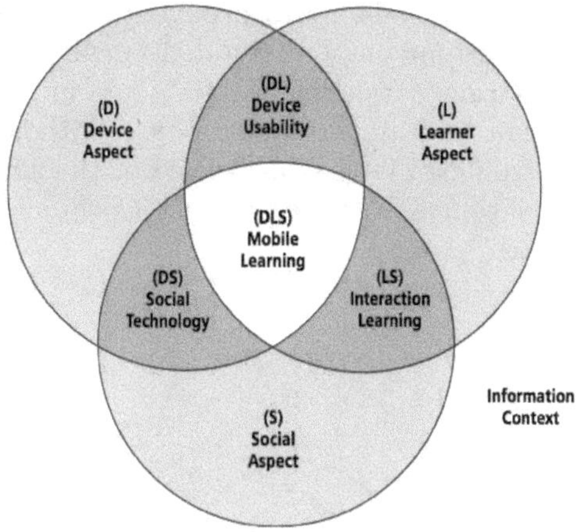

Fig. 1. FRAME framework (Koole, 2009)

in a situation of conflict between the technological and the semiotic subject. In other words, and considering our mLearning vision, there must be a convergence between the smartphone user (technological subject) and the learner (semiotic subject) in order to enable a mLearning opportunity.

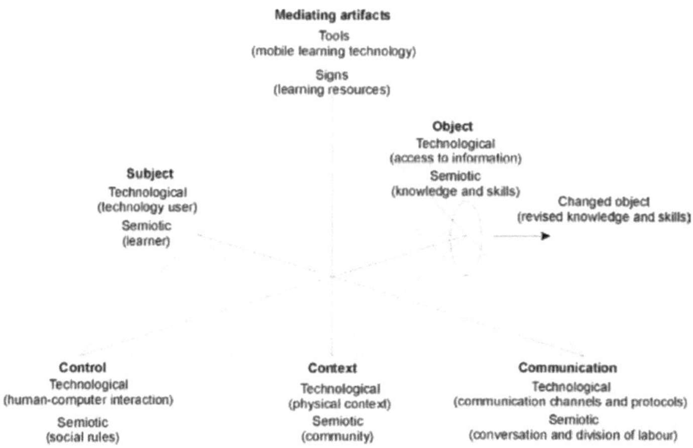

Fig. 2. Adapted framework for analysing mobile learning (Sharples et al., 2007)

The third framework that held our attention, in Figure 3, is from Kearney et al. [15]. Its interest lies in the pedagogical dimension, which shows the convergence of three important aspects of today's pedagogies: collaboration, authenticity and personalisation. This convergence is enabled through the use of space and time. However, by giving priority to pedagogy rather than technological affordances, the authors are missing out the dialectical relation highlighted in the previous framework. How can learners collaborate in an authentic context with personalised content, in a mLearning situation, without the technological affordances of the smartphone? We cannot leave out the affordances and must try to find in them pedagogical applications, like Cochrane and Bateman [11] proposed in their research.

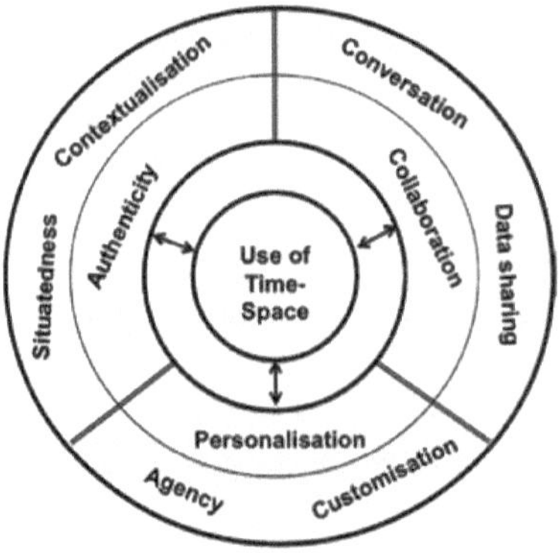

Fig. 3. Pedagogical framework (Kearney et al., 2012)

These three frameworks, as we can see, are very complementary. They show us the intersections, dialectical relations and convergences between contextual, technological, semiotic and pedagogical dimensions. We believe that the mobility should be more central, following the example of the use of time and space in Kearney et al. framework [15]. In many articles the mobility is a polemical concept, the authors wondering if it concerns either the device or the learner. According to Kukulska-Hulme [10], both device and learner are affected by the mobility. As we said in the introduction, a device such as a smartphone represents exactly the convergence between the mobility and the digital. It is also in accordance with Sharples et al. [1] who spoke about "networked mobility". In fact, this expression should be at the heart of mLearning theory, because we are not being mobile alone, we are being mobile in networked spaces and moments, characterised by historical changes and evolutions.

3 Dialogue with Dewey and Freire

John Dewey and Paulo Freire, for their important contributions to the education area over the last century, are often mentioned in educational researches in general, and in mLearning researches more specifically. In this documentary research, our aim was to establish a real dialogue between them and the elements of our literature review. This dialogue sometimes began from the mLearning theoreticians themselves [1].

With this aim in mind, we used the four following books: Experience and Education [17] and Democracy and Education [18], of the American Dewey; Pedagogy of the Oppressed [19] and Pedagogy of Autonomy [20], of the Brazilian Freire. In each of the four books, considering the previous literature review, we attempted to categorise elements that could represent convergences between the mLearning theory and the pedagogical and social ideals of Dewey and Freire. Thus, we could highlight three main dimensions: (i)considerations of everyday life environments, (ii)democratisation and freedom for designing learning experiences, (iii)intercommunication and collaboration in community.

3.1 Considerations of Everyday Life Environments

One of the most obvious convergence between mLearning theory and Dewey and Freire ideals for Education is the importance of considering the learner everyday environments. These environments, according to Dewey [18] can be either social or natural. In the early twentieth century, John Dewey already asked for a more progressive educational system in which a concrete application of knowledge in ordinary life was essential. To him, it was more a question of doing than learning, a question of bridging school aims with ordinary life experiences in natural and social environments.

In Freire's work, the social environment contains undoubtedly a political dimension of education. Like Dewey, he suggested that we should integrate the learner daily life and environment into the school system [20]. Learning outside of school, according to him, represents informal experiences that are important to be considered in the educational process. We learn from the reality of our ordinary life, and we can carry elements of this reality to the learning community. The balance between informal and formal learning is also an important criterion for a new philosophy of education according to Dewey [18].

The consideration of daily environments for the learning process involves spaces and moments. We should take into account these elements when defining aims, subject matters and activities. Effectively, "[t]here is also an inclination to propound aims which are so uniform as to neglect the specific powers and requirements of an individual, forgetting that all learning is something which happens to an individual at a given time and place" [18] (p. 8).

The different environments involved in the learner's mobility are a part of the contextualisation and personalisation of learning. It is a mobility in moments and spaces, hence the importance of history and geography [18]. In this sense, Freire affirmed that "the world is not, the world is being" [20]. We are not in a static reality, but in a reality in transformation, in which we are immersed in a "here and now" dimension as historical human beings [19].

3.2 Democratisation and Freedom for Designing Learning Experiences

Democratisation and freedom are others key components in the design of learning experiences. In the mLearning theory, learners should be active in this design. Freire [19] elaborated a pedagogy of the oppressed in opposition to a banking concept of education, in which the learner is passive. According to the author, in a more democratic and emancipating education, human beings should be creative, active and inventive individuals.

With democratisation and freedom comes the question of content creation. By content, we refer to input as much as output and outcome. Freire gave importance to the input. He showed that in the banking education, it is determined by the educator and only received passively by the learners. This approach minimises the creative power of learners. On the contrary, a pedagogy of freedom would let the learner propose his own input to the community directly from his daily environment. We are talking about participatory education in which the learners could collaborate in the design of their learning process [17]. In mLearning theory, some authors refer to this as "ownership" [13,6].

Creating material and content means, consequently, praxis. It is a key concept in the works of both Freire and Dewey. The American thinker connected the experiences of ordinary life with the possibility of resources, essential input and material of thinking [18]. An intimate relation with the input would lead to outputs and outcomes, through decisive elements in an educational process: aim, subject matter and activity. According to the author, learners are persons concerned in the outcome, so we must be "partakers in the process which produces the result".

In his democratic vision of education Dewey made experience a central point. However, depending on its quality, all experiences are not educative, according to him [17]. Experience is constituted by two fundamental principles in Dewey's work: interaction and continuity. Besides, he assumed that education is a social continuity of life, so it makes sense to connect education with experience. It is also a question of interacting with the environment. "The environment, in other words, is whatever conditions interact with personal desires, purposes, and capacities to create the experience which is had" (p. 44) [17].

Learning by doing means also doing with things. Dewey's words about acquaintance and familiarity with objects are indicative of the potential of today's smartphone affordances: "Having to do with things in an intelligent way issues in acquaintance or familiarity. The things we are best acquainted with are the things we put to frequent use - such things as chairs, tables, pen, paper, clothes, food, knives and forks on the commonplace level, differentiating into more special objects according to a person's occupation in life. Knowledge of things in that intimate and emotional sense suggested by the word acquaintance is a precipitate from our employing them with a purpose" (p. 185).

3.3 Intercommunication and Collaboration in Community

In considering the natural and social environments of learners, in which they have experiences and a certain freedom for acting in a democratic space, we cannot forget the concept of community. Indeed, both Dewey and Freire gave importance to the collective dimension of learning. Men possess things in common by communicating, and it is through these things that they shape a community. It is a question of common aims, knowledge, beliefs. [18].

The authors highlighted the intercommunication between men in an educational process. Intercommunication is a way of mediating authenticity from reality, according to the Brazilian thinker. Human beings educate themselves together, in communion, through the mediation of the world [20]. The intercommunication is, in Dewey's vision, an active connection between members of a community, and from which one has the opportunity to learn from others. Besides, through the social environment, therefore the community, activities are shared between people [18].

Today, the Dewey and Freire ideals that we are pointing out in this documentary research are even more important than at the time of their expression. In fact, we are in a society more and more mobile, more and more connected, and people belong to several communities. We are increasingly active, creative and involved as people, and it is logical that we cannot stay in the traditional concept of education. However, it is hard to leave this old model.

It is obvious that there are several convergences in the dialogue between mLearning theoreticians and Freire and Dewey. Ideals and convictions of last century remain the same today, and mLearning community is clearly influenced by them. Until now, the learning theories based on the use of digital technologies did not succeed in the great mission of changing educational paradigms. However, we saw in the literature review that, from the convergences with twentieth century ideals, there could emerge a new ergonomic alternative with the "new" devices that are the smartphones. Its affordances could allow more free actions to the learners in their ordinary life environments and in their community, to enhance learning opportunities.

4 Contextual and Technology-Oriented Framework

The framework represented in the Figure 4 is the result of the dialogue established earlier, through the documentary research, between contemporary researchers in the mobile learning area, and John Dewey and Paulo Freire. Is is a contextual and technology-oriented framework that situates learning opportunities according to experiences and interactions in specific spaces and moments, enabled by people's networked mobility.

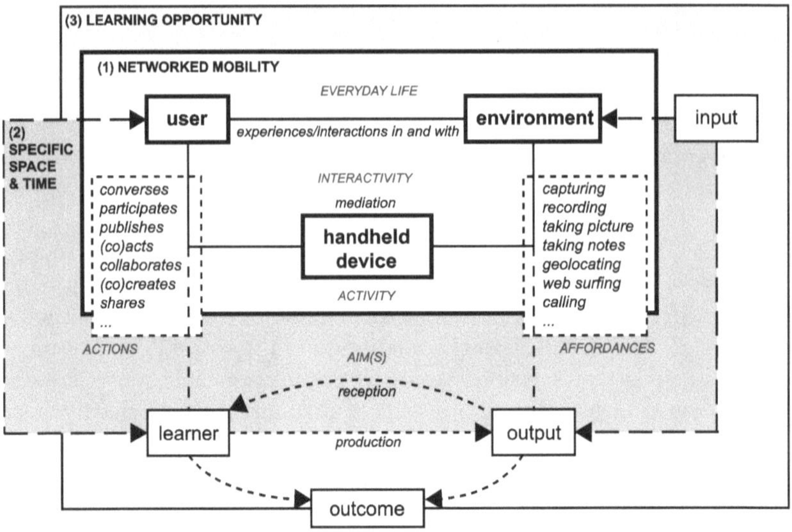

Fig. 4. Contextual and technology-oriented framework for mobile learning

4.1 Networked Mobility

The first layer represents the networked mobility, fundamental element of the framework. That means literally that, without networked mobility, there will not be situations of mobile learning. We want to demonstrate that mLearning is a new and unique learning method, which is not at all the same as studying with an exercise book in a park or at the library with a laptop. Having a networked mobility situation involves three basic elements: the user, which owns a handheld device and interacts with his - natural or social, physical or virtual - environment.

In covering the three aspects of the FRAME [16], we open the environment out to more than a purely social aspect. Moreover, we show that the user cannot be separate from his handheld device, considered as a familiar thing on the commonplace level, according to Dewey. The device - which is not only mobile but above all handheld - is, besides, in the center of this technology-oriented framework. Nowadays, the handheld device at the same time more evolved and more democratic would be a smartphone, reason why we focused on this technology in this paper. But we preferred the term "handheld device" for the framework, because finally the most important is the affordances it allows; and it is possible that in few years many people will be using other kinds of handheld device, such as digital eyeglasses, offering even more affordances.

The handheld device, then, is a mediation tool for the user's experiences and interactions in and with the environment. From this interactivity, in an everyday life situation, the user realises acts according to the affordances. We suppose that the convergence between affordances and actions, in addition to take place in this whole context, is also in relation with a specific activity, either formal, non formal or informal.

4.2 Specific Spaces and Moments

The second layer completes the first one in order to make the framework contextual. Indeed, even though the network mobility is essential, we cannot forget that a specific space and moment is needed to create a learning opportunity. Furthermore, it is difficult to learn while being physically in motion [4].

This layer represents, above all, the converging point between the technological and the semiotic subject [1]. Actually, it is the specific space and time in which the smartphone user will be able to become a learner. In that sense, the environment becomes a learning resource (input) mediated by the use of smartphone. As Parsons underlined, it is a question of capturing a learning experience in the field [4].

To this input will correspond an output produced by a user/learner with specific aim(s), in a community, from an experience in or with the environment, in specific time and place, and through an action enabled by the smartphone affordances. Laurillard [6], in her Conversational Framework, emphasises the peer collaboration, showing that it can be motivating for a learner to share his output and to reflect on it with peers.

We can observe that three aspects - learner, input and output - are situated at the limit between the second layer and the learning opportunity layer. After all, like in the FRAME, it is a question of intersections [16]. The dialectical relation between the two first layers generates a social learning environment, which represents a considerable help for the learner [6]. In this environment where networked mobility and specific space and time merge, a learning opportunity can take place.

4.3 Learning Opportunity

In this layer, we prefer the term "learning opportunity" to "learning" because we cannot affirm that learning will be effective. We agree with Dewey who said that not every experience is educative.

The learning opportunity involves a user/learner belonging to a community. Thus, as we discussed before, he can share or receive output(s). Moreover, the initial input coming from the environment can be integrated by the learner and lead to an output, but it can also be directly the output, depending on the activity and the interactions in and with the environment and the community.

In concrete terms, anyone in the community can ask me to explain what a specific foreign word means. My answer can be a concrete element of my environment. I can take a picture of this element that would be an input coming from my surroundings turned directly into an output for my colleague. In that case, it would mean that people can learn from the context of mobility of other people. Hence the importance of networking, and not only mere mobility. In other cases, following the language learning example, I can be the one who has a doubt about what a specific thing of my environment is. I can make pedagogical use of this input, and maybe share it (taking picture of this thing), as a starting point that will lead me to produce an output (a question, for example) to be shared by the community.

In this layer there is another aspect, the outcome, which is also at the limit of our contextual and technology-oriented framework. Indeed, we consider that this outcome of a mobile learning opportunity can be situated at the intersection with other learning situations. If the outcome/result of our framework, for example, is from an informal situation, it could be integrated later into a formal environment like school. Besides, the framework could actually take place in the school, which is, after all, a everyday life context, part of students cultural practices [13].

The continuity of education and of mobile learning allows precisely the transition between formal and informal contexts [6]. It is the reason why we did not include in this framework the formal and informal aspects. Formality or informality can concern either the context or the activity, but are not determinant in making mLearning a different theory. Moreover, we agree with Sharples et al. [1] when they say that "education in the mobile age does not replace formal education" and that "rather it offers a way to extend the support of learning outside the classroom, to the conversations and interactions of everyday life" (p. 23). We cannot deny, actually, that mLearning is a good possibility to finally integrate the non-formal and informal learning experiences in the formal education.

5 Issues for Mobile Learning: A Theory Facing Ahead and Focusing on New Practices and Technological Perspectives

This documentary research showed us that mobile learning is not really a theoretical novelty in Education, and that the innovation does not really consist in this point. Several convergences were found between main pedagogical ideals of 21st century mLearning theoreticians and 20th century educational thinkers, John Dewey and Paulo Freire.

Since the beginning of the twentieth century, traditional and progressive education have been in a perpetual conflict. The integration of digital technologies in learning and teaching contexts, especially the computers, until now did not lead us to the development of new practices. On the contrary, these technologies are often used in a traditional way, which shows us that the method necessarily should progress together with the medium.

With the mLearning method, we are facing a major new opportunity for change. It is not a question of the integration of technology in the learning process anymore; it is about integrating learning process in the daily use of technology. We can take advantage, maybe for the first time, of a networked mobility, based on the relationship between the user, the smartphone and the environment, to promote the educational continuity in everyday life experiences.

The insertion of mLearning method in formal education is an attractive idea, but it also represents a challenge. We know from Dewey that all experiences are not educative, and Pachler et al. [13] rightly reminded that everyday life is structured by elements like marketing, entertainment or fashion, that don't really belong to the formal education world. In order to make informal everyday life experiences worthwhile, we should try to focus on non-formal learning activities, which envolve intentionality, awareness, and more control and more ownership among learners.

Starting from our contextual and technology-oriented framework, we argue that our efforts in the future should be concentrated on new practices through the development of technological support. From this documentary research, we conclude that mobile learning is an ergonomic alternative for long-awaited educational changes, and that actual handheld devices affordances are showing us the way to these changes.

Future works should contemplate the development of smartphone applications that could fit in the specific space and time layer, in order to bridge networked mobility with learning opportunities. Actually, which applications allow us, today, in an appropriate time and space, to create a learning opportunity, in community, from the smartphone affordances? Social networking applications seem to be the few ones available.

In order to enable and, perhaps, bridge non-formal and formal learning opportunities, we need applications which are more pedagogical-oriented, that could benefit from affordances of today and future handheld devices. Ergonomic studies in specific didactic areas should contemplate this issue, with a specific attention paid to interactivity, central point in a networked mobility situation. A new kind of education is undoubtedly possible and we believe that a mobile learning theory looking ahead could have a role to play in the current transition.

Acknowledgements. The authors would like to acknowledge Professors Marguerite Koole, Matthew Kearney and Mike Sharples for allowing the use of figures from their works. The authors are also grateful to the reviewers, Anna King Murdoch and Eve Demeocq, for their careful reading of this paper.

References

1. Sharples, M., Taylor, J., Vavoula, G.: A Theory of Learning for the Mobile Age. In: Andrews, R., Haythornthwaite, C. (eds.) The Sage Handbook of Elearning Research, pp. 221–247. Sage, London (2007)
2. Unesco: Policy guidelines for mobile learning. Unesco, Paris (2013)
3. Endrizzi, L.: Éléments pour une cartographie de la recherche sur les apprentissages nomades en Europe. In: Poyet, F., Develotte, C. (eds.) L'éducation á l'heure du Numérique: état des Lieux, Enjeux et Perspectives, pp. 167–188. INRP, Lyon (2011)
4. Parsons, D.: The Future of Mobile Learning and Implications for Education and Training. In: Ally, M., Avgoustos, T. (eds.) Increasing Access Through Mobile Learning, pp. 217–229. Commonwealth of Learning and Athabasca University, Vancouver (2014)
5. Patten, B., Arnedillo Sánchez, I., Tangney, B.: Designing collaborative, constructionist and contextual applications for handheld devices. Computers and Education 46(3), 294–308 (2006)
6. Laurillard, D.: Pedagogical forms for mobile learning. In: Pachler, N. (ed.) Mobile Learning: Towards a Research Agenda, pp. 153–175. WLE Centre, loE (2007)
7. Martin, F., Ertzberger, J.: Here and now mobile learning: An experimental study on the use of mobile technology. Computers and Education 68, 76–85 (2013)

8. Kukulska-Hulme, A.: Language learning defined by time and place: A framework for next generation designs. In: Díaz-Vera, J.E. (ed.) Left to My Own Devices: Learner Autonomy and Mobile Assisted Language Learning, pp. 1–13. Emerald, Bingley (2012)
9. Al-Shehri, S.: Context in our pockets: Mobile phones and social networking as tools of contextualizing language learning. In: mLearn 2011 Conference Proceedings, Beijing, pp. 278–286 (2011)
10. Kukulska-Hulme, A.: Will mobile learning change language learning? Re-CALL 21(2), 157–165 (2009)
11. Cochrane, T., Bateman, R.: Smartphones give you wings: Pedagogical affordances of mobile Web 2.0. Australiasian Journal of Educational Technology 26(1), 1–14 (2010)
12. Ally, M., Prieto-Blázquez, J.: What is the future of mobile learning in education? RUSC 11(1), 142–151 (2014), doi:10.7238/rusc.v11i1.2033
13. Pachler, N., Cook, J., Bachmair, B.: Appropriation of mobile cultural resources for learning. International Journal of Mobile and Blended Learning 2(1), 1–21 (2010)
14. Petit, T., Lacerda Santos, G.: A aprendizagem não formal da língua estrangeira usando o smartphone: Por que voltamos a metodologias do século XIX? In: Anais do 5° Simpósio Hipertexto e Tecnologias na Educaç ão, Recife (2014)
15. Kearney, M., Schuck, S., Burden, K., Aubusson, P.: Viewing mobile learning from a pedagogical perspective. Research in Learning Technologies 20 (2012), doi:10.3402/rlt.v20i0.14406
16. Koole, M.L.: A model for framing mobile learning. In: Ally, M. (ed.) Mobile Learning: Transforming the Delivery of Education and Training, pp. 25–47. AU Press, Edmonton (2009)
17. Dewey, J.: Experience and Education, 16th edn. Collier Books, New York (1973)
18. Dewey, J.: Democracy and Education. Free Press, New York (1966)
19. Freire, P.: Pedagogia do Oprimido, 56th edn. Paz e Terra, Rio de Janeiro (2014)
20. Freire, P.: Pedagogia da Autonomia, 47th edn. Paz e Terra, Rio de Janeiro (2013)

Declarative Approach for Adaptivity and Personalization in Mobile Learning: An Algerian Perspective

Nadia Khalfallah, Abdellah Boukerram, and John Traxler

Science Faculty, Computing Department, Sétif University1, Algeria
Computing Department, Bejaïa University, Algeria
Institute for Education, University of Wolverhampton, UK
nadia_khalfallah@yahoo.fr, boukerram@hotmail.com,
john.traxler@wlv.ac.uk

Abstract. In this paper we present how mobile learning (mLearning) can be personalized/adaptive via the use of the declarative approach in order to be adapted with the Algerian context.

The background to this work is mLearning progress in Algerian society, based on several areas such as the socio-cultural, institutional, historical, epistemological, pedagogic, linguistic, infrastructural and demographic context.

This context allows us to think about introducing the mLearning concept to Algerians, especially students, by developing a mLearning platform after the arrival of 3G technology in 2013. We have chosen PBL as an active pedagogical strategy to be implemented with a mLearning platform.

In Algeria, due to the diversity known in many fields, the personalization of the PBL model has become indispensable and vital. That is why we propose in this study an architecture that shows the relationship between the adaptivity, mobile ontology (mOntology) and the reuse of declarativity approach results in order to design an Adaptive/Personalized Virtual Document (AVD) for an institutional strategy model (the PBL model) based on the diversity of the Algerian context (learners, learning needs, languages, cultures, society constraints, infrastructure needs…).

Keywords: Algerian context, Adaptivity, Adaptive/ Personalized Virtual Document, mLearning, mOntology, Declarativity.

1 Introduction

Currently, the use of Information Technologies (IT) in the education environment has become one of a means to providing a more flexible learning experience in both advanced nations or underdeveloped ones.

The Algerian education system is trying to enhance learning quality by introducing some mechanisms and providing a robust infrastructure that allows a major innovation in the learning/teaching processes compared to the traditional one, mainly Internet-based technology applications.

eLearning was the first trend of the modern education environment with the development of various eLearning platforms that: *1)* encourage the collaboration between

M. Kalz et al. (Eds.): mLearn 2014, CCIS 479, pp. 15–28, 2014.

didacticians and computer scientists; *2)* give the opportunity to create a support system for several instructional approaches, such as the Project-based learning as one of the active instructional strategies; *3)* can also be designed for confronting complex issues, which require students to investigate for a more thorough understanding [1].

In addition to the extension of eLearning into wireless, mobile, portable and hand-held computing devices with the help of mLearning, which encourages researchers and developers to integrate the pedagogical strategies models with mLearning platforms, the personalization/adaptivity has become a vital and critical concept in the competitive world of Internet-based application.

In Algeria, the recent arrival of 3G technology, to be followed by 4G, is timely and a significant enabler for a new phase and a major qualitative leap aimed at overall reform in the use of Internet based-technologies and the benefit of the progress that knows the telecommunication market. However, such a technological leap faces a number of barriers and obstacles.

It is in this context that this work investigates the level of preparedness of Algeria to succeed in meeting this challenge. Worth noting however that it was not easy to carry out such an investigation due to the scarcity of related information. We did indeed find that the diversity is the general character of the Algerian society in almost all fields of life (historical, geographical, demographical, social-cultural, epistemological, linguistic, institutional…). This diversity led us to think seriously about the concept of personalization/adaptivity of the various Internet based-applications and not just those destined for mobile devices.

Currently, we identify the need for a mobile ontology that describes semantic models of the domains, respects the good management of tasks sequence and allows the development of a mLearning platform characterized by its flexibility in consulting and managing documents.

Based upon the existing work such as CANDLE[1] and ICCARS[2] that relies on a declarative approach that consists of *ontologies* and *composition engine* (Selection, Organization and assembly Processes), we have reused these projects results, made changes if necessary in order to propose a mOntology which is composed of four main subOntologies: domain ontology; application ontology; user model and metadata model and finally described the specifications of the Adaptivity process (composition engine & Adaptivity Strategies).

First of all, a background study of Internet and mobile technologies used in Algeria is presented. Secondly, the combination Adaptivity & mLearning is the focus of the study including the adaptivity principles and the concept of Adaptive Virtual Document. Thirdly, we focus on the relationship between declarative approach and ontology. Fourthly, a description of our proposed architecture will be presented, *which is somehow simplified because of the page limit condition.* We start as claimed by our ontology design which is consisted of four subOntologies (domain ontology; application ontology, user model and metadata model). Thereafter, the specifications of the Adaptivity process is described including the composition engine which is composed of three main composition mechanisms: logic, semantic and layout composition. Next, the different adaptivity

---

[1] Integrated and Cooperative Computer Assisted Reporting System.

[2] Environment Integrated and Cooperative Computer Assisted Reporting System.

strategies (annotation, direct guidance, tri and masking strategy) are mentioned with some examples. Finally, we will conclude by some perspectives.

2 Research Background

2.1 Algerian Context

The arrival of 3G technology gives an opportunity to Algerian researchers and developers to think about new research topics and applications which are adapted with such technology. But before that, they must take into consideration some constraints that can be the basic of an investigation about mLearning in Algeria. However, we have to mention that it was not easy to do such study because of the lack of information; nevertheless, we have tried to give a general vision of the mLearning situation in Algeria.

Algeria as the Africa's largest country by area (2,3 millions km^2), with its diversity of landforms (coastal, internal, mountain and desert cities…) may have some difficulties to construct a strong infrastructure that provides a good Internet connection mainly in the desert cities (about 84% of Sahara) and rural mountain cities.

Nonetheless, the level of ICT integration is still ongoing and at its early stage. For example the program aiming at providing access to ICT through the Computer for Every Home Initiative was launched in 2003 [2].

Algeria has a population of more than 34 million (69.5% aged 15 to 64 years) with a gross domestic product (GDP) per capital of USD 3,968 in 2007 [4].

It turned out through a specific study that participants (Internet and ITC users) enjoy a higher level education. The majority holds undergraduate degrees, with a rate of 64.79%, followed by holders of university Magister and PhD with a rate of 16.9%, followed by lower degrees (Higher National Diploma/Certificate level) with a rate of 9.86% and end, the students and others with a rate of 8.45%. Internet is therefore accessed by all classes of different educational levels. [4]

Another study on ITC and Internet using, demonstrated that officials have largely a rate of about 46.48% due to the availability of Internet access points in their workplaces. It is important to know that education in Algeria comes in second position with a rate of 35.21%, of which 21.13% are teachers and 14.08% are students. This category uses the Internet in search of the literature. [3]

On other hand, regarding the Internet connection methods in Algeria, a study shows that there are three crucial categories: *1)* the classical and least costly one, which leads with a rate of 78.87% of users. It is usually done by the bias of a telephone line; *2)* the connection lines connected to specialized CERIST, with a rate of 11.27% of users. It is generally used by the public sector; *3)* the use of both connection types at the same time, with 7.04% of users.

Table 1 provides a snapshot of the state of Internet access by all social classes about their age in Algeria[3]. [4]

---

[3] In Algeria, a study revealed that most of the participants (ITC and Internet users) whose age is between 16 and 55 are male (82% men 18% women). The average age is 33 years.

Table 1. Internet Access by Users Age in Algeria

Age	Rate
15 to 20	8%
21 to 26	14%
27 to 32	28%
33 to 38	25%
39 to 44	15%
45 to 50	7%
51 to 56	1%

As a snapshot of the socio-cultural state, Algeria is known by its cultural diversity, which caused languages diversity: 80% to 90% of the population speaks Arabic language as a mother tongue; part of the population (Kabyle, Chaoui, Mezab and Touareg) speaks Amazigh. Besides that, on account of some historical raisons (French colonization of Algeria), the French language has a large part in Algerian daily life (communication, administration, learning, media,..), without forgetting English as a science language which knows a surge in using and learning over the last years. Studies about the linguistic preferences of Algerian Internet users shows that the French language ranks first with a rate of 60%, followed by English with 34% and then 6% for other languages, with rates approximately close. [3]

But in our opinion, the most targeted category will be the illiterate one, because they can neither read nor write. Other category is people that are fluent in Amazigh language for example, they could not use it in their SMS because the mobile phone is not equipped with *Tifinagh,* the Berber alphabet, and this makes them feel isolated. Here, we have to mention that "the Berber language has not been written - until fairly recently - except as short inscriptions on monuments" [5], that is why many Amazigh speakers use Latin or Arabic alphabetic as a transliteration. The Amazigh language researchers are progressing in their project to make Amazigh dialect into a complete language which can be learned in all educational institutions (it has been learned in primary school since 1995). This progress must be accompanied with a parallel one of technologies that allow the easy use of Amazigh language (keyboards with Amazigh letters, applications with language preferences (Arabic, French, Amazigh and English); development of Amazigh learning platforms…).

As solutions, we suggest the following ones:

- To use the Wi-Fi Internet connection of mobiles: in order to solve the infrastructure weakness due to the difficulty of topography. (3G technology)
- To bring education to home without moving to institutes: in order to help learners (mainly girls, workers, special needs categories) to complete their education and learning process whatever the complication of their situations.
- To encourage the evolution of the integration of mobile applications culture in Algerian social life (mLearning, mBusiness,…), especially that the majority of society categories possesses a telephone mobile even illiterate ones.

- To enrich the applications with specific functionalities such as: voice-based applications, videos, voice-mails, in order to allow illiterate enjoying using their mobile.
- To benefit from the experiences of countries that have preceded us in this field; For example some neighbor countries have making a first step in developing and using mobile applications in the people's daily life such as Morocco which has interest in building some mLearning projects as 'MobiMOOC' with a wide range of interests and approaches, resulting in collaborations across countries and even continents [6]. In South Africa, 'Mxit' is an instant messaging servicing available on 3,000 different cell phones including relatively low cost feature phones; 'Dr Maths' is one of the most innovative applications available on Mxit [7]; 'DFAQ'[4] tool with its SMS version that was first used in 2005; The Nokia Mobile Mathematics project is possibly one of the most thoroughly researched mEducation tools in South Africa [7]. In Asia, China is making the largest mLearning market; In India, there is the opportunity of mobile applications market emerging according to the growth of many industries including banking, commerce, entertainments, information, mEducation…[8].

Showing how Mobile technologies are developed in such sub Saharan countries as South Africa and large in area as China or India, gives us hope to success in increasing the robustness of our infrastructure in order to contribute in real mobile technology revolution in Algeria, mainly the new trend which is the mLearning.

2.2 Why the mLearning?

In Algeria, the Distance learning is the most popular form of learning, but recently, the eLearning is taking part in Algerian educational culture which encourages the ***CNIIPDTICE***[5] to announce that the year 2011 was the one of digitisation for the sector of Education mainly in the primary school.

In Algeria the program of ICT training (*Basic ICT, Intermediate and Advanced training*) for teachers has been limited to basic information, with most receiving 30-60 hours of training. Although 100% of secondary teachers and 60% of middle school teachers received the basic ICT training, this has to date very little impact on the quality or method of delivery of education in the classroom. [2]

In 2005, the creation of CERIST[6] project with a Web Review as the only provider (ISP) before market Liberalization, but now there are 71 ISPs and 11 providers of Voice over Internet Protocol (VoIP) services. [9]

Other projects were lunched to enhance the ITC technologies in Algeria, such as:

The education project Academic Research Network (ARN) which connect more than 75 institutions; Virtual Library for Human and Social Sciences provides 30 Algerian university libraries with human and social sciences information; Sidi Abdellah Cyber Park as IT nodes to provide technological support; Wikaya Net as a portal which offers alerts about viruses and warms [9].

---

[4] The Dynamic Frequently Asked Questions.

[5] National integration center for educational innovation and development of information technology and communication in education.

[6] Scientic and Technical Information Research Centre.

On the other hand, the users (98% institute students have a telephone) are increasingly mobile [10] and with the arrival of 3G technology, the problem of Internet connection will be solved, so thinking in other new learning forms is more possible.

And moreover, more than 50% students pass more than 40% of their time apart from the classes or lecture theaters and want to be able to reach their information systems whatever the place where they are [10], for example, the students have a dead time due to university transport which must be taken advantage of.

The mLearning as an educational strategy can solve many of those issues, as it provides a learning process via handheld devices, anywhere and anytime, moreover it facilitates the teaching task and may change the behavior of some teachers towards the use of mobile device at courses (generally they think that it is a lack of respect and sometimes the user will be punished).

Recently, the national market for mobile telephony is growing steadily every year. According to the latest statistics provided by the Regulatory Authority for Post and Telecommunications (ARPT), Algeria intended to the end of November 2011 some 35.2 million subscribers.[11] For raising awareness, some special events have been organized mainly after the arrival of 3G in order to encourage the development of mobile applications such as: the International Exhibition of Information Technology 'Med-IT'[12] which has started from 2004, but this year, in its 11th edition, it organizes a competition that will promote the development of mobile applications in Algeria, in order to encourage the integration of this new trend of technology in both Algerian social and market.

Finally, The mLearning with its characteristics (anyone, anything, anytime and anywhere principles) may be an ideal solution for all this several issues. We have to notice that Algeria may have the potential to create a strong market for mLearning, mainly after the arrival of 3G and soon the 4G, which heats up the Algerian mobile services market. This is an indication of its growth despite not at a very rapid pace, but from now one, the mobile devices technologies, applications, platforms will be also quite strong in Algerian society with its diversity in many life field.

That was a short description of the causes that led us to decide the fact of necessity to collaborate the adaptivity concept with mLearning world, described in the following section.

3 Adaptivity and Personalization in Mobile Learning

Mobile learning can have as a definition: learning delivered or supported solely or mainly by handheld and mobile technologies such as personal digital assistants (PDAs), smartphones or wireless laptop PCs [13]. The mobile learning has several features such as: it enables the convenience since it is accessible virtually at any time and from anywhere with wider access; it offers various technologies for special needs in a special education process, it allows learners to achieve collaborative tasks and build knowledge in different context; it encourages learners to be more courageous, engaged, autonomic and having more fun in their learning process.

Adaptivity and personalization concept in mobile learning systems refers to the process of enabling the system to fit its behavior and functionalities to the educational needs (such as learning goals and interests), the personal characteristics (such as learning styles and different prior knowledge) and the particular circumstances (such as the current location and movements in the environment) of the individual learner or a group of interconnected learners [14]. Next, we list some adaptivity principles proposed by Garlatti and Iksal.

4 Adaptivity and Personalization Principles

The general framework proposed by Garlatti and Iksal is based on these principles:

The user should make a request to the software module through several tasks, such as filling out a form or clicking on a hyperlink for example, which enables the software module to extract retentive data characterizing the user (knowledge, preferences, goals, interests etc...), called user profile. Using it, this module could select a subset of resources and then compose them in order to provide a document or a web service as a response to the user request.

In this context, the semantic web and in the framework of the adaptivity and personalization can help the user and this can be done using steps 1, 2 and 3:

1 *The user request* must be adapted to his needs by proposing appropriate terms to the context of the user task, for example;

2 *Search and filtering relevant resources from the user request and profile*, by reducing the resources space to those, which are appropriate to the request constraints or user needs including his profile of course;

3 *Composition of Resources:* this step should be released after being selecting the relevant resources, which must be organized to be presented to the user.

This organization is the result of the application of rules of composition that can take the form of an *implicit structure* (e.g a calculation result) or an *explicit structure* (e.g a task model) to allow a good document/service organization and orientation that serves as a support for the navigation and understanding. At this level, it is possible to adapt/personalize the organization, content, modes of interaction and presentation of services or documents.

Adaptivity/personalization in the Semantic Web can support the following areas: adaptive hypermedia, user modeling and adaptive virtual documents that we will define in the next part.

5 The Personalized/Adaptive Virtual Document

A virtual document can become adaptive virtual document in case a user can adapt / personalize the real document produced in accordance with its needs.

One of the several existent definitions of AVD is: an adaptive virtual document consists of a set of information: fragments, ontologies and a composition engine which is able to select the relevant information fragments and to organize them according to an overall document structure by adapting various visible aspects of the document delivered to the user. [15]

We should note here that a fragment can be either atomic or abstract, whereas the first type is non-decomposable information unit, which may be a constitute element for the second type, in addition to other abstract fragments and one or several organizations for these fragments.

What we use in the process of the content adaptivity, is the variant fragments that can be defined as follow: two fragments that have the same content description but other different characteristics, either the type (abstract or atomic), the target audience type or the media used (video, audio or text...).

Further, the set of fragments, which are used to be chosen by the user, will constitute the document information space. This one can be organized depending on reading strategies (Author-oriented strategy[7] or reader- oriented strategy[8]).

All of this relies on a general structure that allows the reuse of generic documents and it builds on a declarative approach which is defined next.

6 The Declarative Approach

A declarative approach allows the reuse, by other systems, of abstract fragments and generic documents, designed by the authors. Also, it enables the use of models as operating parameters. Its principle is to facilitate the creation, modification, sharing, reuse and exchange of all or part of generic documents, because it allows the author to modify the documents specification without questioning the other steps of generating the real one.

This approach is based on the three different views of a digital document: semantic, logical and layout, whereas each view has a specific structure organizing it.

The semantic structure of a document conveys the organization of the meaning of the document content. The logical structure reflects the syntactic organization of a document (for example books are organized into chapters and sections). The layout structure describes how the documents have to appear on a device. [15]

Indeed, an ADV is based on two inseparable main steps: the specification and the composition of the actual document, whence the ADV specification is set of necessary information for the system to compose the document and reciprocally, that step obviously depends on the principles used in the composition which is intended for the construction of a real document from a specification and personalization constraints

Consequently, specification and composition are somehow two sides of the same coin, and they cannot be conceived without each other and form an indivisible whole.

For that reason, a composition engine was proposed by Garllati and Iksel with an architecture based on three processes: selection, organization and assemblage process, and it relies on the combination of the precedent three structures (semantic, logical and layout). The composition engine has as main feature, the possibility to be reusable in various contexts, by making just some modifications to change the context.

---

[7] A narrative structure defined by the author in order to present a point of view, or to shed a special light on a set of fragments.

[8] An organization of all fragments defined by the author and calculated according to the reader objectives.

The aim of the semantic composition engine is to compute on the fly an adapted document. The adapted document consists of a set of selected fragments linked by semantic relationships. The logical composition engine browses the adapted document and computes for each node an XML web page with content and navigation tools. In navigation tools, links have properties for managing adaptation. The layout composition engine generates an HTML page from the XML web page. Adaptation processes take place in the three engines [15]. In the next part, the relationship between the two concepts declarativity and ontology is described.

7 Declarativity and Ontology

As we have noticed, the composition architecture is based on ontologies, so the ontology may be considered as a support of the declarativity concept. This architecture is composed on four ontologies: application ontology, domain ontology, user model and metadata schema. Thus, its principle is as follows:

a. via the document model which represents a narrative structure defined by the author and the information space (fragments types and their relationships), the composition engine generates a real document;

b. from the application ontology, we are able to regroup the document model plus the author know-how (fragments types and their relationships),

c. from semantic metadata, some of which take their value in either the domain or the application ontologies, it will have an indexing information fragments placed in the narrative structure, which is represented by a directed graph whose nodes contain specifications to find these pieces of information,

d. from a user model, a personalization of content is made by means of filtering process. The user model allows the system to know the preferences and knowledge of the latter, expressed in terms of the domain model. In what follows, we describe the outline of our architecture.

8 The Architecture

In our framework, we are interested in reusing the results of Garlatti and Iksal proposed approach, by changing the ontologies according to our context (based on: *1)* language diversity mainly Amazigh languge; *2)* learning resource diversity as voice based applications, videos or voice mails for example to solve the illiteracy issue…).

In our context, the adaptive virtual document is destined for reader-oriented strategy, and our reuse is in accordance with our three main domains, which are mLearning, PBL strategy and semantic web. So our four ontologies will be based on their special features. Besides that, the specification of the adaptivity process will be in conformity with these changes. First, we describe the four ontologies then specify the adaptivity process as follow.

8.1 mOntologies

Domain Ontology. This ontology represents all the concepts of the application domain (education, pedagogy, networks, mobile technologies...) and the relationships between these concepts, as it represents knowledge about a particular domain (such as: *1)* the PBL crucial resources could be courses elements; *2)* users represented either by learner, admin or tutor; *3)* presentation, planning, communication or redaction resources such as the keyboard with Amazigh alphabet) and offers opportunities for these knowledge reasoning.

Application Ontology. This ontology has as a goal to achieve provision of a specific design of user's tasks in a global document structure that organizes the access to information and defines Human Interface. Also, the application ontology is based on describing the author know-how, in order to specify a generic document, which is one of the main declarative approach results.

Metadata Schema. In this schema and through the adaptive virtual documents, the fragments indexation has been able, such as the metadata structuring which makes the information search more efficient and accurate, as it allows a more effective use of the resource. The main class in this model is presented by the Resource class which may include information about the resource in terms of use such as its physical location, file format, etc. It can solve the learning resources diversity mainly those destined for the illiterate categories by the capacity of including several resources with its more effective use.

User Model. This model is a formal representation of the user characteristics to facilitate the task of adapting/personalizing a document that requires the availability of the user information due to the diversity of objectives, preferences and knowledge of each one as well as different expectations of the documents that are provided. The appropriate user model in this study is the stereotype, which is a generic model that contains the most representative characteristics of individuals group or class (e.g language preferences: Arabic, French, English, Amazigh; user type: novice, intermediate, expert). The choice of stereotype can be justified by its ability to model the users and allow them to change their information within a certain limit.

As illustrated in fig.1, the four ontologies can be regrouped by several inter-relationships as follow:

8.2 Specifications of the Adaptivity Process

To adapt a document to a reader means to provide him with adequate navigation marks and a set of information and an organization that are pertinent to him, in order to maintain the semantic consistency of the document. [16]

After the design ontology step has been achieved we can pass to the second one which is the specification process in the semantic engine as illustrated in the figure2:

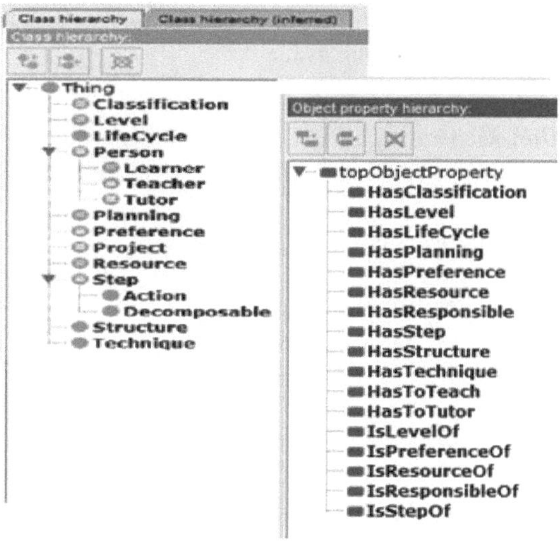

Fig. 1. PBL Ontology Presentation

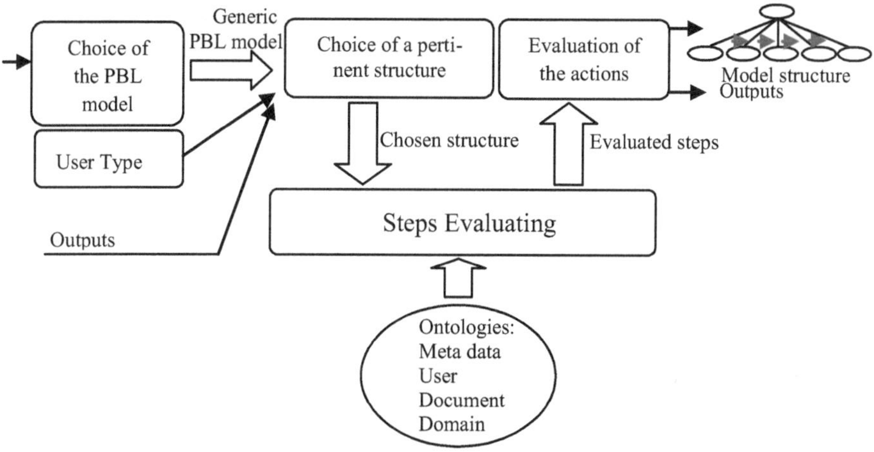

Fig. 2. PBL Adaptive process

This system provides a PBL generic model to the user, according to its type (user model) and some necessary steps inputs, with an adequate structure will be chosen. Via this structure and a control expression, the steps and precisely the actions that must be executed, will be evaluated. At least, via some adaptively rules that allow the system to filter the incoherent steps for the user accompanied by the actions execution results (outputs that may be inputs for other steps), we can have final and pertinent structure for the generic model. Here is an example of the adaptivity rules:

Project Preparation ="((user type = Teacher AND (user type = learner)) AND (Step1 Inputs = Yes)"

⇒ This step could be executed in the case that the user type in teacher and learner *And* the specific inputs at this step are available.

Engine Composition. Generating a PBL model requires the system to search, organize and assemble the relevant steps according to the DVP generating processes:

Semantic Composition. It regroups the two processes: selection and organization, where the result is the instantiation of an adaptive PBL model from a PBL generic model via the semantic composition engine.

Logic Composition. After generating an adaptive PBL model i.e the instantiation of steps, actions, and their execution order. The role of the logic composition process in this stage called the assembly step is to present this PBL model to users (after browsing the adaptive PBL model) in the form of web pages with navigation links for each executable action by two composition mechanisms, the logical one generates, via logical templates, logical pages (expressed in XML, with taking into account user's preferences from user model or the consistent adaptation techniques).

Layout Composition. It is the second assembly mechanism that allows constructing an HTML page from the XML one generated in the logical composition process via presentation templates [17]. It focuses on visual presentation settings of this document according to the following adaptive navigation strategies [18]: annotation; direct guidance; tri; masking strategy.

Adaptivity Strategies. The author may require readers the adaptation strategy which is considered suitable i.e the stereotype associated with one of the strategies appears in user model and this can be a first filter before taking into account the users preferences to be benefited from an adaptation strategy.

For example, we decompose users into three principal categories according to their knowledge: novice, intermediate and expert. We can estimate that the adequate adaptive strategy for the novice users is the direct guidance one as they have to be accompanied in their navigation. Here an example of a direct guidance stereotype:

« (User Type = Novice) AND (Age <20) AND (category = Learner)".

⇒ This grammar represents a stereotype which means that the system will associate typically the direct guidance as an adaptive strategy for users with these characteristics (taking from the user model): a novice learner aged less than 20 years old.

9 Conclusion

In this article, we have presented a proposal of the combination personalization/adaptivity and mLearning with the diversity context in Algeria. In this context, we have chosen project based learning as an active institutional strategy to be personalizaed and implemented with mLearning platform as an important step to enhance the education system especially in our universities and to take advantage of the remarkable progress in the Algerian telecommunication technologies market (3G and 4G technology).

To achieve such a goal, we have reused the results of ICCARS and CANDLE projects represented in a declarative approach that consisted of two crucial elements: ontology and composition engine. The first one was adapted to cover the mOntology concept; it is composed of four main subOntologies as follow: domain ontology; application ontology; user model and metadata model. The second one is based on the composition mechanisms that are the semantic, logic and layout composition, besides the various adaptation strategies such as annotation, direct guidance, tri and masking strategy.

In addition, efforts are also underway to pass to the second step of PBL personalization project, which depends on implementing this adaptive PBL model in our university mLearning platform, soon.

References

1. Milentijevic, I., Ciric, V., Vojinovic, O.: Version Control in Project-Based Learning. Computers & Education 50, 1331–1338 (2008)
2. Hamdy, A.: Survey of ICT in Education in Algeria. Survey of ICT and Education in Africa. 53 Country Reports, vol. 2. info Dev/World Bank, Washington, DC (2007)
3. Le portail de l'actualité des téléphones en Algérie, http://www.mobilealgerie.com/marches/3639-plus-de-35,2-millionsd%E2%80%99abonn%C3%A9s-en-alg%C3%A9rie.html
4. Bekhti, B.: L'Internet en Algerie (2002), http://bbekhti.online.fr/trv_pdf/internet%20en%20algerie.pdf
5. Amazigh, (Berber) the Indigenous Non-Arab Population of North Africa, and Their language, http://phoenicia.org/berber.html
6. Mobimocco 2012 my project draft par Lahmine Sofiane, http://moroccomlearning.blogspot.com/2012/09/mobimooc-2012-my-project-draft.html
7. Thandi, O.: Mobility in Education: Can mobile devices support teaching and learning in South Africa?, http://hsf.org.za/resource-centre/focus/focus-58-september-2010-south-africa-and-the-changing-world/Focus%2058.pdf
8. Patil, P.S.: M-education-Emerging Market Opportunity in India. In: Proceedings of the International Conference on Business Management & Information Systems. K.J. Somaiya College of Arts and Commerce, Vidyavihar, Mumbai (2012)
9. Koubaa, K.: Algeria Arab World Internet Institute, http://www.aw2i.org, http://www.giswatch.org/sites/default/files/Algeria.pdf
10. Kouninef, B., Tlemsani, R., Rebal, S.M., Lotfi, A.: Developing Mobile Learning Approach Platform LMS INTTIC, Oran University. Logy Journal 11(8), 1131–1137 (2012) ISSN 1812-5638
11. http://www.mobilealgerie.com/marches/3639-plus-de-35,2-millionsd%E2%80%99abonn%C3%A9s-en-alg%C3%A9rie.html
12. The International Exhibition of I nformation Technology 'Med-IT', 11th edn., http://www.med-it.com/Alger/
13. John, T.: International Review of Research in Open and Distance Learning. Defining, Discussing, and Evaluating Mobile Learning: The moving finger writes and having writ... 8(2) (2007) ISSN: 1492–3831

14. Wu, S., Chang, A., Chang, M., Liu, T.-C., Heh, J.-S.: Identifying Personalized Context aware Knowledge Structure for Individual User in Ubiquitous Learning Environment. In: Proceedings of the 5th International Conference on Wireless, Mobile and Ubiquitous Technologies in Education (WMUTE 2008), Beijing, China, pp. 95–99 (2008)
15. Garlatti, S., Iksal, S.: Declarative Specifications for Adaptive Hypermedia based on a Semantic Web Approach. In: Brusilovsky, P., Corbett, A., de Rosis, F. (eds.) UM 2003. LNCS (LNAI), vol. 2702, pp. 81–85. Springer, Heidelberg (2003)
16. Iksal, S.: Spécification Déclarative et Composition Sémantique pour des Documents Virtuel Personnalisables. In: IASC-ENST Bretagne, p. 223. ENST Bretagne, Brest (2002)
17. Tanguy, P.: Réalisation d'une plate forme de la composition sémantique de dossiers thématiques de presse (2002)
18. Brusilovsky, P.: Methods and techniques of adaptive hypermedia. User Modeling and User Adapted Interaction 6(2-3), 87–129 (1996)

Using a Commercial Mobile Application as a Mobile Learning Platform

Christodoulos Lytridis and Avgoustos Tsinakos

Department of Computer and Informatics Engineering
Eastern Macedonia and Thrace Institute of Technology
Ag. Loukas, 65404
Kavala, Greece
c_lytridis@gmail.com,
tsinakos@emattech.gr

Abstract. This paper presents an approach in mobile learning in which a commercial mobile application can be used for non-formal heritage education. To demonstrate the feasibility of the proposed approach, an existing travel application for Android devices is adapted in order to demonstrate how educational elements with online interactive material about an archaeological site can be incorporated into the existing framework. The educational material that will be visible to the users of the application will not only be available for reading, but the users will also be able to contribute and enrich the available material by adding more information about the individual artifacts of the site using QR tags.

Keywords: cultural education, Android application, informal learning, user-contributed material.

1 Introduction

The field of mobile learning is one that has attracted much interest during the last few years. The introduction of smart portable devices (smartphones and tablets) in the global market, their increasingly wide availability to the general population, and the continuous improvement in their capabilities in terms of processing power and input/output devices, have created new opportunities in learning (Ally, 2009).

There are several advantages in using mobile devices as means for delivering distance education, such as (Sharples et al., 2009):

- Ease of use: The learner uses a device which he is using in everyday life and is not required to become accustomed with a new tool, thus removing cognitive load and improving the speed at which learners perform tasks.
- Availability of content at anytime: This means that the learner is not limited time-wise to predefined learning sessions, but he is free to choose the time at which he will access the learning materials.
- Portability of the device: Together with the previous point of access to content at anytime, mobile learning allows access to content literally anywhere since the user

M. Kalz et al. (Eds.): mLearn 2014, CCIS 479, pp. 29–37, 2014.

can always carry the smart device with him. Also, it means that situated learning is possible, at the location where training or support is needed.
- Collaborative learning: the communication capabilities such as instant messages, SMS, voice calls, shared calendars, access to forums etc of smart devices through some wireless network, are well suited for cooperative learning activities and sharing content remotely.

These advantages create the potential for developing mobile learning applications which are either web-based or are native applications for the various mobile platforms. Already there have been several studies in recent years which apply mobile learning in real settings, either in the classroom (Markett et al., 2006), (Born et al., 2011) or for fieldwork (Vavoula et al., 2009), (Stead, 2012), (Tsinakos and Ally, 2013).

Especially interesting is the latter case where learning takes place on location and is usually assisted by the location awareness capabilities through global positioning systems (GPS) or network-based location acquisition that is embedded in most of the modern smart mobile devices. In this case, the location awareness feature can allow learning to take place on location (such as a museum, an archaeological site, or any other place of historical interest), and the information that becomes available at that time is directly related to the physical position of the learner (Etxeberria et al., 2012), (Cutri et al., 2008).

Other approaches include applications which use Augmented Reality technology, which takes advantage of the location awareness feature of smart mobile devices. In this context, the experience of the learner in a site of cultural interest is enhanced by moving graphics superimposed on the learner's perception of the real world, that both guide him through the site as well as deliver additional information about the site. For example, in (Gonzàlez et al., 2012) an e-learning project using Augmented Reality for cultural heritage education in various landscape units in Spain is presented.

Motivated by the aforementioned work in the field of mobile learning, this paper proposes a method to enhance an existing commercial mobile application by incorporating mobile learning methods for delivering open educational content. The building blocks of such an approach are presented, as this is part of ongoing development.

2 The Application

The existing Android application is already available through Google Play under the name Thassos-Agenda. It is a travel application which contains tourist information about the island of Thassos, Greece. Its main functions include presentation and categorization of a number of local businesses, customized Google maps with point of interest (POI) overlays, information about the main villages of the island, ferry and bus schedules, and other relevant information such as useful local phone numbers. In addition, there is a section where the user can browse other regional information, including local customs or short descriptions of sites of cultural significance. These are organized in appropriate sub-sections.

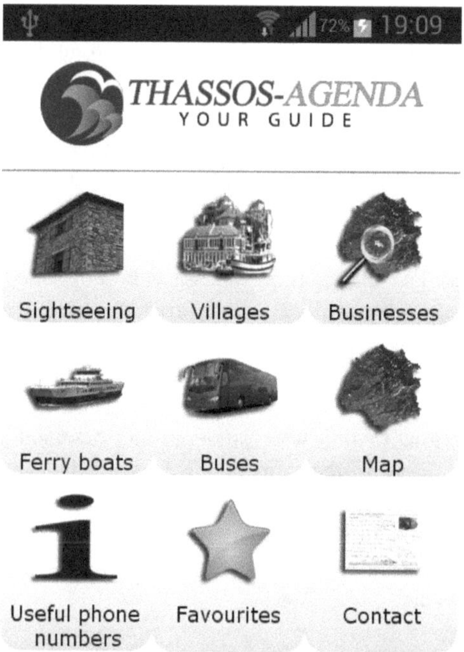

Fig. 1. Screenshot of the original application

However, the content in the application is static and, in this sense, it is limited. Any information material currently present in the application resources is embedded in the application and any additions would require modifying the application and uploading the updated version to Google app-market. Therefore, users would have to download the application again. Obviously, additional material would also affect application size, and this could pose a problem on devices with limited memory capabilities.

Using the section of local points of interest and customs as the basis, another sub-section has been added which deals specifically with the educational material regarding the ancient market of the city of Thassos. The objective is to create a section where the educational material can be accessed and learning will take place. In this sub-section, the users have the opportunity to download and view the educational material regarding the various artifacts in the ancient market, but they are also able to upload their own content. For example, prior of a field trip, teachers will be able to prepare material that will be presented to the students during the visit and the students will be able to access it on location. So the current version of the application mainly focuses on these two processes: delivery of content and adding content.

3 Delivery of Material

Instead of packaging the material within the application, for it to be dynamically changing, it is uploaded to a server where it can be accessed by the application using a wireless or a GPRS network. The content is catalogued based on a serial number which is given to each artifact of interest. All the content will be HTML5 formatted, as such content format is simple to create, even by users with limited technical background. Furthermore, multimedia and scripting such as Javascript can be easily embodied in order to enhance the interactivity of the content.

As mentioned earlier, the information requested and received by the mobile device will be filtered based on tagging of the artifacts. In this stage of development, it was decided that using QR codes was an appropriate method of tagging the various artifacts of the ancient market. The QR codes can be easily created at a number of relevant websites for free. The information that must be contained within each tag is a code which corresponds to an artifact. All the artifact codes are stored on the server. Accordingly, the application is given the capability of reading QR codes using the back camera of the mobile device. QR code detection has been realized using the ZXing library, an open source bar code image processing library, which is incorporated into the application. In this way, students and teachers can easily create and upload material without any further concerns related to the cost or the technology to be used.

When the user navigates into the ancient market information section, the camera is activated and starts detecting QR codes. The following figure illustrates the screen where QR scanning occurs.

Fig. 2. Screenshot of the QR scanning screen

As soon as a QR code of an artifact is read by the mobile device, a request is sent to the server to fetch the content linked to the particular artifact. As more than one source of information can be linked to a single artifact, a list of the relevant webpages appears on screen, together with the corresponding Wiki page, and can provide options for public annotation with other applications such as Foursquare, Pinterest or other social media solutions.

From this list of webpages and options, the user can select the webpage with the content of interest. A WebView is used in order to display the webpage without the user having to leave the application. The WebView is Android's method of properly formatting and displaying the HTML content received from the server, from within an application. It is within this construct that the actual educational content is displayed. In the next section, the format of the content and the process of creating and adding content will be discussed. The following figure illustrates a screenshot of an example web page, loaded immediately after detection of a QR code. The mobile-formatted page contains an embedded YouTube video, demonstrating the ability of the application to display multimedia content.

Fig. 3. An example artifact web page

4 Uploading Material

As discussed in the previous section, the application scans QR codes and displays the educational content that is related to the selected artifact, in the form of a web page. An essential part of the platform is the ability to receive and catalogue user content, at a minimum amount of effort from the part of the contributor. For this reason, a website has been set up, in order to handle user uploads. In order to keep the interface as simple as possible, as student from primary and secondary education level are the targeted users, the file upload user interface of the website contains only the controls to upload a file (Figure 4):

Fig. 4. The file upload interface

The process of uploading material is straightforward and does not require the user to have any special technical skills or training. The user can access the platform either from a desktop computer or a mobile device and simply selects the artifact of interest from a list. This list contains all the artifacts in the site, and is compiled in advance. Each artifact is linked to a QR code, and is tagged on location. Therefore, after the selection of the artifact, the user enters a short description and uploads the corresponding educational content. The limitation at this point is that the content has to be developed in website form, which means that the system only accepts HTML files, with the exception of the wiki pages, which the user can edit in plain text.

The file chosen by the user is uploaded and stored in a main server, and the relevant entry is catalogued in a MySQL database. This entry will be linked to the artifact selected by the user, and will appear in the list of available content materials which is displayed when a user scans the QR code attached to that particular artifact. The titles of the items in that list are created from the content descriptions entered for each webpage during the upload process. Figure 5 illustrates the upload process and how the device accesses the uploaded content.

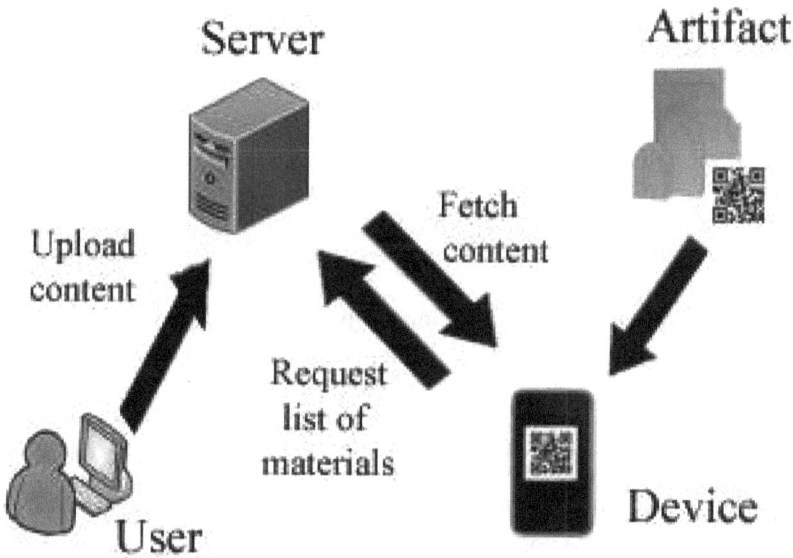

Fig. 5. The content upload process and delivery of material

This process of managing and delivering the content stored on the server, allows multiple sources of information to be available for each artifact. For instance, the local heritage authority can provide the official content regarding the archaeological site and tag the artifact with the corresponding QR. Visitors of the archaeological site will be able to access this content even though they might not be participating in a particular educational activity. On the other hand, it is possible that teachers can deliver their own targeted material for an educational school field trip, by preparing and uploading their customized content to the server beforehand. In this way, the students will access the material that their teachers have prepared by scanning the same QR code with their mobile device and selecting the corresponding webpage based on the description that the teachers have assigned to their material during the upload process. At the same time, students will also be able to access any material previously uploaded for other past activities relating to the particular archaeological site. Using the other open resources offered by the application, students are not limited to their course curriculum or field trip learning objectives, but have the opportunity to expand by further reading from other sources and obtain a broader perspective on the subject.

5 Conclusions

This paper proposes a method of enhancing an existing commercial tourist application to incorporate elements of mobile learning. It is also demonstrated that an Android application containing tourist information can be transformed to serve as a mobile learning tool. The services that are added to the existing platform allow more educational content to be delivered to the users of the application and present the relevant information on location, by simply pointing a mobile device to a QR tag attached to an artifact. In the current development users are enabled to contribute their own content, in order to enrich the information regarding the points of interest in Thassos island, or make the content available in specific events and contexts related to educational activities. The ability of continuously updating the online content by the users, the possibility of easy content personalization of the material by the educators, depending on the target audience, as well as the simple QR tagging mechanism, allow both formal and informal mobile education to happen effectively in a cultural heritage setting at minimum cost.

The next step is to make the application available and test it in real scenarios by involving local schools local schools in the evaluation process. The purpose is initially to test the effectiveness and usefulness of the mobile learning method that is proposed. Using the feedback from the end users (both teachers and students), it will be possible to adapt to the actual educational needs of each school activity, and improve the effectiveness of learning.

It is clear that the educational material which is uploaded and presented in the application is made available not only to students as part of some educational activity, but also to everyone that has installed this application. Therefore, public access to the material immediately points to the open educational sources approach. The Eastern Macedonia and Thrace Institute of Technology is currently involved in a project regarding the creation of Open Courses, delivered via an open source asynchronous e-learning platform. This research is implemented through the Operational Program "Education and Lifelong Learning" and is co-financed by the European Union (European Social Fund) and Greek national funds. The mobile application can be extended to include this type of material, not only for heritage education, but also for a variety of other topics, accessed from the Open Courses repository.

Further research is being carried out so that more mobile learning methods can be implemented and tested on the existing platform. Enhancing the uploading process so that the user can upload video/audio files, Java scripts, cascading stylesheets is one of the upcoming developments. Also, it is the intent of the authors to include Augmented Reality capabilities in the application, so that the benefits of this approach in mobile learning to be also explored.

Acknowledgements. The current research paper is implemented through the Operational Program "Education and Lifelong Learning" and is co-financed by the European Union (European Social Fund) and Greek national funds.

References

1. Ally, M.: Mobile learning: Transforming the delivery of education and training. AU Press, Edmonton (2009)
2. Born, C.J., Nixon, A.L., Tassava, C.: Closing in on vocabulary acquisition: The use of mobile technologies in a foreign language classroom. In: Kitchenham, A. (ed.) Models for Interdisciplinary Mobile Learning: Delivering Information to Students, pp. 195–210. Information Science Reference, Hershey (2011)
3. Cutrí, G., Naccarato, G., Pantano, E.: Mobile cultural heritage: The case study of locri. In: Pan, Z., Zhang, X., El Rhalibi, A., Woo, W., Li, Y. (eds.) Edutainment 2008. LNCS, vol. 5093, pp. 410–420. Springer, Heidelberg (2008)
4. Etxeberria, A.I., Asensio, M., Vicent, N., Cuenca, J.M.: Mobile devices: A tool for tourism and learning at archaeological sites. International Journal of Web Based Communities 8, 57–72 (2012)
5. Gonzàlez, F., Villarejo, L., Miralbell, O., Gomis, J.M.: How to use mobile technology and augmented reality to enhance collaborative learning on cultural and natural heritage? An e-learning experience. In: New Trends on Global Education Conference, North Cyprus (2012)
6. Markett, C., Arnedillo Sấnchez, I., Weber, S., Tangney, B.: Using short message service to encourage interactivity in the classroom. Computers & Education 46, 280–293 (2006)
7. Sharples, M., Amedillo Sanchez, I., Milrad, M., Vavoula, G.: Mobile learning: Small devices, big issues. In: Balacheff, N., Ludvigsen, S., de Jong, T., Barnes, S. (eds.) Technology Enhanced Learning: Principles and Products, pp. 233–249. Springer, Heidelberg (2009)
8. Stead, G.: Towards open formats for mobile learning. In: 11th World Conference on Mobile and Contextual Learning, Helsinki, Finland (2012)
9. Tsinakos, A., Ally, M.: Global Mobile Learning Implementation and Trends. China Central Radio & TV University Press, Beijing (2013) ISBN 978-7-304-06343-6
10. Vavoula, G., Sharples, M., Rudman, P., Meek, J., Lonsdale, P.: Myartspace: Design and evaluation of support for learning with multimedia phones between classrooms and museums. Computers and Education 53, 286–299 (2009)

Creating Library Virtual Learning Spaces Supported by Multiple Mobile Devices

Ieda M. Santos*, Nagla Ali, Gillian Westera, Anthony Hill, and Mubarak Al Jabri

Emirates College for Advanced Education, Abu Dhabi, United Arab Emirates
{isantos,nali,gwestera,ahill,maljabri}@ecae.ac.ae

Abstract. This poster presents an overview of a working in progress funded collaborative action research project being conducted at a higher education institution in the United Arab Emirates. The projects investigates how to create and implement library virtual learning spaces while taking advantage of students' mobile devices to support the activities. The poster also presents preliminary results based on cycle 1 of the action research.

Keywords: virtual learning space, Library, mobile devices, digital students.

1 Introduction

Today's students live in a digital world and technologies such as laptops and smart phones are integrated in their daily lives. Mobile technologies, in particular, are not only becoming ubiquitous in students' lives but are also their first choice to access information, resources and to communicate [1]. Digital students are social, collaborative and multitasking oriented, and generally like hands-on activities [2, 3]. These characteristics overlap with constructivist theory that encourages learning environments that are active, collaborative and student-centred.

Brown [2] stressed that higher education educators must design leaning spaces that maximize the convergence of the digital students, current learning theories and technology. In this respect, higher education institutions are reconfiguring their learning spaces in order to meet the needs and expectations of digital students as well as promote more active student centred learning environments [4]. In particular, the literature shows that many libraries are transforming or reinventing themselves as learning spaces for the digital students [5, 6]. Of the main types of learning spaces at higher education institutions, libraries play a key role in the education of digital students [7]. For example, Bailin's [6] study has provided insights into the needs of students with regard what they expect from 21st library learning spaces. Results showed that collaborative spaces that promote both formal and informal learning equipped with technology were highly valued by participants.

Reconfiguration of learning spaces equally applies to our higher education institution in the United Arab Emirates (UAE) that houses digital students who are arriving

* Corresponding author.

M. Kalz et al. (Eds.): mLearn 2014, CCIS 479, pp. 38–41, 2014.

on campus equipped with personal wireless enabled mobile devices [8]. We have identified a visible need for our library space, in particular, to evolve to avoid disconnection from their 21st century digital students and, quoting Sinclair [9], to "have a place in tomorrow's university… (p. 6). As noted by Brown and Long [10], 21st libraries cannot ignore technology and the fact that students are bringing their mobile devices to university campuses which are broadening the concepts of learning spaces to include anytime anywhere learning. In such virtual learning environments, students might use mobile applications or perhaps will use interactive technologies such as blogs and wikis for collaborative learning. This concept is a growing reality in many library programs. However, according to Oblinger and Oblinger [3] and Temple [11], changes or re-configuration of learning spaces must be accompanied by research to determine whether the planned changes foster one's stated goals.

This poster presentation offers an overview of a working in progress collaborative action research project which aims at investigating how to create and implement library virtual learning spaces while taking advantage of students' mobile devices to support the activities. The poster will also discuss preliminary results based on cycle 1 of the action research. More specifically, analysis of a focus group interview conducted with a group of students will be presented.

The project seeks to answer the following research questions:

1. How can we design and implement library virtual learning spaces that support students' formal and informal learning?
2. What are the perceptions of a selected group of students of the library virtual learning spaces after implementation?

2 Overview of the Project

The project is being conducted at our higher education Institution in the United Arab Emirates and has been approved by our institution's Ethics Committee. The project implementation began in semester 2 of the academic year 2013-2014 and will finish in semester 2 of the academic year 2014-2015. The setting for the study is our library and the virtual space where informal and informal learning can happen anywhere and anytime. A sample of undergraduate students enrolled in different years in the Bachelor of Education program (Years 1-4), a sample of faculty and an external librarian will be invited to participate in the study. A sample of students who participated in the virtual learning spaces during the implementation phase will be invited to take part in the study. Participation in the study is voluntary. All participants will be asked to sign a consent form.

The project adopts an action research approach consisting of two cycles. The action research is most appropriate for this project as it can be conducted collaboratively. The collaborative team include three faculty, our institution librarian and an IT staff member. This multidisciplinary team works on situation, trying to make changes to the library by designing, implementing and evaluating virtual learning spaces.

The literature shows that several action research models are available to guide the research [12, 13]. Although each model explains the research process using different number of steps, most adopt a cyclical and repetitive process of inquiry which begins with identification of a problem, planning, data collection, evaluation [14]. Some models are simpler in design and include common elements of other models [12, 15]. Koshy [16] recommends practitioners to choose an action research model that best suit her/his purpose of inquiry. As some of the co-researchers in this project are new to action research, a simple model proposed by Kemmis and McTaggart [12] is adopted. This model proceeds in a spiral of steps consisting of 1) developing a plan; 2) acting to implement the plan; 3) observing the action by collecting data; and 4) reflecting on findings and develop an action plan to promote changes in future actions[15].

Cycle 1 involves a preliminary phase where we collect qualitative data using focus group interviews conducted with a sample of undergraduate students and faculty. The focus group aims to gather information on participants' perceptions of the design of library virtual learning spaces, formal and informal activities, and technology. In particular, as suggested by Lippincott [7], librarians should consider students' opinions and include them on teams that make decisions about the design and implementation of learning spaces. By considering the perspectives of digital students, it will help us ensure that the virtual learning spaces will meet their needs and expectations. An individual interview will be conducted with an external librarian to gather similar information sought in the focus group interviews. The researchers will use semi-structure interview questions and audio tape.

The information gathered from the interviews in cycle 1 will inform the design of the virtual learning space. Therefore, cycle 2 implements the action plan developed in cycle 1 based on participants' perceptions of the design of library virtual learning spaces. During the implementation, the researchers will write field notes, collect documents and observe the activities. Two focus group interviews with be conducted with a sample of students who participated in the virtual learning space activities. We seek to gain insights into participation, activities, technology and recommendation for improvements.

For both cycles, a thematic analysis technique will be used to analyse the qualitative data. The analysis will be based on procedures suggested by Merriam [17]whereby the researcher codes all the different parts of the data that seem to exemplify similar themes or ideas. The principle researcher will read the documents, make notes on their margins and then create a list of codes in which data will be added later. The next step will consist of looking at the coded data within and across categories for themes. A co-researcher will code the data following same procedure and results will be compared. The coding process will be discussed with other co-researchers to validate interpretations. The interview transcripts will be shared with participants to determine accuracy. The qualitative software NVIVO will be used to assist with data analysis.

The proposed project expects to provide a practical example of how to improve library learning spaces to accommodate students' needs and expectations. It also expects to enhance our library learning spaces to better support our 21st students as well as reflect contemporary learning theories. Outcomes of this project may be significant to other higher education institutions within the UAE that are planning to adapt their library learning spaces to better serve their digital students. Those institutions that have already adapted their library learning spaces can also benefit from this project as we may be developing different virtual activities and deploying other technologies and design.

References

1. Lundin, J., Lymer, G., Holmquist, L.E., Brown, B., Rost, M.: Integrating Students' Mobile Technology in Higher Education. International Journal of Mobile Learning and Organization 4(1), 1–14 (2010)
2. Brown, M.: Learning Spaces (2005), http://www.educause.edu/LearningSpaces/6072
3. Oblinger, D., Oblinger, J.: Is it Age or IT: First Steps Toward Understanding the Net Generation (2005), http://net.educause.edu/ir/library/pdf/pub7101.pdf
4. Frydenberg, M.: Creating a Collaborative Learning Community in the CIS. Presented at IADIS International Conference on e-Learning (2012)
5. Forrest, C.: Academic Libraries as Learning Spaces: Library Effectiveness and the User Experience. Georgia Library Quarterly 46(3) (2009)
6. Bailin, K.: Changes in Academic Library Space: A Case Study at the University of South Wales. Australian Academic & Research Libraries 42(4), 343–359 (2011)
7. Lippincott, J.K.: Net Generation Students and Libraries (2005), http://www.educause.edu/ero/article/net-generation-students-and-libraries
8. Santos, I.M.: Integrating Personal Mobile Devices in Teaching: The Impact on Student Learning and Institutional Support. Learning & Teaching in Higher Education: Gulf Perspectives 10(2), 43–63 (2013)
9. Sinclair, B.: Commons 2.0: Library Spaces Designed for Collaborative Learning. Educause Quarterly 4 (2007)
10. Brown, M., Long, P.: Trends in Learning Space Design (2006), http://www.educause.edu/research-and-publications/books/learning-spaces/chapter-9-trends-learning-space-design
11. Temple, P.: Learning Spaces in Higher Education: An Under-researched Topic. London Review of Education 6(3), 229–241 (2008)
12. Kemmis, S., McTaggart, R.: Participatory Action Research. In: Denzin, N.K., Lincoln, Y.S. (eds.) Handbook of Qualitative Research, 2nd edn., pp. 567–605. Sage, Thousand Oaks (2000)
13. Elliott, J.: Action Research for Educational Change. Open University Press, Milton Keynes (1991)
14. Riding, P., Fowell, S., Levy, P.: An Action Research Approach to Curriculum Development (1995), http://informationr.net/ir/1-1/paper2.html
15. Mertler, C.A.: Action Research: Teachers as Researchers in the Classroom, 2nd edn. Sage, London (2009)
16. Koshy, V.: Action Research for Improving Practice: A Practical Guide. Sage, London (2005)
17. Merriam, S.B.: Case Study Research in Education: A Qualitative Approach. Jossey-Bass, London (1988)

The Tablet Motivating Mathematics Learning in High School

Carlos Fernando Araujo Jr., Eduardo Jesus Dias, and Marcos Andrei Ota

Cruzeiro do Sul University
São Paulo, Brazil
{carlos.araujo,marcos.ota}@cruzeirodosul.edu.br,
eduardo019@gmail.com

Abstract. Using technological devices, particularly the tablet and the collaborative work establishes favorable relationships in mathematics teaching and learning. Being a device with numerous possibilities of interaction, communication, and personal practicalities, the tablet collaborates with people's life in an increasingly globalized world. This research presents the theoretical framework of activity theory as an analytical basis both on uses and interactions of the tablet in a math class in high school. Students' reaction and manifestation in solving proposed problems involving concepts of elementary functions are also examined. Research was conducted in a private high school from São Paulo's east side and entailed three classes from the first year. It was verified that the students organized themselves collaboratively in order to reach the results proposed in the activities. Activity Theory is presented as a structure that reveals the interaction elements between subject and object (students – tablet – application) towards math learning.

Keywords: Activity Theory, tablets in education, mobile learning, math learning.

1 Introduction

In education, tablets appear as a resource of great potential to access didactic material, electronic books, newspapers, magazines, and so on. Combined with Internet access, such devices bring contents to real time. Students making use of it have at their disposal a device for both research and task-solving, although it resembles a conventional notebook. Teachers, in turn, are able to have the same resources offered by a conventional computer without having to take the students to the computer lab, which provides convenience and responsiveness for the pedagogical work.

Tablets are gaining educators' attention regarding their pedagogical use, which means that some educational institutions are testing them as a potentially helpful tool in teaching and learning processes in school context.

One cannot ignore this tool's contributions and applications; hence, researching the use of tablets by high school students in math class must take into account that "the incorporation of all available technology in the world currently is essential to make mathematics a science today" (D'ambrosio, 1996, p. 17).

Considering it, it was noted that this device could be used in order to answer the following question: how a tablet could improve the process of teaching and learning of mathematics?

M. Kalz et al. (Eds.): mLearn 2014, CCIS 479, pp. 42–51, 2014.

In this sense, a starting point will be an activity created and applied in high school classes, more specifically in math lessons of three first year classes.

Thus, by using a tablet, the goal is to design a dynamic and meaningful learning process, so as to allow conditions of graphical and geometric analysis in solving problems and situations in order to provide the students means for them to recognize specific factors of mathematical concepts in the study of functions, trials, investigations, and collaborative research.

This paper is organized as followed: section one presents the context of activity theory in tablet-using context; section two discusses the tablet as a mediating tool for learning; section three examines research organization, methods, and research tool, as well as students target; section four presents and considers the results; section five offers some conclusions.

2 The Activity Theory and the Tablet

Vygotsky (1978), Leontiev (1978), and Luria (1978) started what is currently known as Activity Theory (AT), which conveys a new conception for psychology, known as cultural-historical, created by the trio. AT provides an overview of the phenomenon under study, and analyzes the activity itself, including different aspects of both social and historical activity and relating them to different systems of other activities. According to Kuutti (1996), the AT is a conceptual structure based on studies of different human shapes and practices as mental development processes combined to individual and social spheres.

The first condition for all activity is a need. However, considered in itself, this need cannot determine the actual orientation of an activity, for it is only in the activity object that it finds its determination: it must, so to speak, find it. Once the need finds its determination in the object (it is "objectified" in it), the said object becomes the reason of the activity, stimulating it. (Leontiev, 1978, p. 107-108).

Although it does not constitute neither a methodology (Jonassen and Rohrer-Murphy, 1999) nor a theory, it does comprise a conceptual system that supports different studies related to various forms of human practices, since AT finds its basis in Soviet theorists bound to Marx and Engels' Dialectical Materialism Theory (1984).

In this scenario, the pioneer researches of Vygotsky (1978) and his collaborators Leontiev (1978) and Luria (1978), which were responsible for the continuity of his work on Activity Theory, gains strength.

AT has as basis the interaction subject - device (tablet) - object, that is, it considers that individuals does not simply react to stimuli of the environment they live in, for their actions are mediated by artifacts and object-oriented (Vygotsky, 1991).

The mediation of instruments, tools or devices, that could naturally be a mobile device, such as a tablet, for example, could play the role of a way by which internal psychological operations are directed in order to solve a problem. This vygotskianian theory (1978), named AT's first generation, emphasizes the more individualistic relationship of the Activity Theory (subject – mediator – object).

But Vygotsky's theory (1978) had some limitations as, e.g., the fact it do not associate, in the activities, actions mediated by individuals and collective and social operations, associated to the context of individual action. This first stage of AT, which objective is mediating the process by instruments or devices, is still considered one of the AT pillars.

AT structure has been modified because Vygotsky's model (1978) lacks elements of social correlation such as rules, community and labor division - elements which do not only participate of the subject-object mediation, but also propose interdependency between the established actions.

The emergence of a new structure for the AT was due to the studies of Leontiev (1978), who is considered an anchor of the 2nd generation of the AT, though he has not come to develop such a structure,.

Engeströn (1991), then, based on Leontiev's (1978) ideas, proposed an increase in the vygotskianian model: the activity's social mediators (rules, community and labor division) are actively involved in the mediation between subject and object.

In this system of activity, the subject correspond to the individual, or the communities subgroup; the object is the reason, or the raw material, that accomplishes the activity; the instruments refer to the mediator devices (internal and external) that cooperate and support a transformation in order to achieve the activities' results.

Composed by groups and subgroups that share the same object, the community has the function of situating the subject in the activity's result. The division of labor is structured by a horizontal and vertical division, the first referring to the task between individuals of the community, the second being based on the relationship between power and prestige inside the community.

Lastly, there are the rules that establish the norms to be followed by the individuals, directing the actions and interactions inside the activity's structure.

2.1 The Tablet as a Mediating Tool

A tablet is a clipboard shaped tool that promotes easily Internet access and other resources, like annotations, spreadsheets, games, books, in addition to various applications that can be used both for leisure and work.

One of this tool's possible applications may be, as mentioned before, education. In the United States, e.g., many teaching institutions, considering different areas and educational proposals (Pearson, 2011), are currently researching pedagogical ways and structures for using such a device since the iPad launching (2010).

This device embodies important concepts for education by promoting portability[1], mobility[2], easiness of use (touch screen technology)[3], and technological resources (recording, filming, audio, telephony, photography, etc.).

---

[1] Attribute of a component or computer program that makes it capable of being used in different computer systems.

[2] Attribute of being mobile.

[3] The technology of detecting the presence and the localization of a finger or object touch, inside the screen area, dispensing any input peripheral, like keyboards and mouse. The exhibition area for interaction is a pressure sensitive screen, which various actions can be executed, being possible a direct interaction with exhibited content such as images, numbers, and words. (www.infoescola.com).

Simplicity and practicality enables it to be used both by individuals and by group. An illustration of it is its use in a traditional classroom, promoting interactive activities with wirelessly access to the internet.

Tablets have the possibility of becoming important tools for educators, enabling the exploration of various content such as learning objects, virtual libraries, sites, blogs, social networks, educational applications, among others.

At the same time, it must be mentioned that the increasing use of this technological tool brings gains to our contemporary society, mainly regarding sustainability, for it reduces the consumption of printed matter and, consequently, impacts on the environment.

With the physical reduction of the devices when compared to others, like desktop computers and notebooks, it has emerged today a new conception of their utilization: the m-learning, which might be described as the uses of technology and mobile communication, synchronous and asynchronous, for educational purposes (Freysen, 2004).

Mobile learning or m-learning is a didactic-pedagogical term that labels a new educational "paradigm" based on mobile technology utilization. Largely, it is possible to name m-learning any learning method performed in small sized devices, autonomous to external sources of energy and smaller enough for people to taking it anywhere, anytime (Moura, 2010, p. 39).

There is a great potential in mobile devices (tablets) for education in general since it gives individuals the opportunity of accessing varied content, independent of the environment and the local of use. It works as true portable computers and uses in many cases very user-friendly operational systems, such as Android, Window 8 mobile, and iOS, among others.

3 Research Methodology

This research was based on qualitative and quantitative methods, starting from surveys with closed questions and proposal of classroom evaluations, both in writing and talking (seminar presentations). The closed questions investigation aimed at finding topics that aid pedagogical aspects and the interaction of mathematical objects using tablets, as well as the degree of motivation this tool might have had in math learning.

The proposed activity was applied during two classes of fifty minute each, when, using a tablet and a graphical application (MePlot-Free), students have had to describe the phases for analyzing the graphic behavior of previously studied functions in the Cartesian plan. The application, MePlot, runs on Android 2.1 or newer and has two versions, a free one and a paid one. Due to its educational use in high school, in this research it was used the free version of the application.

This activity aimed at relating some concepts of geometry involving types of graphs in the Cartesian plane exposed on the tablet screen. At the end, all students had to present their analysis of activities and functions observed during class.

In another moment of the activity, after the teacher's analysis presentation, students have had to give a seminar. "The objective and reason of a collective activity are something like a mosaic in constant evolution, a pattern that will never be completely finished" (Engeströn, 1999, p. 20).

The study indicates that, according Leontiev's theory (1978), the object was considered the center of the activity. In this case, the object is engaged to the function analysis observed on the screen device, as well as on the seminar presentation that aimed at discussing the analysis made with the application (MePlot-Free) in connection with collective actions.

Students was the starting point for the activity analysis, that is, as from them it was established contact in order to verify not only whether occurs progression in learning, but also the reaction to this new technology-mediated knowledge that might bring a satisfactory dynamic in teaching and learning processes.

The community or group of individuals sharing the same object was composed by 107 students from the first year of high school and by the teacher (researcher). Labor division refers to horizontal task division among community individuals and vertical based in power relations. In the said activity, it was observed that students divided their presentation of the content studied previously in a sequential approach (point, line, plane, parallel lines, concurrent lines, increasing and decreasing, first degree functions, second degree functions, exponential and logarithm, etc.)

The group established specific points for each of its members, i.e., the first one explained point, line and plane; the second one position of two lines; the third one made simulations with other line examples; the fourth one ended with observations of the applicative manipulation, adding some observed functions. Concerning labor division, it may be verified a hierarchy among them.

Such division was established by the students involved in the activity, which completion was based on established rules that directed the actions and on interactions in an activity system.

Below, a figure shows how this activity was structurally developed according to the referential proposed in Activity Theory (Leontiev, 1978).

The proposed activity was linked to the theoretical content exposed during classes (book) and had as objective making students to report it and to write it in a paper sheet shared by the group. Resolutions and argumentations of students facing the proposed exercises in teacher's activity were justified by observations and by the use of the graphical application (MePlot-Free) installed on the tablet.

Students were oriented to get the application before class. In some devices, the application was not installed; however, using the high school Wi-Fi, the students downloaded it without further problems.

The high school technical team ("Campus Virtual") accompanied the class activity, providing technical support for teacher and students when required. In the first stage, the activity was executed in approximately 70 minutes. In the final 30 minutes, the teacher demanded the students to report positive and negative aspects that might had occurred during the activity.

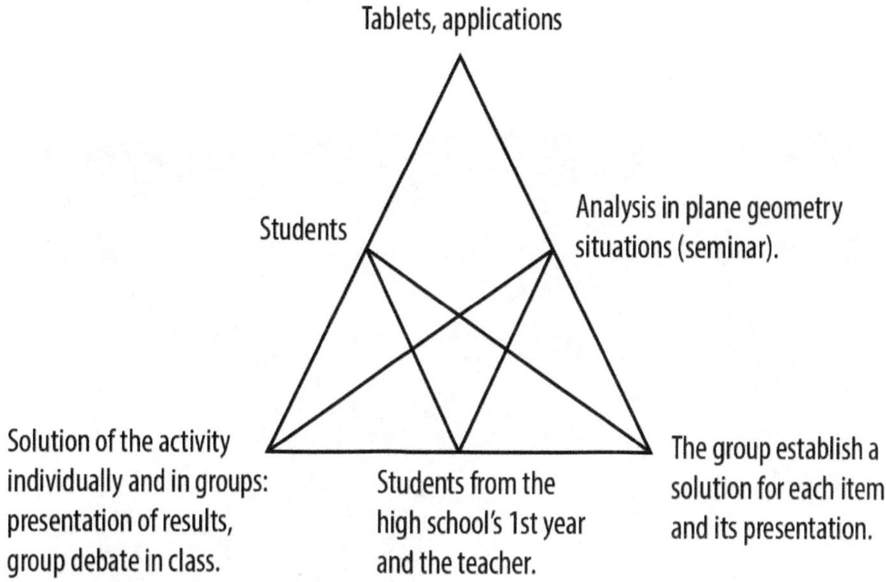

Source: *authors.*

Fig. 1. Schematic *application of class activity*

4 Results and Discussion

Students' answers indicated that the tablet and the applications can provide important interactive analytical meanings of studied content, since 93% of students agreed that the tablet and the application were important in the development of the proposed activity and 98 % reported that the activity became more interactive and dynamic. Regarding the acceptance of the artifact in class, 82% of students said it was great; 8.3 % said it was very good; 3.2% said it was good and 6.1 % said it was regular. The indication of most students is that the tablet provides a new dynamic in the classroom, which contributes to the learning process. According to Dias and Araujo Jr. (2013, p.11): "students understand the simplicity and convenience of mobile technology (tablet) and their potential for interaction and relationship with colleagues". Above, it is shown an excerpt speech of a student characterizing this aspect:

> [...] Using tablets during math class was much more interesting and interactive. Tablets can help us understand better the lesson, the theory and even doing practical exercises. At first, we had some difficulties as to download applications and connect to the Internet, but I think that is a bit lacking usual. Soon we will able to use these devices without any problem (Student A).

Figure 2 shows the graph generated by the application in the activity in the classroom:

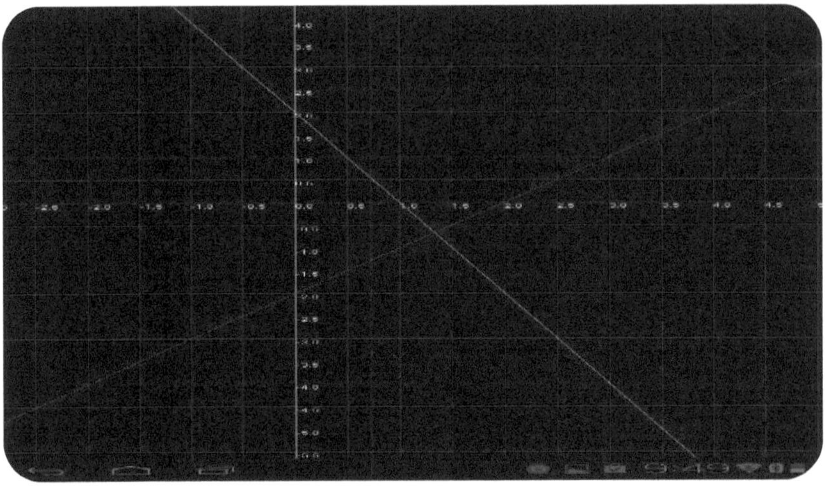

Source: *authors.*

Fig. 2. Meplot-Free | Lesson with the analysis of the function of 1st degree

The image shows a graphical view of the application in which students had to analyze the equations and functions they typed. Additionally, they had to analyze the behavior chart generated by two functions that subsequently they should classify as concurrent or parallel, justifying their answer. Figure 3 shows the image of a student developing the activity in the classroom:

Fig. 3. Student using Meplot-free

Table 1, in which it may be observed before and after the use of tablets in classroom, shows a progression in the indicators of satisfactory grades in math classes for the 2nd and the 3rd quarter. In the school the research was developed, unsatisfactory grades range between 0 to 5.5 and satisfactory grades range between 6-10.

Table 1. Unsatisfactory grades during 2nd and 3rd bimester

Classes	2nd bimester	3rd bimester
1st A	38%	5,8%
1st B	13,51%	2,7%
1st C	27,77%	2,8%

Source: *The research*

Taking this into account, it is possible to consider that the classes involved in the activity responded well to the development of it and were able to significantly improve satisfactory grades immediately. The data is believed to show how a combination of tablet and graphic application with didactic material were able to produce advances in improving the learning of basic concepts in the study of functions planned during the quarters. Thus, according to Prensky (2013) "technology is not something we need , in addition to mental activity ; technology is now part of mental activity . And we need to use it wisely" (Prensky , 2013, p.1).

Consequently, the data indicate the importance of using a device such as a tablet in the development of the proposed activity in math class for it enables an interaction with the content. In the presented observations, which were based on Engeströn's theory of the second generation activity (1999), it may be seen that this tool-mediated relationship (tablet) eased students to solve and analyze exercises. Another pertinent aspect is the interaction of students with the tablet, which provided an opportunity not only of a written speed reading, but also of a manipulation of mathematical concepts.

5 Final Considerations

It is possible to affirm that, associated to the scenario created by a technological tool used on pedagogical conjunction, TICs create, in high school, a technological, argumentative and participatory education culture in students in math class.

The introduction of technology (tablets) in school classes bestows a multifactorial question that embodies the convergence and the integration of aspects related to technology, pedagogy and knowledge of a specific domain. In order to conceptually integrate the presented aspects inside a comprehensive and meaningful organization for students and teachers alike, it was used in this research the Activity Theory (Leontiev, 1978).

Survey results display that the meplot-free application provided an opportunity and a meaningful graphical analysis to the discussion and resolution of exercises of functions by mediating objects and mathematical signs observed in the tablet screen, thus helping students to specifically complete the suggested activities. It also was detected that both the tablet and the application have developed student's mathematical logical decisions with regard to graphical interpretation based on the study of functions.

It is notable that AT provided a more accurate analysis of the work accomplished by students. Labor division and rules established by students in seminar presentations, as well as the resolution of tasks proposed in this research helped to significantly improve the results of assessments in school context regarding mathematics. It is considered that the mediating tool (tablet) has provided not only a dynamism in learning mathematics, but also eased the connection of primary pedagogical actions during the process. Hence, the data demonstrate the possibilities provided by AT in a theoretical analysis of the pedagogical work accomplished by a teacher. Another factor concerning the activities generated with the aid of devices such as tablets is that students identified the relationship between variables and values that appeared in some questions concerning functions, providing then opportunity and condition for specific understanding to occur. We consider that, via interaction and use of mobile technology (tablet), students may have access to the necessary tools to achieve a more significant learning.

References

1. D'ambrosio, U.: Educação Matemática: da Teoria à Prática. Papirus, Campinas (1996)
2. Kuutti, K.: A framework for HCI research. In: Nardi, B.A. (Org.) Context and Consciousness: Activity Theory and Human-computer Interaction. MIT, Cambridge (1996)
3. Leontiev, A.N.: Atividade, consciência e personalidade. Primeira Edição (1978)
4. Jonassen, D.H., Rohrer-Murphy, L.: Activity theory as a framework for designing constructivist learning environments. Educational Technology Research and Development 47(1), 61–79 (1999)
5. Marx, K., Engels, F.: A ideologia alemã. Moraes, São Paulo (1984)
6. Pearson Foundation (2011), Disponível em: `http://www.pearsonfoundation.org/great-learning/research-reports-and-surveys/pearson-foundation-survey-on-students-and-tablets.html` (acesso em: June 18, 2012)
7. Engeströn, Y.: Activity theory and individual and social transformation. Multidisciplinary Newsletter for Activity Theory 7(8), 14–15 (1991)
8. Engeström, Y.: Learning by Expanding: Na Activity – Theoretical Approach to Developmental Research. Orienta Konsultit, Helsinki (1987)
9. Engeström, Y.: Activity theory and individual transformation. Cambridge University, Cambridge (1999)
10. Freysen, J.: M-Learning: An educational perspective. In: Atwell, J.E., Savill-Smith, C. (Org.) Moblie Learing Anytime Everywhere, Mlearn 2004, Londres, UK (2004)
11. Moura, A.M.C.: Apropriação do Telemóvel como Ferramenta de Mediação em Mobile Learning: Estudos de Caso em Contexto Educativo em Contexto Educativo. 601 f. Tese (Tese de Doutoramento em Ciências da Educação, na Especialidade de Tecnologia Educativa) – Universidade do Minho, Instituto de Educação, Lisboa, Portugal (2010)

12. Dias, E., Araujo Jr., C.F.: Mobile learning no ensino de Matemática: um framework conceitual para uso dos tablets na Educação Básica. Encontro de Produção Discente PUCSP/Cruzeiro do Sul 1(1) (2013)
13. Prensky, M.: Our Brains Extender. Tecnologia-Rich Aprendizagem 70(6), 22–27 (2013), Disponível em: `http://www.ascd.org/publications/educationalleadership/mar13/vol70/num06/abstract.aspx#Our_Brains_Extended` (accesso em January 04, 2014)
14. Lévy, P.: As Tecnologias da Inteligência. Editora 34. Rio de Janeiro (1993)
15. Vygotsky, L.S.: Mente na sociedade: o desenvolvimento dos processos psicológicos superiores. Imprensa da Universidade de Harvard (1978)
16. Vygotsky, L.: Pensamento e linguagem, 3rd edn. M. Fontes, São Paulo (1991)

Taking eBook Readers to Prisons: A Tale of Two Projects

Helen Farley, Angela Murphy, Tas Bedford, and Susan Hopkins

University of Southern Queensland, Toowoomba, Australia
helen.farley@usq.edu.au

Abstract. This paper reports on a project that was first introduced to World Conference on Mobile and Contextual Learning attendees in 2012, the PLEIADES project (Portable Learning Environments for Incarcerated Adult Distance Education Students) and discusses how this project evolved into two separate projects; one each from the two technologies originally trialled. PLEIADES introduced a version of an internet-independent version of the Learning Management System (LMS) called Stand Alone Moodle (SAM) and eBook readers to incarcerated students in a correctional centre in Southern Queensland. The Triple 'E' Project (Empowerment, E-Learning and E-Readers) using eBook readers similar to those trialled in the PLEIADES project, were rolled out to a further four correctional centres. This paper explores the issues and challenges involved with deploying eBook readers to incarcerated students through the PLEIADES and Triple 'E' projects.

Keywords: digital inclusion, distance learning, higher education, prisons, eBook readers, eReaders, mobile learning, design-based research.

1 Introduction

The use of mobile technologies has become increasingly essential in the delivery of distance education courses, largely in response to an emerging demand for flexibility in learning [1]. Unfortunately, this increasing reliance on digital technologies is based on the assumption that students have reliable access to the internet. For many students this is not the case, for example forty-four per cent of the students who participated in the University of Southern Queensland's (USQ's) Tertiary Preparation Program (TPP) in Semester 2 2012 claimed they did not have access to the internet (USQ, unpublished data). Consequently, this increasing reliance on mobile devices for learning increases the risk of further excluding disadvantaged students without reliable access to the internet from higher education [2].

Correctional centres are highly reliant on the provision of distance education for offenders who wish to undertake higher education [3]. Higher education institutions that still provide course materials for students without internet access, including incarcerated students, often employ exceptions handling processes, using large volumes of printed copies of the course materials and learning support resources [4]. This is costly for universities to assemble, print and post, is in no way interactive, and cannot incorporate all of the learning support resources of the course. With the emergence of

M. Kalz et al. (Eds.): mLearn 2014, CCIS 479, pp. 52–60, 2014.

e-learning and mobile learning, distance education institutions are becoming increasingly reluctant to print course materials.

Researchers at the University of Southern Queensland (USQ) began to investigate the possibilities of providing electronic access to course materials using eBook readers. These eBook readers would contain all of the readings that a student would need to complete a course, allowing them to highlight text and take notes. This approach would preclude the necessity of printing course materials though the eBook readers would still need to be delivered to the appropriate correctional centre. The researchers began to plan a trial project in conjunction with Queensland Corrective Services and Serco Asia Pacific (private prison providers).

The PLEIADES (Portable Learning Environments for Incarcerated Distance Education Students) project piloted the use of secure e-learning and mobile learning technologies that were independent of the internet, for learning within a correctional centre. Incarcerated offenders are prohibited from accessing the internet and are therefore excluded from accessing course materials electronically. In order to address this exclusion, course materials were loaded onto eBook readers that were incapable of accessing the internet. Additionally, an internet-independent version of the open source LMS, Moodle, was developed and piloted over a seventeen-week-semester period in a Queensland correctional centre. *TPP 7120 Studying to Succeed*, a foundation course in USQ's Tertiary Preparation Program, was modified to provide incarcerated students with access to course materials, learning experiences and assessment activities, without needing access to the internet. At the end of the trial period, additional funding was sourced from the Australian Government's Higher Education Partnerships and Participation Program (HEPPP) to fund the Triple 'E' Project (Empowerment, E-Learning and E-Readers) to deploy eBook readers in five correctional centres. This paper reports on the issues, challenges and lessons learned from deploying eBook readers in correctional centres through the PLEIADES and Triple 'E' projects.

1.1 USQ's Tertiary Preparation Program

USQ delivers programs both face-to-face and via distance education, and has a high proportion of students from low socio-economic status (SES) backgrounds. USQ's Open Access College offers a Tertiary Preparation Program (TPP) which once successfully completed, guarantees entry to a USQ undergraduate degree. The only formal entry requirements to study the TPP program is that students need to be at least 18 years of age. For students who meet the citizenship and residency requirements, the course is fee-free and Australian Government assistance is usually available to cover living costs for eligible students. *TPP7120 Studying to Succeed* is a compulsory course in the TPP and students are required to obtain a passing grade in order to gain direct entry to an undergraduate program. These students usually require significant academic and social support to complete their courses successfully [5]. A significant number of incarcerated students from Queensland enroll in the TPP but with many having limited numeracy and literacy skills, this cohort often struggles. Though the TPP is designed to ready students for higher education, it is unlikely this aim can be achieved when instruction is primarily delivered online and incarcerated students cannot access the internet. The challenge is to develop those digital literacy skills in the incarcerated cohort, while they do not have access to the internet. The PLEIADES project, short for Portable Learning Environments for Incarcerated Adult Distance Education Students, was formulated to try and address this problem while reducing also reducing the costs associated with the supply of hard-copy materials.

2 The PLEIADES Project and Beyond

The PLEIADES project, short for Portable Learning Environments for Incarcerated Adult Distance Education Students, was formulated to try and address this problem while reducing the costs associated with the supply of hard-copy materials. Project resulted from discussions between staff at the Southern Queensland Correctional Centre (SQCC), Queensland Corrective Services (QCS), the Australian Digital Futures Institute (ADFI) and the Open Access College (OAC), the latter two both at USQ. A portable version of USQ's LMS Moodle, called Stand Alone Moodle (SAM), was deployed in an attempt to replicate USQ's online learning environment for incarcerated students enrolled in the TPP. SAM had no capacity for internet connection in any form and was accessed by incarcerated students in either of SQCC's two computer labs which were serviced by a server. Because students typically only spend two hours a week in the computer lab, a second arm of the project facilitated learning by issuing these students with eBook readers loaded with course materials and relevant resources. The project was deployed in Semesters 2 and 3 2012. The initial aim was for seventeen students to participate in the project, but only seven proceeded as participants in the trial. Though SAM and eBook readers were successfully deployed, a number of factors made it extremely challenging to collect meaningful data. These challenges are summarised below:

- The correctional centre relocated from Borallon to another location some 60 kilometres away which resulted in the withdrawal of several key project team members. These people understandably elected not to move their families and could not continue working at the new site;
- High turnover in relevant correctional centre staff to try and cover a number of positions left vacant;
- Insufficient handover procedures in place in the correctional centre to ensure continued operation of the trial;
- Resulting in poor training of the students in the use of the new technologies; and
- Prejudice manifesting in many forms included delayed approvals and purchases.

The project team managed to work its way through these difficulties. There was an enormous amount of good will and belief in the project on the part of all team members and the students in the trial that ensured that the project proceeded. Though it was difficult to collect much meaningful data, the project did demonstrate that it was feasible that eBook readers and SAM could provide a reasonable simulation of an online study environment to help incarcerated students develop the necessary digital literacy skills to successfully complete a higher education program. On this basis, the project team successfully applied for an Australian Government Office for Learning and Teaching grant to further develop SAM. Though a large amount of funding was secured, it was still insufficient to expand the roll out of the eBook readers to other correctional centres. The Open Access College, in consultation with the Australian Digital Futures Institute and Queensland Corrective Services, secured funding from another Australian Government funding scheme aimed at broadening participation in higher education (Higher Education Partnerships and Participation Program or HEPPP) to purchase some 200 eBook readers and chargers.

3 Emergence of the Triple 'E' Project

Though there were many difficulties encountered with deploying eBook readers during the PLEIADES Project, the project team felt strongly that this pilot demonstrated the feasibility of these technologies for use in this context. The Triple 'E' Project (Empowerment, E-Learning and E-Readers) deployed eBook readers at four additional correctional centres beyond the site of the original PLEIADES at SQCC. The correctional centres were chosen on the basis of the number of students enrolled in *TPP7120 Studying to Succeed* and strong relationships between TPP staff with the correctional centre education staff. The additional correctional centres were Brisbane Women's, Wolston, Woodford and Maryborough. These four centres are directly administered by QCS while SQCC is administered by Serco Asia-Pacific for QCS. A training and information session was conducted with the education officers from the relevant correctional centres. These officers were unanimously enthusiastic about the project and the prospect of having eBook readers for the students enrolled in *TPP7120 Studying to Succeed* in their own correctional centres. Though initially, some of the education officers expressed doubts that they were 'tech savvy' enough to use the eBook readers themselves, with some basic instruction they were soon using the devices confidently.

In addition, training sessions in the use of the eBook readers was delivered by an Open Access College tutor at the correctional centres themselves. Inmate tutors were also trained and could act as a 'go to' person should the incarcerated students experience difficulties with the eBook readers. These tutors were given a small amount of money by the correctional centres to fulfil this role. As one peer tutor explained: 'If it comes from prisoners to prisoners they are more likely to pay attention to it.'

The eBook readers needed to conform to QCS security requirements. These include having a non-removable battery, inability to access the internet and no SD card slot. In the PLEIADES trial, Sony PRS-300 eReaders were used which fulfilled these requirements. For the Triple 'E' project Book Pures were used as the Sonys were no longer available. Security clearance was obtained by incapacitating the SD card slot of the BeBook Pure by filling it with 'builder's bog', a timber repair filler.

3.1 The Education Officers

The education officers in the correctional centres generally oversee the provision of education to the incarcerated students. As incarcerated students don't have access to the internet, it falls to the education officers to download lecture materials, assignment coversheets and conduct any internet research that students might need to complete assignments. These resources are downloaded and printed by the education officers. This forms a significant portion of an education officer's workload.

3.2 The Course Materials

The same course materials, but updated with appropriate dates, were used for the Triple 'E' Project as were used in the PLEIADES project. In the PLEIADES project, permission from academic publishers to load course readings onto the eReader was obtained by USQ's Learning and Resources Development Services department and took a period of four months. Course materials written by lecturers from the OAC were converted to ePub format. This format was chosen as it enables text to reflow on a screen. Text within PDFs is not able to reflow and therefore limits text size without the need for scrolling. The eBook readers were preloaded with all the course readings for the *TPP7120 Studying to Succeed* in ePub format prior to sending them to the correctional centres. The eBook readers were handed back to the education officers by the students about once per fortnight for charging and checking to ensure that the devices are not tampered with or damaged.

4 Evaluation

In total, there were 47 students using the eBook readers across five correctional centres (5 at Brisbane Women's, 16 at Wolston, 8 at the Southern Queensland, 8 at Woodford, and 10 at Maryborough).

The projects were conceptualised using a design-based research methodology with iterative cycles of planning, implementation and revision in response to feedback from participants and key-stakeholders. Design-based research is a blend of empirical research with the theory-based design of learning environments [6]. The method centres on the systematic investigation of innovations designed to improve educational practice through an iterative process of design, development, implementation and analysis in real-world settings [7]. A major strength of design-based research lies in its adaptability to adjust the intervention based on on-going findings from participants.

A range of data sources inform the development of the projects as well as inform the on-going evaluation of the project progress and areas of refinement required. Data sources include notes from project team members taken during meetings with stakeholders and during project planning activities, email communications between team members and stakeholders, focus groups conducted with students and education officers at correctional centres, and observations made by project team members during and after the launch of the PLEIADES pilot and during the subsequent deployment of the Triple 'E' Project.

5 Issues and Challenges

The PLEIADES project ran over two semesters in 2012, and the Triple 'E' Project ran in semester 1 2013 (from early March). There were some common problems between the two projects with the eBook readers.

5.1 Technical

In the Triple 'E' project, once the BeBook Pures were distributed to students, a number of issues were encountered. Some of the eBook readers had issues with frozen screens and had to be swapped for others. It seemed to be that this tendency was exacerbated when students did not wait for one action to complete before pressing buttons to elicit another. We believe the most likely cause is the large files sizes of some of the course materials and the limited processing power of the eBook readers. A related issue is that the eBooks can be slow to open. This is almost certainly related to file size. These problems were not previously encountered with the Sony eBook readers used in the PLEIADES project.

Another significant issue is that the standard font size of the text is sometimes too small for easy reading. The students have been shown how to increase the font size but sometimes the text does not reflow correctly making the text too difficult to read. This was also a problem encountered in the earlier PLEIADES project.

5.2 Issues at the Correctional Centres

Despite the security reports and memorandum from QCS head office, each centre General Manager had their own questions and concerns in regards to the eBook readers which unexpectedly further delayed the booking and delivering of training. Coupled with the fact that the first few assessments in *TPP7120 Studying to Succeed* are due in the first week meant that the tutor made the decision to release hard copies of study materials to ensure no incarcerated students were disadvantaged by this delay. This was an issue because often the correctional centres were quite distant from the Toowoomba campus and return visits by the OAC tutor could not be scheduled in a timely manner. Training sessions were sometimes interrupted by random drug searches with sniffer dogs. Unscheduled lockdowns impact on training sessions. A 'lock down' meant students could not come to class unless they were already in the building. Though these lockdowns meant that students missed out in training in some cases, in other situations, the eBook readers enabled students to study during these lockdown periods.

5.3 Student Concerns

During the initial trial in 2012, some students were very positive about the eBook readers. Even so, many students expressed a preference for hard copy materials. One student in the trial refused to use the eBook reader and handed it back to the education officer. A number of students have relayed that they enjoyed using the eBook readers for recreational reading, but for study they found it too difficult.

Some students have asked to use the eBook readers and have the hard copy of the materials as well. They are quite determined to find a way to use the eBook readers but find it too difficult with the eBook readers alone. Education officers concur, they have noted that students are very happy to have access to an electronic device and feel honoured to be part of the trial, yet they are also frustrated that they cannot view multiple books at once. Some prisoners are frustrated because they can only have one book open at once when they really need to be flicking backwards and forwards and between books. Another expressed frustration that windows could not be minimised to allow the quick switching between materials as on a computer. Because the BeBook Pure is true to its name, it doesn't have any capacity for notes to be taken or text highlighted. The Sony eBook readers used in the PLEIADES pilot trial at SQCC did have the capacity for notes and highlighting text but many prisoners found them too complicated.

5.4 Digital Literacy of the Students

During the planning of the projects, it was expected that the digital literacies of students would be low. To overcome this, the project team ensured that the education officers were trained in the use of the technologies. These staff in turn, would train the students in the use of them. The prior levels of experience of the students with mobile devices was mixed, with some students (usually those who had been in prison for a shorter period of time) being more familiar with the technologies than others. The variability in digital literacies of students as well as staff, impacted extensively on the effectiveness of both projects. Students did not make effective use of the technologies as a result of lack of knowledge about how they worked and education officers were often not aware of the necessary features to assist students effectively in spite of training by team members. This variability in digital literacies was also evident in the Triple 'E' Project with students struggling to effectively use the eBook readers.

5.5 Copyright

Despite extensive assistance from USQ's copyright officer, copyright issues were a major hindrance to the success of the PLEIADES project and necessitated some redesign of the course for the Triple 'E' Project. Although most publishers permitted readings to be placed on the eBook readers in the original PDF format, few permitted the conversion to more usable formats such as ePub without large fees. Affordable copyright permissions could not be obtained for two of the articles and the education officer needed to provide these articles in hard copy.

The majority of the course materials were originally only available in PDF format. This format is difficult to read on eBook readers as the text is often small and zooming in only increases the size of the page. Users therefore need to scroll back and forth on the page to read each line of text. Where copyright permitted, documents were converted to ePub format to provide greater functionality and ease of use for students. The process to convert these documents was laborious and a number of technical challenges were experienced. Materials needed to be converted into Rich Text Format first and the software, Apple Pages, was used to convert to ePub. Styles often did not apply consistently within the same documents and tables and images were not converted correctly during this process, requiring extensive reformatting of the documents. A number of the readings did not convert well when saved as rich text format from pdf documents. To overcome this, each reading had to be thoroughly checked, which was enormously time consuming. The aim was originally to combine all course readings into one document for each week, but the file size of these documents was too big for the software to cope with and individual files for each reading were created instead.

6 Conclusion

This paper has outlined many of the challenges faced and contingencies that had to planned for in the deployment of eBook readers across two projects: the PLEIADES Project and the Triple 'E' Project. In both cases these projects deployed eBook readers in correctional centres in Queensland, Australia for students enrolled in *TPP7120 Studying to Succeed*. Challenges could be broadly arranged into a number of groups: technical issues with the eBook readers, issues at the correctional centres, student concerns, digital literacy of the students and copyright concerns.

Of the seven students who participated in TPP7120 in the Southern Queensland Correctional Centre in semester 2 2012, two participated fully, one obtaining a High Distinction and one obtaining an A. Four students in total obtained a passing grade. Two did not participate in the course at all and received a fail grade and one withdrew from the course but re-enrolled for semester 3 and obtained an A. It is difficult to determine whether the participation trends were as a result of the changing technologies or because of the disruptions within the prison environment which were remarked upon by the students in the post-semester 2 focus group in the PLEIADES Project. Though a very small cohort participated in that trial, the results are encouraging. Though many challenges were encountered, these could be overcome with careful planning and goodwill on the part of both correctional centre staff and USQ staff.

The Triple 'E' ran in Semester 1 of 2013 across five correctional centres. As with the PLEIADES Project, there were many unexpected obstacles in deploying the eBook readers across the centres. Most were out of control of the research team. In no case were the delays caused by ill will or were the result of a deliberate attempt to derail the project. Special mention must be made of the students who have endeavoured to continue using the eBook readers even when they struggled to make them work.

Senior management at both the correctional centre and within USQ have acknowledged the importance of supporting further development and advancement of the projects. Although the security of the technologies was a primary consideration in the development of the project, no issues were encountered with the students' use of the technologies during the trial. It is anticipated that with certain modifications the eBook readers could provide a solution for other groups of students without access to reliable internet. These groups would include students from low SES backgrounds, Indigenous communities, rural, regional and remote communities and students in countries with poor ICT infrastructure.

References

[1] West, D.: Digital Schools: How Technology Can Transform Education. Brookings, Washington (2012)
[2] Hancock, V.: Essential, desirable or optional? Making distance e-learning courses available to those without internet access. European Journal of Open, Distance and E-Learning, 2(2) (2010), http://www.eurodl.org/?article=410 (retrieved)
[3] Justice Action: Computers in Cells: Maintaining Community Ties (2012), http://justiceaction.org.au/cms/images/JusticeReformPapers/computers%20in%20cells%20pdf.pdf (retrieved)
[4] Dorman, M., Bull, D.: Aligning educational needs with institutional priorities: Facilitating offender reintegration into contemporary society. Paper presented at the IFECSA Conference: Unlocking Doors - Rebuilding Lives Through Education (2003), http://www.acea.org.au/Content/2003%20papers/Paper%20Dorman_Bull.pdf (retrieved)
[5] Mehrotra, C.M., Hollister, C.D., McGahey, L.: Distance learning: Principles for effective design, delivery, and evaluation. Sage, Thousand Oaks (2001)
[6] Design-Based Research Collective. Design-based research: An emerging paradigm for educational inquiry. Educational Researcher 32(1), 5–8 (2003)
[7] Wang, F., Hannafin, M.J.: Design-based research and technology-enhanced learning environments. Educational Technology Research & Development 53(4), 5–23 (2005)

Exploring Tools to Promote Teacher Efficacy with mLearning

Robert Power[1], Dean Cristol[2], and Belinda Gimbert[2]

[1] College of the North Atlantic-Qatar, Doha, Qatar
robert.power@cna-qatar.edu.qa
[2] Ohio State University, Columbus, Ohio, USA
{cristol.2,gimbert.1}@osu.edu

Abstract. Teachers' perceptions of self-efficacy have been identified as a barrier to widespread integration of mobile learning strategies in teaching practice. This paper describes the development of the Collaborative Situated Active Mobile (CSAM) learning design framework and the mobile Teacher's Sense of Efficacy Scale (mTSES) survey. The CSAM framework aims to help with making instructional design decisions for using mobile reusable learning objects. The mTSES survey is designed to measure teacher's perceptions of self-efficacy with mobile learning. This paper describes how the CSAM framework and the mTSES survey were used to develop a professional development course to help teachers to become more confident with their ability to integrate mobile reusable learning objects into their teaching practice.

Keywords: CSAM framework, efficacy, mobile learning, mLearning, mobile RLOs, mTSES, professional development, reusable learning objects, self-efficacy, teacher training.

1 Introduction

The range of technical solutions available for the implementation of mobile learning strategies is expanding rapidly. At the same time, there has been an increase in calls for scholarly discourse around mobile learning to shift away from technical deployment issues, and towards pedagogical strategies. The urgency to bring mobile learning pedagogy to the forefront of discourse was highlighted by the panel discussion on tablet deployment initiatives at the 12th World Conference on Mobile and Contextual Learning (mLearn 2013) in Doha, Qatar [2]. When asked what he viewed as the single greatest barrier to wider adoption of mobile learning strategies, Athabasca University researcher Dr. Mohamed Ally cited the human elements of teachers' confidence in both mobile technologies and their own skills with mobile learning pedagogy. Ally and Prieto-Blázquez [3] point to one of the reasons why teachers lack confidence with pedagogies that employ mobile technologies:

> The current educational model is outdated because it was developed before the advent of information and communication technologies. The current model, based on classroom-based face-to-face delivery, is geared towards educating a certain

M. Kalz et al. (Eds.): mLearn 2014, CCIS 479, pp. 61–68, 2014.

> segment of the population. Also, teachers are being trained for the current model of education, and will therefore continue using the model when they become teachers. Teacher training must be re-invented to prepare teachers for the technology-enhanced educational system (pp. 145-146).

Lack of training in the pedagogical considerations for the integration of a specific type of technology can have a negative impact upon teachers' perceptions of self-efficacy. However, Kenny et al. [11] note that:

> While a significant body of research exists on learners' feelings of self-efficacy concerning computer technology, online learning, and even podcasting... this concept does not yet appear to have been examined in any detail in a mobile learning context (p. 2).

This paper describes a partnership to explore potential tools and professional development strategies to help increase perceptions of self-efficacy for teachers who want to integrate mobile learning strategies into their teaching and learning practice. The following section describes the development of the Collaborative Situated Active Mobile (CSAM) learning design framework and the mobile Teacher's Sense of Efficacy Scale (mTSES) survey tool. The next section outlines the partnership between researchers from Ohio State University and a doctoral student from Athabasca University to develop a professional development course integrating CSAM and the mTSES. The impact of the professional development on perceptions of self-efficacy amongst participants in the professional development course is the focus of an ongoing doctoral dissertation research project. However, early results from that study can be used to discuss the effectiveness of the training, and to begin exploring the next steps in the partnership to develop an open educational resource for teacher professional development based upon CSAM and the mTSES.

2 Developing Tools to Promote Teacher Efficacy with mLearning

This section describes the development of two tools that formed the core of a professional development course designed to help increase teachers' perceptions of self-efficacy with mobile learning strategies. The first tool is the Collaborative Situated Active Mobile (CSAM) learning design framework. The second tool is the mobile Teacher's Sense of Efficacy Scale (mTSES).

2.1 The Collaborative Situated Active Mobile Learning Design Framework

The CSAM framework has been developed to highlight the primary pedagogical components of instructional design for mobile reusable learning objects (RLOs) used to facilitate collaborate learning. A mobile RLO is a "digital object that can be [accessed via a mobile device and] reused to facilitate and support learning activities"

[14], [22]. Collaborative learning refers to learning activities in which participants are mutually engaged towards achieving a shared goal [8].

An analysis of examples of mobile RLOs presented in recent mobile learning literature reveals four primary pedagogical elements addressed by their instructional design. These pedagogical elements are a focus on Collaborative learner interactions, Situation of the learning in an authentic or meaningful context, Active engagement in learning and the production of artefacts of learning, and the affordance of Mobility in terms of time, space, and learning strategies [15, 16]. The key elements of the CSAM learning design framework are depicted in Figure 1.

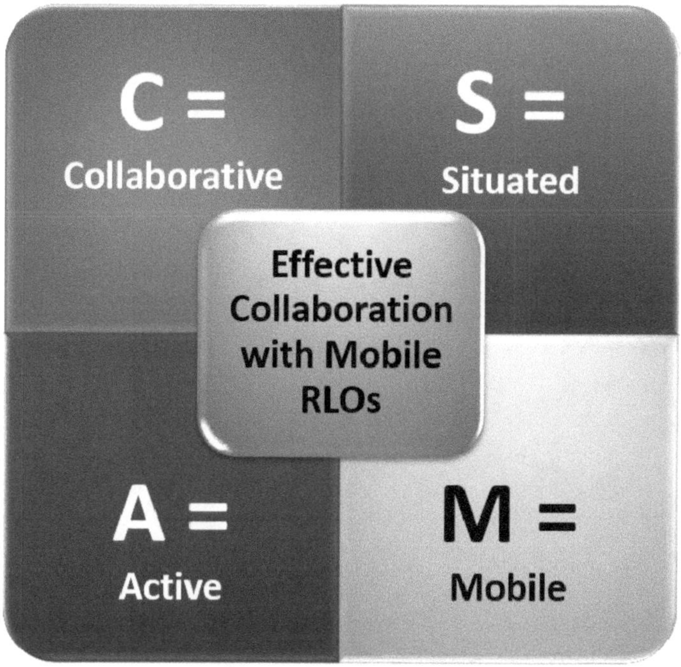

Fig. 1. The CSAM learning design framework

A more detailed qualitative meta-analysis was conducted on 403 recent articles and chapters on mobile learning. The samples were from textbooks and conference proceedings publications between the years 2009 and 2014 [1], [4], [5], [7], [12], [13], [17], [19], as well as the first five volumes of the *International Journal of Mobile and Blended Learning* [10]. Over one quarter of all of the sample publications either focused on mobile RLOs, or provided a mobile RLO or a suite of mobile applications used as supporting examples in the article or chapter. Of the publications identified as RLO examples, 86% focused on collaborative learning strategies. Of the RLOs identified as intended for collaborative learning, 96% explicitly addressed all four of the CSAM pedagogical framework components. The prevalence of CSAM-compliant RLOs in recent mobile learning literature is depicted in Figure 2.

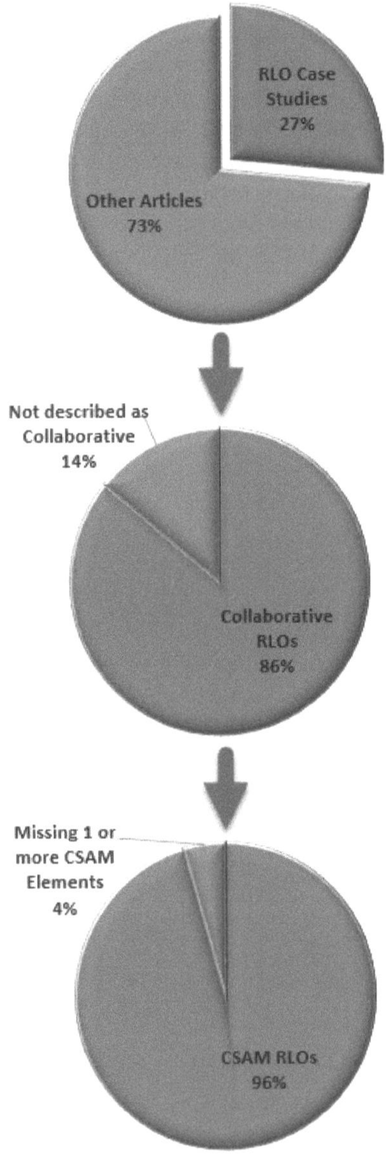

Fig. 2. Prevalence of CSAM-compliant RLOs in recent mLearning literature

2.2 The Mobile Teacher's Sense of Efficacy Scale

The mTSES survey tool is designed to measure teacher's perceptions of self-efficacy with respect to integrating mobile RLOs into their teaching and learning practice. The survey was adapted from the Ohio State Teacher's Sense of Efficacy Scale (TSES) [20, 21]. The TSES survey was selected for this research study and professional development resource because its reliability and construct validity have been previously established through comparison with the Rand scale and the personal teaching efficacy and general teacher efficacy factors of the Gibson and Dembo instrument [20].

The original TSES instrument consists of 24 questions using a nine-point Likert-scale to rate perceptions of self-efficacy in the areas of student engagement, instructional strategies, and class management skills [20, 21]. Minor changes were made to the wording of some questions from the original TSES for contextualization purposes. Those questions that were modified were added to the original set of TSES questions, resulting in a new 38 question survey. The combination of the TSES and mTSES questions into a single instrument allows for measurement and cross-comparison of teachers' perceptions of self-efficacy for teaching practices and general, as well as the use of mobile learning strategies. Benton-Borghi [6] used a similar strategy to adapt the TSES for a new instrument called the Inclusion Teacher's Sense of Efficacy Scale (I-TSES). The I-TSES was developed in the context of measuring teacher's perceptions of self-efficacy with using technology to facilitate the inclusion of students with disabilities. The dissertation research study being undertaken by one of the authors [18] will use the procedures outline by Benton-Borghi [6] to conduct statistical analyses to determine the actual construct validity and reliability for the mTSES. It is anticipated that the wording changes for the mTSES will have a similar effect upon both construct validity and overall instrument reliability when compared to the TSES as those observed by Benton-Borghi [6] for the I-TSES.

3 Exploring Strategies to Promote Teacher Efficacy with mLearning

The CSAM learning design framework and the mTSES survey instrument form the core of an online professional development course called Creating Mobile Reusable Learning Objects Using Collaborative Situated Active Mobile (CSAM) Learning Strategies. The course was developed in partnership between the authors as part of a doctoral dissertation research study [18], and was conducted in the May of 2014. Participants in the course included approximately 70 teachers and graduate-level education students affiliated with Ohio State University, Athabasca University, College of the North Atlantic, and College of the North Atlantic-Qatar.

Participants in the professional development course were introduced to key mobile learning concepts, including the CSAM framework. Learning activities included using the CSAM framework as a lens for exploring the instructional designs of mobile RLOs presented in recent literature, using CSAM to guide participants' own instructional design plans for mobile RLOs, using free online tools to develop mobile RLOs, and using the CSAM framework to guide reflective practice after building a mobile RLO. Free online tools were used by course participants to build their mobile RLOs so that they could focus on pedagogical instructional design decisions. The mTSES survey was used at the beginning of the course to measure participants' pre-course perceptions of self-efficacy with mobile learning. Participants were asked to reflect upon their initial mTSES scores, and to contribute a discussion forum posting about what their scores told them about their confidence with mobile learning. The mTSES survey was used administered again in the final module of the course, and, again, participants were asked to reflect upon their scores and how their scores might have changed since the beginning of the course.

The Creative Mobile Reusable Learning Objects Using Collaborative Situated Active Mobile (CSAM) Learning Strategies course was developed using the Canvas open learning management system. The Canvas LMS is free for use by individual teachers, and can also be licensed for use by educational institutions [9]. The Canvas LMS was selected because it allows for participant self-enrollment in courses, and is optimized for access using either a personal computer or a mobile device. The course consists of six modules, as outline in Table 1:

Table 1. Course modules for Creating Mobile Reusable Learning Objects Using Collaborative Situated Active Mobile (CSAM) Learning Strategies

Module	Topic
Module 0	Getting Started
Module 1	Exploring Your Skills with mLearning
Module 2	Introduction to Using Mobile RLOs for Collaborative Learning
Module 3	Planning to Use Mobile RLOs
Module 4	Creating Mobile RLOs
Module 5	Reflective Practice

Data collection for the doctoral dissertation study included the two in-course mTSES survey administrations, follow-up interviews with selected course participants, and a third mTSES survey administration to be conducted four months after the completion of the professional development course. The two in-course mTSES survey administrations will be analyzed to show changes, if any, in teachers' perceptions of self-efficacy between the beginning and the end of the course. The forthcoming third mTSES survey administration will be used to determine if any changes in perceptions of efficacy resulting from the training are sustained over time. The follow-up interviews will be qualitatively analyzed to provide further insights into participants' perceptions of self-efficacy with mobile learning, the value of the course as a professional development tool, and the utility of the CSAM framework as a guide for instructional design decision-making.

4 Early Results and Next Steps

Data collection for the dissertation research study is ongoing. Early results of the analyses of the first and second mTSES survey administrations are expected to be available by November 2014. Analyses of the follow-up interview data and the third administration of the mTSES survey are expected to be available by January 2015.

The intent of the partnership for the dissertation research study and the development of the training course with Ohio State University is to determine how useful the CSAM framework, the mTSES survey, and the training course itself are in promoting increased teacher self-efficacy with mobile reusable learning objects. The findings of the research study will be used to guide iterative improvements to the design of the Creating Mobile Reusable Learning Objects Using Collaborative Situated Active Mobile (CSAM) Learning Strategies course. Course instructional design improvements will be integrated into the development of a stand-alone reusable learning object that can be incorporated into professional development programs by schools and teacher-training institutions. A preliminary version of this RLO has already been developed for use with the Canvas learning management system. Canvas users can download a version of the original professional development course that has been exported as a data-pack file. The data-pack can be imported into a new course using Canvas, and can then be adapted to meet the training schedule needs of the users' institution.

5 Conclusions

It is premature to draw conclusions about the effects of the CSAM learning design framework, the mTSES survey, or the online professional development course, on teachers' perceptions of self-efficacy with mobile learning strategies. However, the partnership that has led to the development of the Creating Mobile Reusable Learning Objects Using Collaborative Situated Active Mobile (CSAM) Learning Strategies course has verified the prevalence of the CSAM pedagogical components in mobile instructional design for collaborative learning. It is hoped that the CSAM framework can help more teachers to become comfortable and confident with making decisions about how and why to integrate mobile RLOs into their teaching and learning practice. It is also hoped that the resources developed for the professional development course discussed in this paper will help increase perceptions of self-efficacy by allowing teachers to focus on pedagogical decision-making without needing to master complex web and mobile application authoring tools.

References

1. Ally, M. (ed.): Mobile Learning: Transforming the Delivery of Education and Training. AU Press, Edmonton (2009)
2. Ally, M., Farias, G., Gitsaki, C., Jones, V., MacLeod, C., Power, R., Stein, A.: Tablet deployment in higher education: Lessons learned and best practices. Panel Discussion at the 12th World Conference on Mobile and Contextual Learning (mLearn 2013), Doha, Qatar, October 22-24 (2013)
3. Ally, M., Prieto-Blázquez, J.: What is the future of mobile learning in education? RUSC 11(1), 142–151 (2014), `http://doi.dx.org/10.7238/rusc.v11i1.2033`

4. Ally, M., Tsinakos, A.: Increasing Access Through Mobile Learning. AU Press, Edmonton (2014), http://www.col.org/resources/publications/Pages/detail.aspx?PID=466
5. Proceedings 10th World Conference on Mobile and Contextual Learning (mLearn 2009), October 18-21. Beijing Normal University, Beijing (2011)
6. Benton-Borghi, B.: Teaching every student in the 21st century: Teacher efficacy and technology (Doctoral dissertation, Ohio State University) (2006)
7. Berge, Z., Muilenburge, L.: Handbook of Mobile Learning. Routledge, New York (2013)
8. Boyinbode, O.: Towards an interactive mobile lecturing model: A higher-level engagement for enhancing learning (Doctoral dissertation, University of Cape Town) (2013), http://hdl.handle.net/11180/6383
9. Canvas: About Us, http://www.instructure.com/about-us
10. IGI Global: International Journal of Mobile and Blended Learning (2014), http://www.igi-global.com/journal/international-journal-mobile-blended-learning/1115
11. Kenny, R.F., Park, C.L., Van Neste-Kenny, J.M.C., Burton, P.A.: Mobile self-efficacy in Canadian nursing education programs. In: Montebello, M., Camilleri, V., Dingli, A. (eds.) Proceedings of mLearn 2010, the 9th World Conference on Mobile Learning, Valletta, Malta (2010)
12. McConatha, D., Penny, C., Schugar, J., Bolton, D.: Mobile pedagogy and perspectives on teaching and learning. IGI Global, Hershey (2014)
13. Palfreyman, D. (ed.): Proceedings of the Mobile Learning: Gulf Perspectives Research Symposium. Abu Dhabi, UAE (April 25, 2013)
14. Polsani, P.: Use and Abuse of Reusable Learning Objects. Journal of Digital Information 3(4) (2013), http://journals.tdl.org/jodi/article/viewArticle/89/88
15. Power, R.: Collaborative situated active mobile (CSAM) learning strategies: A new perspective on effective mobile learning. Learning and Teaching in Higher Education: Gulf Perspectives 10(2) (2013), http://lthe.zu.ac.ae/index.php/lthehome/article/view/137
16. Power, R.: Collaborative Situated Active Mobile (CSAM) learning strategies: A new perspective on effective mobile learning. Presentation at the Mobile Learning: Gulf Perspectives Research Symposium. United Arab Emirates, Abu Dhabi (April 25, 2013)
17. Power, R. (ed.): QScience Proceedings: 12th World Conference on Mobile and Contextual Learning (mLearn 2013), vol. 2013 (2013), http://www.qscience.com/toc/qproc/2013/3
18. Power, R.: A framework for promoting teacher self-efficacy with mobile reusable learning objects (dissertation proposal candidacy exam presentation) (2014), http://www.slideshare.net/RobertPower1/a-framework-for-promoting-teacher-selfefficacy-with-mobile-reusable-learning-objects
19. Specht, M., Sharples, M., Multisilta, J. (eds.): Proceedings of the 11th Annual World Conference on Mobile and Contextual Learning (mLearn 2012), held in Helsinki, Finland, October 16-18 (2012), http://ceur-ws.org/Vol-955/
20. Tschannen-Moran, M., Woolfolk Hoy, A.: Teacher efficacy: Capturing and elusive construct. Teaching and Teacher Education 17(7), 783–805 (2001)
21. Tschannen-Moran, M., Woolfolk Hoy, A.: Teacher's sense of efficacy scale (2001), http://people.ehe.osu.edu/ahoy/files/2009/02/tses.pdf
22. University of Wolverhampton: Reusable learning objects (RLO) and open educational resources (OER), http://www.wlv.ac.uk/default.aspx?page=26086

Students' Use of Mobile Phones for School Work

Torbjörn Ott, Therése Haglind, and Berner Lindström

University of Gothenburg, Department of Applied IT, Gothenburg, Sweden
{torbjorn.ott,therese.haglind,
berner.lindstrom}@ait.gu.se

Abstract. Findings from a student survey at a Swedish upper secondary school class concerning the use of mobile phones for school work are presented in this paper. A previous study indicated that a majority of the students did not regard the mobile phone as an appropriate tool for school work at school. However 56% of the students stated that they used mobile phone for school work at home every week (Haglind, 2013). In relation to the previous study this paper explores the students' perception of the mobile phone as a tool for school work in school and the students' use of it for learning at home. The results indicate that the mobile phone can be described as a boundary object between the students' social worlds of home and of school. The results also show that the students use their mobile phones for school work related tasks, when the task is suitable for the mobile phone format.

Keywords: Boundary objects, mobile phone, school work, usage, 1:1, upper secondary school.

1 Introduction

Our modern way of life is highly dependent on connected technologies and activities mediated by those technologies e.g. laptop computers, tablets and mobile phones. This affects several layers of the society including school (OECD, 2012).

In Sweden, like in many other countries, several schools are 1:1 schools. In a report to the Swedish Municipalities and County Councils Hylén (2013) states that many municipalities have on going or are planning 1:1 projects. This means that the school equips every student with a digital tool, often a laptop or a tablet. However, the devices provided by the school, are not the only technology students have at hand. Today, most students also have a mobile phone in their possession (Norris and Soloway, 2010). In Sweden in 2013, 99% of the population in the age span 15-24 years in their homes had access to some kind of mobile phone. 90% had access to smartphones (NORDICOM, n.d.). Throughout the society outside of school mobile phones are important tools for communication. In school the use of mobile phones are more scarce. Students' use and even possession of mobile phones in classroom context is contested (Ott, 2014). However, as a report from the European Commission (2013) concludes, students bring their mobile phones to school no matter if the school supports a use or a ban of mobile phones.

M. Kalz et al. (Eds.): mLearn 2014, CCIS 479, pp. 69–80, 2014.

The question might not be *if* but *how* mobile phones are to be used for learning in school. "[...] we may feel that mobile learning is no longer an innovation within institutional learning but a reflection of the world in which institutional learning takes place" (Traxler and Vosloo, 2014, p. 21). It is then interesting to find out more about how students use their private mobile phones for school work at home. It can be uses that might transfer to the school work in school. The students might be used to tackle problems in certain ways that involve usage of mobile phones. It might also be that the students have competences for using mobile phones for problem solving relevant to school work. These habits and competences are not left at home when entering school. This paper reports of a small-scale study of this issue, conducted in a 1:1 learning environment.

2 Problem Area

The problem presented in this paper originates from different ways to perceive various modern technologies as tools for learning. Technologies like mobile phones, laptops and tablets are mobile, connected and personal. These devices can be utilized for learning when moving between formal and informal learning settings, in and out of school, and between various communities (Chan et al., 2006). Mobile learning as an approach to learning entails learning mediated by mobile technologies, making the learner mobile (Sharples et al., 2007). Mobility can be understood as temporal, spatial and contextual (Kakihari and Sørensen, 2002). One of the two most active areas of research within the field of mobile learning at present is the use of portable devices to support the curricular learning (Sharples, 2013). Mobile technologies, e.g. mobile phones, are now ubiquitous in most parts of the society including school (Norris and Soloway, 2010). According to O´Bannon and Thomas (2014) mobile phones can support creativity, collaboration, connectivity, reflection and instruction in the classroom. But there are also stumbling blocks to mobile phone usage in the classroom e.g. disruption, cheating, texting, sexting, cyber bullying and poor technological infrastructure (ibid). And there are many factors affecting the learning outcome, for example learners' motivation and attitudes towards using technologies; their skills in using technologies; access to technologies; and the nature of technological use (Adhikari et al., 2012). When it comes to the nature of technological use, use in informal settings – in students' everyday life in non-schooling context – is different from use in formal educational settings.

Mobile phones typically do not belong to curricular activities. They are predominantly tools used in non-curricular and personal activities. Although students appreciate some benefits of technology for making their school work more productive, they do not expect a teaching radically changed by technology. However OECD (2012) suggests this could change, if the experiences of using technology are more rewarding.

In order to understand the premises for using mobile phones as tools for school work it might be productive to consider the boundaries between school and non-school activities and mobile phones as boundary objects (Akkerman and Bakker, 2011;

Star, 2012). Students in school can be regarded collectively or as individuals. They are brought together in various constellations by school activities and by school organization. In the wording of Wenger (1998), they may share repertoire and may have a mutual engagement when collaborating, but basically they are their own enterprises. Their success or failure is personal. Phelan et al. (1991) suggest that students are participating in several social worlds with stronger or weaker boundaries between them. These worlds can be family, school, and peers. Students have to handle and to move between these social worlds, which do not always share a common ground of values and norms.

Akkerman and Bakker (2011) state that all learning involves boundaries. Students develop competences on different levels in crossing boundaries between different social worlds. These competences are of importance for the students' capacity to use the school system as stepping stone for a successful life (Phelan et al., 1991).

Boundaries are comprised by sameness and continuity as well as socio cultural differences that generate discontinuity in action and interaction (Akkerman and Bakker, 2011). The boundary entails a shared space where the sense of the own and the others' practices are being mixed up (Star, 2010).

Crossing boundaries could be done either by brokering, people who transfer elements of different practices between practices, or by boundary objects (Wenger, 1998). Boundary objects are objects of interpretive flexibility and objects of action. A boundary object resides between social worlds where it is poor structured. It is an object that is perceived more as private than as shared. Further it is an object that in the absence of consensus between the cooperating groups causes the subjects to wobble between the local and shared perception of the object (Star, 2012).

There is a contradictory relation between an interpretive flexibility and common recognition, "[...] boundary objects have different meanings in different social worlds but at the same time have a structure that is common enough to make them recognizable across these worlds" (Akkerman and Bakker, 2011, pp.140-141). Boundary objects have a strong structure in their own social world. In the boundary the structure is weak. If a strong common structure is developed around the object, it is no longer a boundary object.

In a study of use of ICT in school and out of school, carried out in 2013, 56% of the students stated they use their mobile phone for *school work at home* at least once a week or more often (Haglind, 2014). When the students were asked to rank how they appreciated different technologies in relation to their *functionality regarding school work*, mobile phones (Android and iPhone Smartphones) were by most respondents *not* regarded as a suitable alternative (Haglind, 2013). In group interviews the students expressed that they did not think of it as a good idea to let them use their own technology in school (Haglind, 2013).

This study shows that there is a contradiction between the students' use of mobile phones for school work at home and their perception of it as a suitable tool to be used for school work in school. The study presented in this paper intends to focus this ambiguity by unpacking the perceptions of the tool. We connect this ambiguity to the crossing into the boundaries between different social worlds. The boundary for the students resides in the area where social world of school and social world of the home are confounded. Certain tools can be linked to that area as being boundary objects. The objects both bridge and separate the social worlds. We identify the mobile phone as being one such object. How this is enacted in the group of students is what we examine. The relationship between the students' use of the mobile phone for school work at home and their perceptions of school work related use of mobile phones in school is in focus.

3 Method

A survey was distributed to an upper secondary school class in the Business Management and Economics program in a Swedish midsized city. Several studies regarding 1:1 initiatives in Sweden have been conducted during the past years (Grönlund et al., 2011 & 2013; Håkansson Lindqvist, 2013; Samuelsson, 2014).

The school in this paper is one of four upper secondary schools that have been subject of a larger study. The study is an evaluation of an on going 1:1 initiative, conducted over two years in the municipality. The evaluation indicates that the implementation of ICT in the municipality shares common features with the utilization of ICT in other Swedish municipalities. ICTs are mainly use for utilization of the LMS, for writing texts, for seeking information, for taking notes and for communication. The evaluation in which the school in this paper were part indicates that the school does not deviate from the other schools and can be considered an ordinary Swedish school (Player-Koro et al., 2014a).

The school has both vocational program and preparatory program for higher education. The study was conducted at the Business Management and Economics Program and this program is a preparatory program for higher education. The survey involved one school class of thirty-one students in their first year at the upper secondary school. The class can be considered as being a quite regular class.

To reach an extended understanding of the ambiguous results of the 2013 study (Haglind, 2013) a second data collection was carried out in 2014 with the same students. In the 2014 study the students were asked questions about their use of mobile phones for school work. The small sample size of the study raises questions concerning the conclusions that can be drawn from the data and about generalization of the results. However, since the study presented in this paper is carried out within a larger study of an ongoing 1:1 initiative, we argue that our data add important information on students' use of ICT for school work.

The survey was distributed in the classroom during a lesson and answered on paper. Twenty-eight students of which fourteen were males and fourteen were females answered the survey.

The questions in the survey were:

1. Are you a male or a female?
2. Do you own a mobile phone, if so what kind; a smartphone or a regular mobile phone?

3. Do you carry your mobile phones with you to school every day?
4. Regarding digital technology, what of a user are you; beginner, normal, habituate or expert?
5. Below there are a number of activities listed where you use the mobile phone for school work at home. Specify how often these activities occur in your school work? (for the categories see table 1).
6. What do you think is suitable areas of use of the mobile phone for schoolwork in school? (for categories see table 2).
7. What obstacles do you experience for using the mobile phone for school work in school?
8. How would you like to use the mobile phone for the school work in school?

Question one, two, three and four were used as background variables. Question five, six, seven and eight were analysed in relation to these. The methods of analyses were descriptive, using frequency analyses and crosstabulations. Non-parametric testing procedures (chi-square tests) were applied. Answers on open questions were analysed based on their content and categorized.

In gathering data in the Player-Koro et al. (2014a) evaluation study, a survey was distributed to the upper secondary teachers in the municipality's schools during the spring of 2013. One of the questions the teachers were to answer was to what extent they allowed students to use their mobile phones in their lessons. On the Likert scale value 1 was not at all and value 5 was to a very high extent. Six teachers on the Business Management and Economics Program answered the question and the mean value was 1,83. The teachers on the Business Management and Economics Program at the school were not particularly allowing of mobile phones in their lessons. However this does not mean that they were negative to using ICT in their instruction. Results from the 2013 survey (Haglind, 2013) show that a majority of the teachers teaching the class have a positive attitude towards ICT.

4 Results

Twenty-six students stated that they owned a smartphone and two owned a regular mobile phone. All of the students stated that they brought their mobile phones to school every day. Twelve students regarded themselves as *expert users*, twelve regarded themselves as *habituate users* and four regarded themselves as *normal users*. No student choose the lowest user category *beginner*.

All of the students stated that they use the mobile phone for school work at home *for some purpose* at least once a month. 93% of the students stated they use it for school work at home *weekly*. 61% use it more than three times a week and thirty-two percent use the mobile phone for school work at home 1-3 times a week.

Table 1 presents the statistics of the students' answers regarding their specific uses of mobile phones for school work at home.

Table 1. Frequencies of students' use of mobile phones for school work at home

	Never	At least once a month	1-3 times/ week	3 times/ week<	Daily
I cooperate with my classmates by texting	11	50	25	0	14
I cooperate with my classmates by talk or video calls	54	25	7	11	4
I cooperate with my classmates by social media	11	47	25	14	4
I communicate with my teachers	61	32	4	4	0
I use the mobile phone to access the LMS	43	29	21	0	7
I browse the Internet for information	21	11	32	18	18
I take pictures for school assignments	36	29	29	0	7
I revise lesson notes	50	21	25	0	4
I look at pictures	26	30	22	0	22
I record audio and video	61	29	0	4	7
I edit audio and video	64	14	4	7	11
I use the mobile phone to connect the computer to the Internet	39	50	11	0	0
I watch informational films on the Internet	61	39	0	0	0
I watch material produced by the teacher	55	30	7	4	4
I write texts for school assignments	78	11	4	7	0
I use the mobile phone as calculator	4	11	52	30	4
I use the mobile phone for translation of words.	11	11	53	18	7

(Numbers are in percentages. N=28.)

The most common uses among the whole class were cooperation with the classmates by texting; cooperation with the classmates by social media; browsing the Internet for information; looking at pictures; calculation; and translation of words.

One question is if there are any differences between the expert user group and the group of normal and habituate users regarding frequency of specific type of use. Performing chi square-tests, the only significant ($p=0,015$) difference found concerned the students' use the mobile phone to access the LMS. It was more common among the expert users.

Another matter of interest is what students' appreciation of the mobile phone as a tool for school work in school are. The students were asked to consider various uses and decide on one of two alternatives, appropriate or not appropriate. Table 2 displays the percentage of students stating a particular type of use of the mobile phone as appropriate in school settings. Values <50% are highlighted in grey.

Table 2 indicates that in school the students regard the mobile phones as an *appropriate tool* for most uses except for revising lesson notes, watching material produced by the teacher and for writing school assignments. The usage areas that the students find most appropriate to use the mobile phone for school work at home for are typically valued as highest with respect to appropriateness for school work in school, e.g. browsing the Internet for information; calculation and translation of words; cooperation with the classmates by texting and by social media; looking at pictures; and taking photos for school assignments.

Table 2. Proportion of students positive to different types of use of the mobile phone for school work in school

Type of use	Normal and habitual users*	Expert users**	All**
For cooperation with my classmates by texting	88	67	79
For cooperation with my classmates by talk or video calls	69	42	57
For cooperation with my classmates by social media	88	58	75
For communication with my teachers	69	33	54
To access the LMS	69	75	71
For browsing the Internet for information	88	92	89
For taking pictures for school assignments	88	92	89
For revision of lesson notes	50	50	50
For looking at pictures	75	75	75
For recording audio and video	81	75	79
For editing audio and video	63	25	46
To connect the computer to the Internet	63	58	61
For watching informational films on the Internet	56	42	50
For watching material produced by the teacher	50	42	46
For writing texts for school assignments	38	17	29
For calculation	100	100	100
For translation of words.	81	100	84

*N=16 **N=12*** N=28

Comparing the expert users with the normal and habituate users a difference in judgement is revealed. The students that classify themselves as expert users tend to be more sceptical than the group of normal and habituate users. More specifically, expert users are more sceptical regarding communicative use areas.

Another question is what *obstacles* for using mobile phones for school work in school the students identify. Using content analyses, four themes of obstacles could be identified from the students' answers; size, cheating, teachers and distraction (table 3). Some students identified several obstacles.

Noteworthy is that the students tend to state *cheating* and *teachers* separate. The two words do not always occur together in the students' answers.

Having identified obstacles for utilization of mobile phones for school work in school it is also of interest to investigate what uses of mobile phones for school work in school the students might *desire*. Some desirable usage areas mentioned by the students were; for calculation twenty-five percent, for browsing information eighteen percent, for translation of words fourteen percent for communication eleven percent as a calendar and for taking pictures eleven percent. 46% of the students however stated that they either do not know, or that they do not want to use their mobile phone for school work. The reason why is because they have a computer. This is

Table 3. Obstacles for use of mobile phones for school work in school

Theme	%	Quotation
Size	39	"It is too small. It is more flexible and faster to use a computer. Also the mobile phone has smaller screen (S22)."
Cheating	21	"Cheating. If there is a test you can use information on the phone and check. [...] (S10)."
Teachers	21	"That teachers do not want the mobile phone to be used during lesson (S23)."
Distraction	21	"Distractions for most (S26)."

exemplified with the following two answers from the survey: "Rather not, prefer using the computer, there you have a bigger screen and it is easier to have several tabs", and "Ideally nothing, since it is my private thing. But it facilitates little now and then. Boring when you have to share your Internet".

5 Discussion

The results presented indicate that the mobile phone is not an integrated tool for school work in school, but yet the students use it for school work at home. All of the students bring their mobile phones to school every day. But from the students' answers regarding desirable usage areas and obstacles for using the mobile phones we can assume that the mobile phone is not an integrated tool for school work in school. The results reveal that the mobile phones in the class examined is functioning as a boundary objects as defined by Star (2010). The boundary objects are unstable and ill structured in the boundaries. Crossing the boundaries is done, but it is not done without hesitation. The school work related areas and purposes the students find the mobile phone appropriate for in school are also the areas in which they tend to use them at home.

The students identify several obstacles for their use of the mobile phone for school work. The students that describe themselves as expert users are more restrictive and selective regarding appropriate school work usage areas of the mobile phone. It might be that the more demanding the use of the mobile phone is, the more evident are its shortcomings. Almost half of the students stated that they did not know what to do with the mobile phone, or that they did not want to use it for school work. Plausible explanations for this could relate to how school tasks are constructed; how the students are used to solve tasks in school and the students' teachers attitudes for using the mobile phone for school work. The structure of school tasks and the tools that are supposed to be used in working with them are set. The mobile phones does not belong the primary repertoire of tools to be used. Also, as one student stated, the mobile phone is a private thing. It belongs to the social worlds outside school. And the students to some extent use their mobile phones for school work at home.

Mobile phones are ubiquitous in society. Access to mobile phones in the age group is high. The data indicates that for solving school tasks at home the students develop certain patterns of use that involve mobile phones. The students do however not fully acknowledge that these patterns of use is appropriate in the school setting.

The mobile phone is crossing the boundaries between the social world of home or, more generally, everyday life and the social world of school. The students do not in unison perceive the mobile phone as a tool desirable to utilize for school work. Quite a few of the students claim that they do not know what to use the mobile phone for, or that they do not want to the mobile phone for school work in school. Still they do bring the mobile phone to school and identify several possible school work related usage areas. Moving into the curricular social world of the school the mobile phones is not left in the social world of everyday life outside school. One way to look at this is that the students construct the mobile phone as a boundary object. School and the teachers are mostly reactive to this process, even though some of them acknowledge the mobile phone as a potential resource. And as a private tool for students to be used outside curricular activities.

They also regard the computer as a more functional tool for school work in school. Most of the students regard the mobile phone as a compliment but not as a substitute to the computer.

The limitations that the students experience with the mobile phones could be grouped into four categories: Limitations in the devices; perceptions of the devices; distraction and the nature of school work in relation to the potential of the device.

When the students in the survey of 2013 are asked whether they think that the mobile phone is a suitable tool for school work, many of them answer no. But when "unpacking" the question and asking them about specific range of use, we see another pattern in the answers. Students are answering in positive terms when presented with specific potential use areas.

Studies of 1:1 initiatives show that student use their laptop computers in school most for writing texts, taking notes and downloading material from the LMS (Skolverket, 2013; Player-Koro et al.; 2014a). Mobile phones are less optimal for these types of use. Some tasks are more conveniently carried out with a computer (write texts, take lessons notes, download big files) and some things are easier to do with the mobile phone (e.g. taking pictures, recording audio, and to communicate). Both technologies can, however, be regarded as mobile devices, but they have different interfaces supporting different uses.

Player-Koro et al. (2014b) show that although a school has integrated a new digital tool this does not have to mean that the way of organizing school has changed. In their study, the way schooling is carried out is still to be considered traditional. However, within frames of a traditional way of organizing school activities, there are activities where students (and teachers) use technology in a way that are renewing the work of instruction and learning. Even though the school might be traditional in some sense, students in this study identify uses supporting collaboration and communication in school. For tasks such as writing texts for school assignments, the students do not find the mobile phone a suitable tool. But for other tasks the digital development has changed its conditions, e.g. information seeking.

Students generally expect their education to be traditional, but they appreciate technology if it enhances their productivity OECD (2012). To the students it is not rational to use a highly mobile device to solve traditional tasks designed to be performed sitting in the classroom. In accordance with O'Bannon and Thomas (2014) the results indicate that there are both benefits and obstacles for using mobile phones in school. The results show that all of the students use their mobile phones at home for school work related usage areas that they think the devices can handle; cooperation with the classmates by texting, cooperation with the classmate by social media, browsing the Internet for information, looking at pictures, using the mobile phone as a calendar, using the mobile phone for translation of words and calculation. A lower frequency of use at home is reported concerning activities that the students do not think the mobile phones is a proper tool for. There is also a lower frequency of use in usage areas that the mobile phone potentially could be a proper tool for; recording audio and video and to some extent taking pictures for school assignments. These areas of usage coincide with activities that the students state that they also seldom use their Chromebooks for (Haglind, 2013). The introduction of Crombooks and iPads, or other mobile surfing devices, is a contemporary development that points at the importance of considering also the mobile phone as a tool in school. Especially since these kinds of devices essentially are upsized mobile smartphones.

Boundary objects are unstable and the mobile phone is mainly residing in the social world of home (or everyday life outside school). Future research will have to examine whether it will remain a boundary object or if school will annex the mobile phone into its digital toolbox.

The students' usage of their mobile phones for school work at home is dependent on the task. If solving of the task benefits from the use of the mobile phone the students see no hindrance for using it.

To summarize the two surveys, the picture that arise from the questions concerning mobile phones and school work are ambiguity. We explain that ambiguity with the mobile phone being a boundary object between the students' social world of home and students' social world in school, at least concerning school work. The students use the mobile phone for school work at home, they identify several usage areas for which the mobile phone could be a suitable tool, still they identify severe hindrances. As a boundary object, the mobile phone also has affordances that potentially can be picked up more broadly by students and teachers and in the end serve school work in productive ways.

References

1. Adhikari, J., Parsons, D., Mathrani, A.: Bridging Digital Divides in the Learning Process: Challenges and Implications of Integrating ICTs. In: Proceedings of the 11th International Conference on Mobile and Contextual Learning 2012, pp. 224–227 (2012)
2. Akkerman, S.L., Bakker, A.: Boundary Crossing and Boundary Objects. Review of Educational Research 81(2) (2011)
3. Chan, T.W., Roschelle, J., Hsi, S., Sharples, M., Brown, T.O.M., Pattob, C., et al.: One-To-One Technology-Enhanced Learning: An Opportunity for Global Research Collaboration. Research & Practice in Technology Enhanced Learning 1(1) (2006)

4. European Commission. Survey of schools: ICT in Education. Benchmarking access, use and attitudes to technology in Europe's (2013), https://ec.europa.eu/digital-agenda/sites/digital-agenda/files/KK-31-13-401-EN-N.pdf (accessed October 5, 2013)
5. Grönlund, Å., Englund, T., Andersson, A., Wiklund, M., Norén, I.: Unos uno årsrapport 2012. Örebro universitet, Örebro (2013)
6. Gröndlund, Å., Englund, T., Andersson, A., Wiklund, M., Norén, I., Hatakka, M.: Årsrapport Unos uno 2011. Örebro universitet, Örebro (2011)
7. Haglind, T: A report about an evaluation of Sandagymnasiets (an upper secondary school in Sweden) Chromebook project fall semester 2013 (2013), Provisional edition. Jönköpings municipality (unpublished)
8. Haglind, T.: Student Use of Digital Technologies for Schoolwork in Upper Secondary. In: Searson, M., Ochoa, M. (eds.) Proceedings of Society for Information Technology & Teacher Education International Conference 2014, pp. 2123–2129. AACE, Chesapeake (2014)
9. Hylén, J.: Lägesrapport om it i skolan. Nationellt forum för it i skolan (2013), http://www.skl.se/MediaBinaryLoader.axd?MediaArchive_FileID=b691ea66-750c-4e9d-a862-8c6ad9d40a34&FileName=Nationellt+forum+f%C3%B6r+digitaliseringen+av+skolan.pdf (accessed August 5, 2014)
10. Håkansson Lindqvist, M.J.: Possibilities and challenges for TEL from a student perspective through the uptake and use of digital technologies in a 1:1 Initiative. Education Inquiry 4(4), 629–647 (2013)
11. Kakihari, M., Sørensen, C.: Mobility: An extended perspective. In: Proceedings of the 35th Hawaii International Conference on Systems Sciences, pp. 1756–1766 (2002)
12. Nordicom: Tillgång till medier i hemmet 2013 (procent) (n.d.), http://www.nordicom.gu.se/sites/default/files/medieforskning-statistik/10410_tillgang_i_hemmet_i_grupper_2013.xlsx (accessed August 5, 2014)
13. Norris, C., Soloway, E.: Why Is Mobile Technology Different From Other Technology? Among many other reasons, students will be using their own devices (2010), http://www.districtadministration.com/article/why-mobile-technology-different-other-technology (accessed April 11, 2014)
14. O'Bannon, B., Thomas, K.: Teachers perceptions of using mobile phones in the classroom: Age matters! Journal of Computers & Education 74, 15–25 (2014)
15. OECD. Connected minds: Technology and today's learners, Educational Research and Innovation. OECD Publishing (2012)
16. Ott, T.: A Historical Materialist Analysis of the Debate in Swedish Print Media on Mobile Phones in School Settings. International Journal of Mobile and Blended Learning (2014)
17. Phelan, P., Davidson, A.L., Cao, H.T.: Students' multiple worlds: Negotiating the boundaries of family, peer, and school cultures. Anthropology and Education Quarterly 22, 91–129 (1991)
18. Player-Koro, C., Karlsson, N., Ott, T., Tallvid, T., Lindström, B.: Utvärderingsstudie av Jönköpingskommuns satsning på IKT för att utveckla elevers lärande I de kommunala gymnasieskolorna. University of Gothenburg (2014)
19. Player-Koro, C., Tallvid, M., Lindström, B.: Traditional teaching with digital technology. In: Searson, M., Ochoa, M. (eds.) Proceedings of Society for Information Technology & Teacher Education International Conference 2014, pp. 947–951. AACE, Chesapeake (2014)

20. Samuelsson, U.: Digital (o)jämlikhet? IKT-användning i skolan och elevers tekniska kapital (Doctoral dissertation). School of Education and Communication, Jönköping (2014)
21. Skolverket: It-användning och It-kompetens i skolan. Skolverket, Stockholm (2013)
22. Sharples, M., Taylor, J., Vavoula, G.: A Theory of Learning for the Mobile Age. In: Andrews, R., Haythornthwaite, C. (eds.) The Sage Handbook of Elearning Research, pp. 221–247. Sage, London (2007)
23. Sharples, M.: Mobile learning: Research, practice and challenges. Distance Education in China 3(5), 5–11 (2013)
24. Star, S.L.: This is not a boundary object: Reflections on the origin of a concept. Science, Technology & Human Values 35(5), 601–617 (2010)
25. Traxler, J., Vosloo, S.: Introduction: The prospects for mobile learning. PROSPECTS (2014), doi:10.1007/s11125-014-9296-z
26. Wenger, E.: Communities of practice: Learning, meaning, and identity. Cambridge university press, New York (1998)

Students' Perceptions of Mobile Devices Usage during Class Time and Policy Suggestions for Appropriate Practices

Ieda M. Santos[1,*] and Otávio Bocheco[2]

[1] Emirates College for Advanced Education, Abu Dhabi, United Arab Emirates
isantos@ecae.ac.ae
[2] Instituto Federal Catarinense Campus Rio do Sul, Santa Catarina, Brazil
o.bocheco@ifc-riodosul.edu.br

Abstract. This paper reports preliminary findings based on a survey that explored students' perceptions of mobile device usage in the classroom and policy suggestions for appropriate practices to use the devices during class time. The survey was administrated to undergraduate students enrolled in two courses in different countries namely Brazil and United Arab Emirates. Key results suggested that the Emirati and Brazilian students tended to rare use of mobile devices to engage in content and non-content related activities during class time. However, both groups are bringing their mobile devices to the classroom and have suggested policies for appropriate practices to use those devices in the classroom.

Keywords: Mobile devices, classroom, student perceptions, policies for appropriate practices.

1 Introduction

Mobile technologies such as mobile phones have evolved rapidly to become lightweight, more powerful and small enough to fit in one's pocket [1,2]. Smart phones and tablets, in particular, will continue evolve in capacity and functionality tending to replace laptops [3]. We are now living in the mobile age [4] where individuals use Internet-enabled mobile devices anytime and anywhere to perform a variety of activities. Inevitably, Higher Education (HE) institutions are experiencing an influx of students' personal mobile devices on their campuses.

Mobile technology ownership among university students is widespread, with predictions to increased ownership including among low income students [5, 6]. Students' personal mobile devices, as noted by Parsons [7, p. 223] "…[open] up major new opportunities for digital learning in the classroom, since the old constraints of having to provide all learning technologies from central resources gradually fade away." Research has shown that mobile devices can potentially support classroom

* Corresponding author.

M. Kalz et al. (Eds.): mLearn 2014, CCIS 479, pp. 81–91, 2014.

activities [8, 9, 10]. However, mobile technologies students bring to the classroom can disrupt lecturers [11, 12]. Universities are faced with challenges on how to implement effective policies regarding mobile device usage in the classroom [13]. Additionally, many instructors may not know how to deal with the disruptions and some may not even be aware there is a problem [14].

As observed by Parsons [7], it would be unwise to focus only on the positive aspects of mobile devices. In this respect, researchers have started acknowledging the importance of investigating the use of mobile devices in the classroom aiming to better understand disruptions caused by those devices as well as explore policies for appropriate practices [13], [15] ,[16]. Further research is needed to advance understanding on disruptions caused by mobile devices and inform policy development [16]. In addition, Tindell and Bohlander [13] recommended researchers to conduct studies with different groups of students and institutions to advanced knowledge in this field of inquiry. In line with the need for more research and taking into account the widespread ownership of mobile devices in our institutions, we explored student and faculty perceptions of mobile device usage in the classroom and policies for appropriate practices. The study was implemented in three undergraduate courses taught in three different countries namely Brazil, Egypt and the United Arab Emirates (UAE) during the academic year 2013-2014. The study adopted multiple methods of data collection including survey, focus group and individual interviews. This paper presents preliminary results based on the survey results administered to the students enrolled in the UAE and Brazilian courses only.

2 Literature Review

Students' personal mobile devices can potentially enhance classroom learning experience [10], [17]. However, they can also disrupt the lessons [12], [18]. Research has reported common disruptions caused by mobile devices during class time including phone ringing or vibrating, text messaging, sending or reading emails, and using social media such as Facebook and Twitter [15], [19], [20], [21]. These disruptions may impact on instructors' decision on whether they would allow personal mobile devices in their classes. For example, Thomas and Bannon's [11] research explored pre-service teachers' perceptions of using mobile phones in the classroom. The majority of the participants suggested class disruption and cheating as main barriers to using phones in class. The authors stressed that these concerns expressed by participants equally apply to other mobile technologies used in class such as iPads.

Researchers like Campbell [19] studied both faculty and student perceptions of mobile devices in the classroom as a source of disruption. Through a survey delivered across disciplines, Campbell [19] reported phone ringing as a source of in-class distraction for both faculty and students and support for policies restricting phone usage in class. However, further analysis suggested that, in comparison with older participants, the younger counterparts reported significantly less support for policies banning phone usage in class and more tolerance for phone ringing. Similarly, Burns and Lohenry [20] found that most of the faculty and students agreed that mobile phones were a source of disruption in class. Additionally, more than half of the students were aware of policies regarding phone usage in class. Yet, 76.3% of the students suggested that their phone had rung during class time.

The above studies were more concerned with studying disruptions caused by mobile devices in the classroom, in particular, mobile phones. Others have investigated how students use mobile technology in the classroom to provide better understanding of disruptions and explore students' perceptions of policies for appropriate practices in the classroom [13], [16], [22]. Based on a survey administrated to undergraduate students, Jackson [16] found that many of the students used their devices for both academic and non-academic activities during class time. The study further reported that many students considered mobile phones as a distraction to their learning. In addition, most of the students suggested that instructors should have clear policies regarding mobile technology use in the classroom. A survey conducted by Tindell and Bohlander [13] indicated that the majority of students admitted sending or receiving text messages during class time, with some even indicating texting every day. Students also mentioned they had observed their peers performing a similar activity during class time.

Baker, Lusk and Neuhauser [15] used a comprehensive survey to investigate both student and faculty perceptions regarding use of mobile phones and other devices in the classroom at three public universities in the United States. Contrary to other studies [19], Baker, Lusk and Neuhauser found significant differences between faculty and student perceptions of whether mobile devices disrupted lectures. The authors stated: "In every instance in which we found a difference, students perceived electronic devices to be more appropriate, and less disruptive, than faculty members did" (p .285). For example, both faculty and students stated that phone ringing bothered them; however, the former viewed phone ringing as more disruptive. In addition, students tended to favor democratically decided policies while faculty favored more university-wide policy regarding use of mobile devices in class. The survey also sought information on students' frequent use of mobile devices to perform certain activities. For example, on a typical semester, 71% of the students said they had never taken a call during class time. Many of the students were engaged in text messaging, checking emails or text message; however, the frequency varied. For example, a quarter of the students stated sending text messages in almost every class while 15% sent 5-10 text messages per week.

While the above studies relied more on surveys, Gikas and Grant [8] adopted a qualitative small scale study to investigate student and instructor perceptions of implementing mobile technology in their courses at three different universities in the United States. The study reported that some instructors from the three universities perceived the devices as inappropriate and asked students to store them away when entering the classroom. Findings further suggested that the average students at one university felt that, at times, mobile devices can be disruptive. The results indicated that social networking applications used for non-academic activities potentially threatened their concentration. In contrast, older students from one of the universities felt the devices were not disruptive.

With regard to policies to guide appropriate practices in the classroom, McCoy [22] found that the majority of the participants were in favor of policies to manage disruptions caused by mobile devices in the classroom. However, Baker, Lusk and Neuhauser [15] suggested that administrators should consider both student and faculty opinions before creating mobile device usage policies for the classroom. In a similar vein, Jackson [16] reported that students would like instructor to consider their opinions when suggesting policies to manage mobile devices in the classroom.

3 Context and Participants

This paper is based on two case studies that explored students' perceptions of mobile devices usage in the classroom and policy suggestions for appropriate practices. Study 1 was conducted at a Higher Education institution in the United Arab Emirates (UAE). A 15-week course offered within the Bachelor of Education program in Year 1 was selected for the study. Nineteen out of 20 Emirati female students enrolled in the course gave their consent to participate in the study. More than half of the Emiratis (53%) were under 20 years old. Thirty five percent were within the age group 20-25. One student was in the 26-30 age group, while another was in 41-45 age group. Study 2 was conducted in a 20-week course within the Physics Education program offered at a Federal Institute located in a regional area in the South of Brazil. Of the 26 students enrolled in the course, 19 (73%) gave their consent to take part in the study. The sample consisted of 53% females and 47% males. More than half (58%) were under 20 years old, 26% were within the 20-25 age group and 16% were in the 26-30 age group. Both courses were offered face-to-face in the first semester of the 2013-2014 academic year. The first author of this paper taught the UAE course and the other was taught by the second author. These two courses were chosen because the authors had easy access to the students.

4 Data Collection and Analysis

The project adopted multiple methods of data collection including survey, focus group, and individual interviews. This paper is based on the survey delivered to students enrolled in the two above courses. The survey aimed at obtaining information on demographics, students' perceptions of the use of mobile devices in the classroom and appropriate practices for using the devices during class teaching. The survey was considered the most appropriate approach because it allowed convenient data collection. It had a mixture of closed and open ended questions, multiple choices (a series of statements), ticking boxes, yes and no type of questions, and a 4-point rating scale consisting of frequently (1), occasionally (2), rarely (3) and never (4). The authors of the project developed an anonymous survey based on the literature on mobile learning [15], [19]. Some of the survey questions were adapted from Tindell and Bohlander [13] and McCoy [22] survey. A colleague in the field reviewed the survey questions for content validity. The survey was piloted with four students outside the study. Based on feedback received, some questions were revised to improve clarity. The first author translated the survey from English into Portuguese which was later reviewed by a Brazilian colleague for accuracy. We delivered a printed-based questionnaire to students. Eighteen Emirati students (95%) and all 19 Brazilian students taking part in the study completed and returned the questionnaire.

We examined the returned questionnaires for completion of all questions and accuracy such as checking the right box. Two Brazilian students skipped questions 6, 7, 8 and 9 in the questionnaire. The questionnaire was printed as double size and these students may have not noticed. We decided to include these two questionnaires in the analysis as other questions were completed. Next, and guided by Cohen, Manion and Morrison [23], analysis of closed questions from the two questionnaires consisted of assigning a code to each closed question (e.g. yes=1; no=2). On an Excel Spreadsheet (for each study), we used one column for participants' names and a column for each question where the coded information was added. For each column, percentages were calculated. We also performed mean calculation for questions 4, 5 and 7 consisting of a 4-point scale (frequently, occasionally, rarely and never). One of the questions in the questionnaire allowed multiple responses (a series of policy statements for appropriate practices). We organized students' responses which had similar options under categories (e.g. category 1, 2) and counted the number of students who fell under each category. Later, we calculated percentages of students choosing each category (Appendix A). The preliminary results presented in this paper are based on the analysis of the closed questions from the questionnaire.

5 Preliminary Results

The questionnaire sought information on which mobile device students usually brought to the classroom. Results show that the majority of Emiratis and all Brazilian students brought mobile phones or smart phones to the classroom (94% and 100% respectively). One Emirati usually brought an iPad to class. In comparison, the Brazilian students had more Internet access on mobile devices they brought to the class (79%) than did the Emiratis (61%). As seen in Table 1, the majority of the Brazilians had their devices set to either vibrate or on silence mode during class time. The striking difference is that 22% of the Emiratis had their devices set to ring during class time.

Table 1. Students' mobile devices mode during class time

When you are in class, your mobile device is:	UAE Class (N=18) %	Brazil Class (N=19) %
Turned off	0	5
Set to vibrate	67	42
On silence mode	11	53
Set to ring	22	0

Table 2 indicates the frequency of students' use of mobile devices to perform content and non-content related activities during class time. Generally speaking, both groups tended to rarely use their mobile devices to perform content and non-content related activities. There is, however, a slightly tendency for the Emiratis to occasionally access WhatsApp or similar software to engage in content and non-content related activities as well as surf the web for content. Table 2 also suggests a slightly tendency to occasional use of SMS to send and receive text message related to non-content within the Brazilian and Emirati group. As displayed in Table 2, there is also a slightly tendency to occasional use of SMS to access content by the Emiratis.

Table 2. Use of mobile devices during class time to perform content and non-content related activities (Scale 1-4)

Statement	Content related activities		Non-content related activities	
	UAE Class (N=18) Mean	Brazil Class (N=19) Mean	UAE Class (N=18) Mean	Brazil Class (N=19) Mean
Send and receive phone calls	2.83	3.26	2.78	3.00
Send and receive text (SMS)	2.67	2.89	2.61	2.68
Send and receive emails	2.83	3.26	3.17	3.42
Access WhatsApp or similar software (e.g.Viber, Blackberry Messenger)	2.44	3.05	2.56	3.00
Access Twitter or Facebook	2.89	3.21	2.94	2.89
Web surfing	2.56	2.68	2.89	3.11
Other activities	3.22	3.05	3.17	3.16

Students were inquired how often they had observed their peers using mobile devices during class time. Within the Emirati group, Table 3 shows a clear tendency to occasional observations of peers using their mobile devices to perform the activities stated in Table 3. In contrast, there is a slightly tendency to frequent observations of peers using text message and social media during class time within the Brazilian group.

Table 3. Observation of peers using mobile devices in the classroom (Scale 1-4)

Statement	UAE Class (N=18) Mean	Brazil Class* (N=19) Mean
I have seen another student sending or receiving text message during class time.	1.83	1.47
I have heard mobile phones ringing or vibrating during class time.	1.89	2.18
I have seen another student using social media (e.g. Twitter, Facebook, WhatsApp) during class time.	1.89	1.35

*Two students did not answer the question

Students were asked to state how much distraction from learning mobile devices cause when used during class time to perform non-content related activities. Table 4 indicates differed opinions between the two groups. The majority of the Brazilians felt that mobile devices cause more than little distraction, with 56% choosing they cause a big distraction from learning. In contrast, many of the Emiratis seemed less distracted by mobile devices in class, with only 28% feeling they cause more than a little distraction from learning.

Table 4. Distraction from learning caused by mobile devices

Statement**	UAE Class (N=18) %	Brazil Class* (N=19) %
No distraction from learning	16	0
Little distraction from learning	56	6
More than a little distraction from learning	28	38
Big distraction from learning	0	56

*Three students did not answer the question
** Statements adapted from McCoy [22]

Students were asked to choose policies for appropriate practices regarding the use of mobile devices during class time. Analysis did not suggest a strong pattern within the two groups (see Appendix A for categories and policy statements under each category). Within the Emirati group, five categories emerged in which two included a single policy statement. For example, 28% of the Emiratis indicated that mobile devices may be used to send and received text messages during class time as long as they do not distract other students. Across the five categories, four included the policy statement that mobile devices must be placed on vibrate or silence mode during class time. None of the categories included the policy statement that mobile devices must be turned off during class time.

Within the Brazilian group, six categories emerged in which one included a single policy statement (Appendix A). For example, 35% of the students suggested three policies which include 1) mobile devices must be placed on vibrate or silence, 2) they may be used to send or receive SMS as long as the devices do not distract others, and 3) the devices should be allowed to use for class related activities. Across the six categories, five included the policy statement that mobile devices should be allowed to use for class related activities. In contrast to the Emiratis, one of the categories included the policy statement that mobile devices should be turned off during class time.

To complement to above information, students were also asked to choose which policy the teacher should apply in a hypothetical case a student caused class disruption by using his or her mobile device to perform non-content related activity. Table 5 shows that in both groups, less than half of students had a similar opinion that the device should be turned off. Comparatively, a higher percentage of the Brazilians felt the teacher should ask the student to leave the class.

Table 5. Teacher's application of policy

During class time, if a student causes disruption by using a mobile device for non-class activities, the teacher should*:	UAE Class (N=18) %	Brazil Class (N=19) %
Talk to the student	39	26
Ask the student to leave the classroom	17	32
Turn off the student device	44	42

*Adapted from McCoy [22]

6 Concluding Remarks

This paper reported preliminary findings based on a survey that explored students' perceptions of mobile device use in the classroom and suggestions for appropriate practices. It was found that all Emirati and Brazilian students usually brought a mobile device to the classroom. In both groups, evidence suggests an overall tendency to rare use of mobile devices to perform content and non-content related activities during class time. However, one could observe within the Emirati group, a slightly tendency to occasional use of a few applications to perform content and non-content activities. Within the Brazilian group, there was a slightly tendency to occasional use of SMS for non-content related activities. The overall findings differ from others who reported more frequent use of mobile devices during class time [16], [22]. A closer look at Table 3 suggests that both groups tended to occasionally observe their peers performing specific activities using their mobile devices while in class. Others have reported similar findings [13]. When looking across Tables 2 and 3, one could argue that some students within the two groups might be using more often their devices in class without being aware. The interviews conducted with students and instructors might shed light on student use of mobile devices in class.

Overall, findings suggest that both groups seem to favor policies for appropriate practices that do not involve turning off the devices during class time (Appendix A). However, a high percentage of the Emirati and Brazilian students agreed if a student disrupted the class, the teacher should turn off the device. Results suggest that the two groups might not accept classroom policies that ban or limit mobile devices in class. This contrasts with Campbell [19] who reported students' support for policies restricting mobile phone use in the classroom. Furthermore, the two groups differed in their perceptions of how much distraction from learning mobile devices cause and the status of their mobile devices while in class, with 22% Emiratis having their devices set to ring. Although this requires further investigation through the interviews, perhaps the differences in perceptions might be related to student age, students' characteristics, culture and context. This may imply that instructors need to consider these issues when creating classroom policies for mobile usage.

Within the Brazilian group, a common policy statement found in most of the categories (Appendix A) indicates that mobile devices should be allowed in the classroom to perform content related activities. Table 2 suggests a slightly tendency to occasional use of some applications for content related activities within the Emirati group. This may send a message to instructors to explore those devices to support teaching and learning as well as include as policy for appropriate practices.

References

1. Crompton, H.: A historical Overview of M-learning. In: Berge, Z.L., Muilenburg, L.Y. (eds.) Handbook of Mobile Learning, pp. 3–14. Routledge, London (2013)
2. Kukulska-Hulme, A.: Introduction. In: Kukulska-Hulme, A., Traxler, J. (eds.) Mobile Learning: A Handbook for Educators and Trainers, pp. 1–6. Routledge, London (2005)
3. Ally, M.: Mobile Learning: From Research to Practice to Impact Education. Learning & Teaching in Higher Education: Gulf Perspectives 10(2), 10–22 (2013)
4. Ally, M., Palalas, A.: State of Mobile learning in Canada and Future Directions (2011), http://www.rogersbizresources.com/files/308/Mobile_Learning_in_Canada_Final_Report_EN.pdf
5. Dahlstrom, E., diFilipo, S.: The Consumerization of Technology and the Bringing your Own Everything (BYOT) Era of Higher Education, Education Report (2013), http://net.educause.edu/ir/library/pdf/ERS1301/ers1301.pdf
6. Kobus, M.B.W., Rietveld, P., van Ommeren, J.N.: Ownership versus On-campus Use of Mobile IT Devices by University Students. Computers & Education 68, 29–41 (2013)
7. Parsons, D.: The Future of Mobile Learning and Implications for Education and Training. In: Ally, M., Tsinakos, A. (eds.) Increasing Access through Mobile Learning, pp. 217–229. Commonwealth of Learning and Athabasca University, Vancouver (2014)
8. Gikas, J., Grant, M.M.: Mobile Computing Devices in Higher Education: Student Perspectives on Learning with Cell Phones, Smartphones & Social Media. Internet & Higher Education 19, 18–26 (2013)

9. Lundin, J., Lymer, G., Holmquist, L.E., Brown, B., Rost, M.: Integrating Students' mobile Technology in Higher Education. International Journal of Mobile Learning and Organization 4(1), 1–14 (2010)
10. Santos, I.M.: Integrating Personal Mobile Devices in Teaching: The Impact on Student Learning and Institutional Support. Learning & Teaching in Higher Education: Gulf Perspectives 10(2), 43–63 (2013)
11. Thomas, K., O'Bannon, B.: Cell Phones in the Classroom: Pre-service Teachers' Perceptions. Journal of Digital Learning in Teacher Education 30(1), 11–20 (2013)
12. Sharples, M.: Disruptive Devices: Mobile Technology for Conversational Learning (2002), https://www.tlu.ee/~kpata/haridustehnoloogiaTLU/mobilesharples.pdf
13. Tindell, D.R., Bohlander, R.W.: The Use and Abuse of Cell Phones and Text Messaging in the Classroom: A Survey of College Students. College Teaching 60(1), 1–9 (2012)
14. Dodds, R., Mason, C.Y.: Cell phones and PDA's Hit K-6. Education Digest 70, 52 (2005)
15. Baker, W.M., Lusk, E.J., Neuhauser, K.L.: On the Use of Cell Phones and Other Electronic Devices in the Classroom: Evidence from a Survey of Faculty and Students. Journal of Education for Business 87(5), 275–289 (2012)
16. Jackson, L.D.: Is Mobile Technology in the Classroom a Helpful Tool or a Distraction? A Report of University Students' attitudes, Usage Practices, and Suggestions for policies. The International Journal of Technology, Knowledge and Society 8(5), 129–140 (2013)
17. Scornavacca, E., Huff, S., Marshall, S.: Mobile Phones in the Classroom: If You Can't Beat Them, Join Them. Communication of the ACM 52(4), 143–146 (2009)
18. Fang, B.: From Distraction to Engagement Wireless Devices in the Classroom. EDUCAUSE Quarterly 32(4) (2009), http://www.educause.edu/ero/article/distraction-engagement-wireless-devices-classroom
19. Campbell, S.W.: Perceptions of Mobile Phones in College Classrooms: Ringing, Cheating, and Classroom Policies. Communication Education 55(3), 280–294 (2006)
20. Burns, S.M., Lohenry, K.: Cellular Phone Use in Class: Implications for Teaching and Learning a Pilot Study. College Student Journal 44(3), 805–810 (2010)
21. Geist, E.: The Game Changer: Using iPads in College Teacher Education Classes. College Student Journal 45(4), 758–768 (2011)
22. McCoy, B.: Digital Distractions in the Classroom: Student Classroom Use of Digital Devices for Non-Class Related Purposes (2013), http://digitalcommons.unl.edu/journalismfacpub/71
23. Cohen, L., Manion, L., Morrison, K.: Research Methods in Education, 6th edn. Routledge, London (2007)

Appendix A

Category	Brazil Class (N=19)*		UAE Class (N=18)	
1	1. Mobile devices must be placed on vibrate or silence mode. 2. Mobile devices may be used to send and receive text messages during class as long as they do not distract other students. 3. Mobile devices should be allowed to use for class related activities.	35 %	1. Mobile devices may be used to send and receive text messages during class as long as they do not distract other students.	28 %
2	1. Mobile devices must be placed on vibrate or silence mode. 2. Mobile devices must not be used to access social media during class time to discuss non-related class activities. 3. Mobile devices should be allowed to use for class related activities.	24 %	1. Mobile devices must be placed on vibrate or silence mode. 2. Mobile devices must not be used to access social media during class time to discuss non-content related activities. 3. Mobile devices should be allowed to use for class related activities.	22 %
3	1. Mobile devices must be placed on vibrate or silence mode. 2. Mobile devices may be used to send and receive text messages during class as long as they do not distract other students.	18 %	1. Mobile devices must be placed on vibrate or silence mode. 2. Mobile devices may be used to send and receive text messages during class as long as they not distract other students.	22 %
4	1. Mobile devices must not be used to access social media during class time to discuss non-related class activities. 2. Mobile devices should be allowed to use for class related activities.	12 %	1. Mobile devices must be placed on vibrate or silence mode. 2. Mobile devices may be used to send and receive text messages during class as long as they do not distract other students. 3. Mobile devices should be allowed to use for class related activities.	11 %
5	1. Mobile devices must be turned off during class time. 2. Mobile devices must not be used to access social media during class time to discuss non-related class activities. 3. Mobile devices should be allowed to use for class related activities.	6 %	1. Mobile devices must be placed on vibrate or silence mode.	17 %
6	1. Mobile devices should be allowed to use for class related activities.	6 %		

*Two Brazilian students did not complete the question.

In the Eye of the Beholder: Promoting Learner-Centric Design to Develop Mobile Games for Learning

Birgit Schmitz, Marco Kalz, Roland Klemke, and Marcus Specht

Welten Institute – Research Centre for Learning, Teaching and Technology
Faculty of Psychology and Educational Sciences
Open University of the Netherlands
NL-PO Box 2960, 6401 DL Heerlen, The Netherlands
{birgit.schmitz,marco.kalz,roland.klemke,
marcus.specht}@ou.nl

Abstract. Out of the project EMuRgency a game-based learning environment evolved, which trains school children in providing reanimation and cardiopulmonary resuscitation (CPR). The application gets players to act as if they were in a real case of emergency. This paper reports on a formal usability study conducted with two different groups of learners, regular learners and learners with special educational needs (SEN). With the study we compared the two groups of learners with regard to game usability and effectiveness of the intervention. Our intention was to better understand the different needs and requirements to learning materials that game designer need to take into consideration in order to make the learning experience successful for both groups. A total of 89 children played the game simulation. Results showed differences in perception and effectiveness of individual mechanisms for the two groups with regard to usability or switching between tasks and mobile device.

Keywords: games for health, mobile learning, serious games, learner centric design.

1 Introduction

The need for constantly applying new motivational strategies and providing innovative learning scenarios is comprehensive and not restricted to a particular target group. Mobile technology is currently regarded one foothold to react to this demand and bring sustainable change to the classroom. The interactive nature of mobile media enhances the learning experience by augmenting objects, places and activities thus adding context to learning objects and inherently supports the concept of student centred, active learning approaches [27] such as inquiry-based learning or discovery learning. Still, educational practitioners seem to be reluctant in using this technology. It is arguable that beyond learning theory und motivational aspects, e.g. school children's attraction for the device, its perceived potential to deliver relevant knowledge is not consistently clear. While more mobile applications are being developed for educational settings, little is known about how the learning process can be transformed

M. Kalz et al. (Eds.): mLearn 2014, CCIS 479, pp. 92–106, 2014.

when using them [20]. A particular challenge is to find adequate curricular functions in school where the inclusion of these new cultural resources can and should be introduced [7]. Also, most of the devices used for learning were not designed with educational applications in mind and usability issues are often the case. In order to support the use of mobile devices for educational purposes, usability issues should be tracked over a longer period, from initial use through to a state of relative experience with the technology [12]. Still, many digital tools for learning do not take this into consideration.

In order to provide effective tools for learning, it is important to consider learning over technology, i.e. considering how technology can be used to provide different groups of learners with better opportunities and enhanced learning outcomes. The need to align learning materials to its target audience seems to go without saying. There has been common understanding of learning and motivation as 'natural processes that occur when the conditions and context of learning are supportive of individual learner needs, capacities, experiences and interests [16, p. 5]. Along these lines, the process of developing digital learning offers needs to focus on the tasks learners must undertake, the tools they can use to deal with those tasks and the interface for those tools [28]. As suggested in the learner-centred framework [16], learners must be supported in their diverse needs and capacities. This is especially true for people with disabilities who can benefit from the successful integration of technology based learning offers in the educational scenario, creating a safe and personalized educational environment for the students [22]. Brown, McHugh, Standen, Evett, Shopland, and Battersby [6] in their article on the design of location-based learning experiences for people with intellectual disabilities and additional sensory impairments provide concrete examples of evidence for the advantages that computer games provide for this particular target group. They argue that via computer games, learning can take place in a context similar to that in which it is required. This for example is important for a target audience described as 'concrete thinkers' whose performance is characterized as rigid and context dependent. Especially the notion of 'context-aware learning', which is of eminent importance for this target audience, finds translation in mobile devices. This technology is especially well suited to context-aware applications simply because it is available in different contexts [6]. Even though scientists have increasingly started to consider how technology can support the needs and capacities of learners, there still is a surprising lack of systematic evaluations that investigate the benefits of the 'new' communication technologies for SEN learners with more complex and severe communication and language needs at different stages of their education [8], [34].

This study investigates how SEN learners compared to average pupils engage with the mobile learning game in order to evaluate its design and implementation. In order to further specify the SEN group of learners we apply the definition of Davis and Florian [8]. In their study, several groups of children were described, among them children with speech, language and communication needs (SLCN). They are referred to as having specific speech and or language impairments (SSLI), or as children with specific speech and or language difficulties (SSLD) who are unable to express themselves in the normal effortless way, and where the difficulty cannot be attributed to physical or sensory impairments, [3], [1] as cited in Davis and Florian [8]. They often have difficulties in learning to read, write and spell, in processing information and in sequencing and organizing activities. Research has provided evidence that educational software is capable to countervail this potential for development by providing chances to experience everyday situation, to explore curriculum related topics such as mathematics or reading and vocabulary, to support attention or to promote problem solving skills [10], [35]. Also, Smith [26] in her study emphasizes that mobile technology is effective at facilitating and improving digital communication skills for students with intellectual disabilities, this way helping them to develop self-determination skills and increased autonomy.

Our research aims at educational interventions for improving school children's' understanding of cardiopulmonary resuscitation (CPR) and engage them in practice. Addressing school children is regarded one toehold to enlarge the number of trained laymen who can provide CPR in case of emergency. Plant and Taylor [17] in their review point out diverse methods of first aid training that have been successful with children. They state that especially the use of "virtual worlds and multiplayer online simulation could be an attractive training and/or retention tool to use in this age group" (p. 3). Building on this evidence, we set up the mobile game simulation *HeartRun*. It uses smartphones as a location-based gaming platform to enable an authentic, simulation like environment. Based on different roles it prompts learners to help the victim or to get an automated external defibrillator (AED) in order to help. Using mobile phones allows us to augment the situation with location dependent information as well as instructive, situation-dependent educational materials, to structure the process by sending out notifications monitor and to record user behaviour, which can be reused for later reflection and feedback sessions. A recent review by Zhu, Hadadgar, Masiello and Zary [36] corroborates this and in the context of medical education stresses the potential of augmented reality (AR) to improve health care learning from different educational objectives. Their work stresses AR to be "not only useful for healthcare learner understanding the spatial interrelationships and concepts, acquiring the skill and knowledge, strengthening cognitive psychomotor, shortening learning curve and prolonging learning retention, but also is subjective attractiveness for student because it provided them the authentic simulated experiences" (p. 15).

With this paper we further contribute to research in the field. We present an AR learning game environment, which trains and increases procedural CPR knowledge, and evaluate its use for two different groups of learners, regular learners and learners with special educational needs (SEN). The studies assessed game usability and effectiveness of the intervention, i.e. knowledge gain. The mobile simulation game in use was not explicitly designed as a learning tool for SEN students but was used to train them in the course of the general CPR school trainings. We report results from the formal usability studies and direct to the differences that surfaced in the perception of the educational mobile game by these two groups. Our focus was to better understand the different needs and requirements to learning materials that game designer need to take into consideration to make a learning experience successful for both groups of learners. This paper comprises a description of the research design and summary of the main data collecting methods that we adopted. This is followed by a description of the game-concept. We conclude by discussing and highlighting design implications, which we derived from the research data.

2 Related Work

For the past decade, the field of health education and health care has started to investigate how digital games can assist in teaching first-aid concepts and cardiopulmonary resuscitation (CPR). The range of games currently available provides vital functional elements to address the problems related to health emergency training [18]. *Staying alive*[1], for example, is a video-based 3D game experience that requires players to apply their CPR knowledge to a virtual person suffering from cardiac arrest. Other examples are *LIVESAVER*[2], a crisis simulator that fuses interactivity with live-action film to show how to respond to heart attacks or *Relive* [23], a first person 3D adventure to increase kids and young adults' awareness on the need of instant CPR. The game prompts to attend CPR classes and prepares for intervention in case of cardiac arrest.

By now, research offers little information on the outcomes of using games for teaching CPR. Marchiori, Ferrer, Fernandez-Manjon, Povar-Marco, Suberviola, and Gimenez-Valverde, [15] evaluated the *First Aid Game*[3], a video-based game commissioned by the Aragonese Center for Educational Technologies (CATEDU) that teaches high school students basic first-aid and reanimation procedures. Their study results indicate that unsupervised use of the video game in a single 45-minute session significantly improved their knowledge of the action protocol and the procedures involved (p. 435). Also, Tüzün [30] investigated the use of video games to teach first-aid knowledge. The study points to their potential in yielding positive pedagogical changes, e.g. more active learning modes.

With regard to mobile game-based learning, several approaches have come into existence that aim at improving the performance of an advanced life support provider in a simulated emergency situation, for instance [14]. Other examples are *Icpr*, [24], an iPhone application designed for laypersons and healthcare professionals. It is able to detect the rate of chest compressions performance by using the built-in accelerometer. *Viva! CPR*[4] [23] is a smartphone application designed by the Italian Resuscitation Council that provides real time feedback on chest compression quality. Recent innovative developments are more comprehensive such as the approach by Wattanasoontorn, Magdics, Boada, and Sbert, [33]. They describe a Kinect-based system for *LISSA*, (*LI*fe *S*upport *S*imulation *A*pplication), which is able to provide feedback on the performance of specific parameters of the CPR procedure, i.e. chest compression rate and correct arm position. Results showed that using *LISSA* in combination with mannequin-based simulation led to improved achievement of teaching and learning

[1] `http://www.stayingalive.fr/`

[2] `https://life-saver.org.uk/`

[3] `http://first-aid-game.e-ucm.es/`

[4] `http://www.viva2013.it/vivacpr`

outcomes [32]. To the best knowledge of the authors, none of the existing applications to train resuscitation and CPR is based on a mobile simulation game approach.

3 Instructional Design

Drawing on previous results in the field of mobile learning and game-based learning for health education we set up the mobile game simulation *HeartRun*. We took the decision to base our CPR training on a mobile platform because we wanted to have a scenario that:

a) is authentic, i.e. prepares learners to react adequately in a closely related situation. Within the game just as in reality, notification systems are the base for first responders. Dispatch centres send them in case of emergency.
b) includes different locations and different roles. Within the game just as in reality, different roles are involved in case of emergency. While one person gets the nearest automated external defibrillator (AED), another person already starts providing CPR to help the victim. A mobile game can include different locations and different roles.
c) allows for situation-depended educational materials. In case of emergency several steps need to be taken that are spatially distributed and that need to be accessed at different times. Simple augmentations of physical objects, places and situations can provide necessary information at the time needed.

In order to realize features frequently associated with mobile games, i.e. cooperative action between team players that have different tasks or roles and time-critical orientation in physical space [4], we based our development on the game engine ARLearn. It almost comprehensively reflects the required features. ARLearn is an open source tool suite for educators and learners [29]. Its android client allows playing a run with mobile users. By building on the Google App Engine (GAE) stack, the server architecture offers a scalable web service for content and notification management as well as game state persistence [29].

Comparable to an unexpected emergency, our training environment involves instant decisions on what to do and the recall of CPR knowledge under unexpected circumstances involving time pressure and stress. This way we intend to enhance psychological preparedness of the rescuer and thus achieving a more prompt and appropriate response. Social psychology presumes that 'the more realistic the situation is made, the closer it is to reality, the better able the subject should be to imagine what he would actually do if he were in the real situation' [9; p. 108]. The game is laid out to complement traditional resuscitation training.

HeartRun is played in teams of two. Every team player is randomly assigned to one of the roles (player role AED and player role CPR). Opening the game, the first message shows. According to the operating mode of ARLearn, opening a message automatically triggers a new message. The set-up of the individual messages and the corresponding learning content is related to the *Chain of Survival*, i.e. (a) to prevent cardiac arrest, (b) to buy time, (c) to restart the heart and (d) to restore quality of life.

The line of action described there is reflected in the sequence of messages, e.g. making sure that the victim and any bystanders are safe, checking the victim for a response by gently shaking his/her shoulders and asking loudly: "*Are you all right?*"

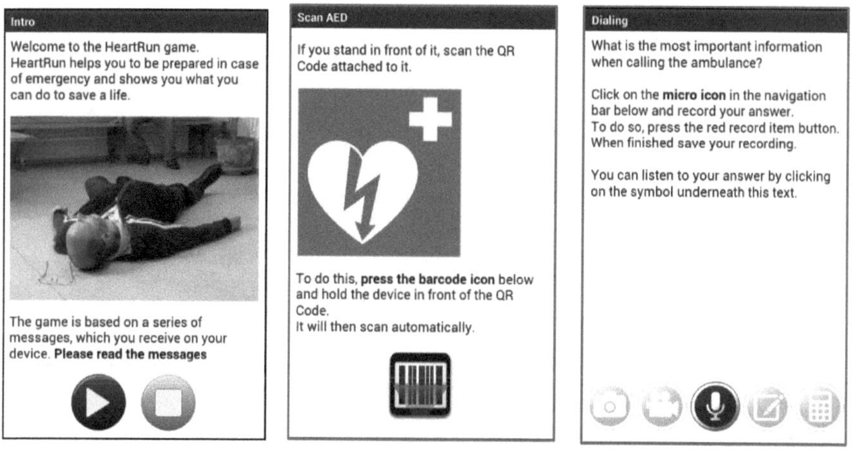

Fig. 1. Game screens *HeartRun*: first message task, and instruction screen

Time and location play an important role in the scenario. While Player A (role AED) heads for the AED, player B (role CPR) runs to the victim to provide CPR. At the scene of emergency a manikin is provided with which player B interacts. Meanwhile, player A searches the next AED. As soon as s/he found it and scanned the QR code attached to it, information on the AED is provided and the game requires him/her to bring the AED to the victim. At the scene of emergency another QR code is placed, which payer A has to scan. This synchronizes the players. Henceforth, both roles get the same information on how to correctly apply the AED. Both players follow the instructions on the screen and apply the AED to the manikin until the ambulance arrives (last message).

HeartRun can be played several times, with participants switching roles. This way, the game allows students to perceive the emergency situation from different perspectives. By "putting oneself into other's shoes" they have the chance to experience and control both feelings of panic and fear of emergency.

4 Method

The two studies under analysis were conducted between September 2013 and February 2014. Both studies applied the same method and measures to catch positive and negative aspects of the activity. They similarly involved (a) an introduction phase where players were presented a short introduction to the game, e.g. how to read QR codes with a telephone and (b) the game phase with pupils playing *HeartRun*. For the gaming sessions, we relied on a number of smart phones we brought along and randomly assigned pupils to one of the two roles involved in the teams (player role AED and player role CPR). While playing, intervention was kept to a minimum. The researcher's role during the case study was participant observer. Participants had to fill in a form after they had played the game.

With the studies we aimed at obtaining information on (a) how different learners perceive the game simulation, i.e. system usability, game-play experience, attitude towards the use of educational games in general and (b) their performance (knowledge gain) as well as their self-assessed learning outcomes.

4.1 Participants

A total of 89 children played the game simulation. The sample size comprised pupils from two different schools with two groups of learners respectively: Group A consisted of 53 subjects from a regular school in Landgraaf. Pupils were between 10 and 15 years of age (M = 13 years, SD = 1.109). Group B encompassed 36 learners from a school for pupils with special educational needs in Stolberg. Pupils were between the ages of 12 and 18 years (M = 15 years, SD = 1.499). None of the pupils had played *HeartRun* before.

4.2 Measures

For the data collection we mixed qualitative data, such as videos and field notes of researchers' observations and focus groups with quantitative data coming from a questionnaire, which was largely based on the standardized system usability scale (SUS). Quantitative data was analysed with Excel and the SUS guide and calculator package [21].

Learners' Perception of the Game Simulation. To measure learners' perception of the game simulation, we assessed system usability, learnability, game-play experience and attitude towards the use of educational games in general. In order to assess system usability and learnability, we chose the SUS because it is an accepted measure that makes results easily comparable [13]. We used the standard Overall SUS score, instead of the 8-item Usability scale, because the Overall SUS allows estimating perceived Learnability along with perceived Usability [13]. The items had to be rated on a 5-point Likert-scale ranging from 1=strongly disagree to 5=strongly agree.

In order to further specify usability aspects from SUS feedback, we added four bound questions and one open text question in order to further improve the design of the simulation game as well as event log files generated automatically by the game (archived voice recordings and pictures taken). The bound questions provided a range of five to eight possible answers and learners were asked to mark any answer they considered appropriate.

Learners' Performance. To measure learner performance, we distributed a knowledge test after the intervention in order to evaluate CPR knowledge. The test comprised four question items that were introduced and validated by medical experts in the course of the project EMuRgency [11]. Pupils had to mark their answer in a box. Furthermore, we added a bound question to contrast results from the knowledge-test with pupils' self-assessed learning outcomes. The increase in active CPR knowledge, i.e. depth of pressure or frequency of pressure rates, was not assessed.

5 Results

In the following we present the results from our study. This section is structured according to the two different points of view we focus on: learners' perception of the game simulation indicating system usability, learnability, game-play experience and attitude towards the use of educational games, and learners' performance indicated by their knowledge gain.

5.1 Learners' Perception of the Game Simulation

From the studies it showed that pupils from both groups readily engaged with *HeartRun*, and found it to be a rewarding and stimulating experience. Children felt engaged to learn by playing the mobile game application with the phones, and 67 out of 89 children claimed they would like to use such systems in classes more often.

With regard to system usability and learnability, i.e. the ease of getting used to the application, the *HeartRun* questionnaire reveals different values for the two studies. The overall SUS for Group A shows a mean SUS score of 53.3 (SD=17.5). The internal reliability as measured by Cronbach's alpha is calculated at 0.754. Results indicate a mean SUS score of 52.8 for usability and 55.0 for learnability. A SUS score between 51.2 and 62.4 could be considered grade D, i.e. a marginal level and on an adjective rating scale could be described as OK [2]. Results from Group B show an overall SUS of 69.6 (SD=22.0). The internal reliability as measured by Cronbach's alpha is calculated at 0.879.

Results reveal a mean SUS score of 72.9 for usability and 56.4 for learnability. According to standardized interpretation, a SUS score between 64.6 and 70.9 equals grade C, i.e. an acceptable level and on an adjective rating scale could be described as Good [2]. A mean SUS score above 72.4 equals grade B and on an acceptable level could be described as good to excellent (Bangor, et al., 2009). Table 1 compares the average values per questionnaire item from the two studies.

Table 1. Comparison of SUS scores between group A and group B

Items	Group A M n=53	Group B M n=36
I think that I would like to use this game frequently to train BLS.	2.68	3.89
I found the game unnecessarily complex.	3.06	1.92
I thought the game was easy to use.	3.42	3.81
I think that I would need the support of a technical person to be able to use this game.	2.92	2.72
I found the various functions in this game were well integrated.	3.34	3.97
I thought there was too much inconsistency in this game.	3.15	2.16
I would imagine that most people would learn to use this game very quickly.	3.38	4.12
I found the game very awkward to use.	3.06	2.03
I felt very confident using the game.	3.17	3.65
I needed to learn a lot of things before I could get going with this game.	2.66	2.76

Quantitative feedback analysis shows that the inclusion of physical tasks was motivating and enabled participants to enhance their skills. Both groups of pupils rated the integration of time-critical physical tasks, i.e. running to the victim and saving the victims life by interacting with the manikin and providing CPR, to be strong motivational factors when playing the game. Table 2 compares results between the two groups. From this, it showed that pupils from study two were more engaged in game related activities than pupils from study one. Game elements such as *looking at the videos* were rated less important by children of group B (SEN) for their involvement in the game and considered them rather disturbing.

Table 2. Importance of topics related to resuscitation and CPR as a percentage of the sample, comparison of results between group A and group B

	Group A (n=53)			**Group B** (n=36)		
What was particularly important to you when playing the game?	**M**	**No CPR** (n=43)	**CPR** (n=10)	**M**	**no CPR** (n=13)	**CPR** (n=23)
Fulfilling all the tasks	28.30	32.56	10.00	55.56	69.23	47.83
Acting quickly	62.26	60.47	70.00	80.56	69.23	86.96
Reading the text messages completely	33.96	34.89	30.00	30.56	30.77	30.44
Looking at the videos	33.96	39.54	10.00	25.00	30.77	21.74
Saving the victim's life	60.38	58.14	70.00	77.78	76.92	64.29
Learning how to provide CPR	39.62	41.86	30.00	30.56	46.15	21.74
Learning how to use the AED	49.06	55.81	20.00	44.44	30.77	52.14

S 1: I found the videos disturbing.
S 2: If such a video shows and you look at it. I stood there, for example, round about 15 minutes and did not know what to do.
S 1: That is almost redundant [video]. If I am supposed to save a life and then, I have to learn something. [...] Researcher: Even though the videos were meant to information you what to do next?
S1: Yes even though, I would simply use audio items, because I do not look on the phone anyway.

When asked for the disturbances while playing the simulation game, a different picture shows. Pupils mostly stressed the several technical problems they experienced. The disturbances reported were mainly due to bad Internet connectivity, which is characteristically for many institutions. Bad network accessibility mainly hampered information access and caused a high degree of distraction and frustration. Due to connectivity problems some messages did not load correctly and/or immediately, videos did not show and scanning the barcodes did not always trigger the next item, thus they were stuck and the game did not progress. From table 3 it shows that pupils from group A considered videos and tasks more disturbing than SEN learners. Game data confirm this result. From the event log files and the archived game data it shows that learners from group A frequently skipped the voice recording task.

Table 3. Disturbances when playing the game, comparison between group A and group B

What has disturbed you when playing the game?		
	Group A n=53	**Group B** n=36
graphics/design	0.245	0.027
usablity	0.151	0.027
audios	0.189	0.054
videos	0.264	0.027
tasks	0.208	0.027
technical problems	0.396	0.432
nothing	0.245	0.514

The multitasking, i.e. moving in the physical environment and at the same time processing mobile information, was not considered problematic. When asked if the device was in the way participants from both groups responded along the same lines and one student poignantly put it: *If I can text friends while driving on my scooter, I can easily use a mobile device while running around in the school building.*

When asked for the kind of support they would have wished for in order to make the game work better for them, both groups frequently replied in favour of a usage scenario of 1:2 (as in pair work, sharing a device). They argued that this way: "... they *could compare notes. One deals with the device, the other one acts. This way both players could help each other.*

S 1: The fact that two people can make it together, running and so.
Researcher: Two people with one cell phone?
S 1: Yes S 2: To discuss what to do next.
S 1: Yes, I think so too.

From the qualitative feedback it showed that both groups valued the list of message items because it helped them to structure the process and guided them through the course of action. For the SEN learners, the organisation of the message items was rather problematic. They had problems connecting individual messages and tasks to the overall process: "*The relationship of the messages was not clear. It was difficult to keep the overview. Maybe better use something else, like a list to check off.*"

5.2 Learners' Performance

With regard to learner performance, table 4 shows that pupils assessed their learning high when interaction with tangibles was involved, i.e. *providing CPR* (on the manikin) and *using the AED*. This is true especially for the SEN learners and for learners who had no resuscitation training beforehand. Results suggest that regular learners who already had resuscitation training rather took interest in different process oriented issues than learners who had never had resuscitation training before, e.g. *securing the scene of emergency, calling emergency*, or *opening the airway*. As for the SEN learners, there is little difference between pupils who already had resuscitation training and those who stated that they have never had resuscitation training before.

Table 4. Self-assessed learning outcomes as a percentage of the sample, comparison of results from group A and group B

	Group A (n=53)			Group B (n=36)		
Do you think you have learned something in the field of …?	M	**No CPR** (n=43)	**CPR** (n=10)	M	**no CPR** (n=13)	**CPR** (n=23)
Securing the scene of emergency	46.15	18.18	53.66	27.03	23.08	29.16
Checking the victim for a response	34.62	18.18	39.02	37.83	30.77	41.66
Calling emergency	36.54	36.36	36.59	32.43	30.77	33.34
Checking for normal breathing	25.00	36.36	21.95	32.43	23.08	37.50
Opening the airway	23.08	36.36	19.51	24.32	23.08	25.00
Providing CPR	30.77	9.10	36.59	59.46	61.54	58.33
Using the AED	59.62	54.55	60.98	59.46	53.85	62.50
Rescue Breathing	19.23	18.18	19.51	18.92	23.08	16.67

We added the knowledge test to our study in order to further detail the results and effects of mobile game applications for diverse groups of learners. From group A, a total of 24 pupils completed the knowledge test. With regard to group B, a total of 19 pupils completed the knowledge test. A total of 23 out of 24 pupils from group A already had resuscitation training. With regard to group B, 14 pupils out of 19 already had resuscitation training. Results from the knowledge test confirm the tendency of results pupils' self-assessed learning outcomes. Table 5 shows that pupils from group A attained higher scores than pupils from group B.

Table 5. Correct answers from the knowledge test from group A and group B

Knowledge Test	**Average number of questions answered correctly**	
	Group A (n=24)	**Group B** (n=19)
What is the European emergency number?	1.000	0.684
Where must you place your hands when providing Cardiopulmonary Resuscitation (CPR)?	0.958	0.579
People who have a cardiovascular arrest, do not move and show no respiratory movements.	0.875	0.789
What does this sign stand for? (AED sign)	1.000	0.189

6 Discussion

The primary purpose of this study was to evaluate design and implementation issues of a mobile game for learning that is meant to engage school children in using such an educational tool for learning. Its particular focus was on the possibilities technology offers to support the diverse needs and capacities of learners. By investigating the benefits of a simulation game for both regular and SEN learners it illustrates the differences, which individual design mechanisms have regarding learners' perception of the game simulation, i.e. system usability, game-play experience, and attitude towards the use of educational games in general and their impact on learners' knowledge gain.

Results from the usability study indicated that SEN learners were in favour of the game and considered it less complex. In general, they rated the system usability of *HeartRun* good. It can be argued that the introduction of innovative technology accounts for this result. For the SEN learners, using smartphones for learning was an absolute novelty. It showed that several of the SEN learners did not possess a state-of-the-art smartphone and therefore were very enthusiastic about using the devices for learning, which is reflected in their answers. On the other hand, learners from the group of regular education were not as positive, reflecting on the benefits and drawbacks as well as the appropriateness of the use of mobile technology. Almost all of them possessed a smartphone, many of them were state-of-the-art devices and on regular intervals they had access to new technology for learning.

Learners from both groups perceived physical activities, e.g. *acting quickly* and *saving the victim's life* as the most vital element for engagement. This is substantiated by connatural research. For example, Blum, et al., [4] report research wherein the initial task immediately put players into action and created a physical and emotional peak, which involved players in the simulation game. However, coordinating tasks such as receiving directions on the device while running through the physical environment needs careful consideration. In this context, audio emerged as a core design issue. Setting up the task with audio instructions that direct players through the environment avoids an unnecessary switch of visual focus (looking at the device to read subsequent instructions while players are already running to get to the scene of emergency), which is also a safety issue. This especially is true for children between 11 and 14 years, who quickly became immersed in the game and acted as if they were in a real emergency situation. However, an in depth analysis of qualitative and quantitative feedback also yielded three differences between the two groups of learners.

The first difference concerned the switch between the integrated tasks and the smartphone. Pupils from group A (learners from regular education) rated the videos and especially the tasks disturbing. Game data confirmed this result. From the event log files and the archived game data it showed that learners from group A frequently skipped the voice recording task, which could have enabled them to enhance their skills. Shatte, Holdsworth, and Lee [25] also state that dealing with mobile devices frequently requires multitasking, e.g. processing mobile information and learning, which is often taken as hurdle for learning. However, transcripts showed that pupils from group B took the effort to make use of this learning offer. Switching between device, task and physical environment is sometimes credited with a potential for *sensemaking activities* i.e. the device supports people in finding structure in an uncertain situation through using a combination of information, communication and computation [20]. Future studies should target the perception of task-based interactions in more detail for the reluctance of pupils to carry out tasks, e.g. video or voice recording tasks in the course of mobile learning experiences is a frequently reported phenomenon.

Second, collaboration and co-operation between learners showed to be of importance. From the qualitative feedback it surfaced that pupils strongly supported a *student – device ratio of 1:2*, in order to support discussion and collaboration among team members. The importance of co-operative action for pupils with learning disabilities (children with specific speech and or language difficulties) has been highlighted in several contexts. It allows exploring new technologies with a peer, which makes them feel more confident [5] and provides opportunity to apply interactive communication and language skills, which they have been taught elsewhere [35]. This accords with the ideas of Vygotsky's Zone of proximal development [31], which emphasizes the collaboration with peers as decisive element for developing skills and is corroborated by recent research.

The third difference was related to scaffold learning. The notion of scaffolding in the context of interactive learning environments refers to software as supporting instance for learners that structures activities by organized material and clear instructions for example [35]. Reiser [19] describes software scaffolding as providing "some aspect of support that helps make the learning more tractable for learners." (p. 275). In the context of *HeartRun*, sequencing elements, i.e. text message items, were provided that structured the resuscitation activities involved and reminded learners what steps to take. For SEN pupils this element provided orientation and it was considered helpful. However, for the SEN learners, the organisation of the message items was not intuitive.

7 Conclusion

This paper has reported on a formal usability study comparing SEN learners having specific speech and or language difficulties and learners from regular education. The mobile simulation game in use was not explicitly designed as a learning tool for SEN students but was used to train them in the course of the general CPR school trainings. From this, several design issues emerged regarding their perception of the game, which this paper collocated. Results clearly showed differences in the perception and effectiveness of individual mechanisms for the two groups with regard to switching between tasks and devices as well as physical movement and activities. The purpose of our studies was to turn the attention of educational designers to possible weak-points when designing technology-based learning offers for this particular target group.

From the gaming sessions it showed that mobile devices can effectively support real-life activities and facilitate the switching between such activities and tasks on the mobile, which is substantiated by connatural research [20]. Learners with special needs highly valued the application, even though the handling was hampered and less clear for them. What proved to be essential for them were the activity elements, the structured content and its relevance to real-life situations. Even though there is evidence that SEN learners gain much benefit from using ICT, there has been a lack of research investigating the differences of using ICT and more specific mobile technology for people with SEN [35].

Acknowledgement. This publication was partly financed by the European Regional Development Fund (ERDF), regions of the Euregio Meuse-Rhine and the participating institutions under the INTERREG IVa program (EMR.INT4-1.2.-2011-04/070, http://www.emurgency.eu).

References

1. Adams, C., Byers Brown, B., Edwards, M.: Developmental Disorders of Language. Whurr, London (1997)

2. Bangor, A., Kortum, M., Miller, J.: Determining What Individual SUS Scores Mean: Adding an Adjective Rating Scale. Journal of Usability Studies 4(3), 114–123 (2009)
3. Bishop, D.V.M.: Uncommon Understanding: Development and disorders of language comprehension in children. Psychology Press, Hove (1997)
4. Blum, L., Wetzel, R., McCall, R., Oppermann, L., Broll, W.: The final TimeWarp: Using form and content to support player experience and presence when designing location-aware mobile augmented reality games. In: Proceedings of the Designing Interactive Systems Conference, pp. 711–720. ACM Press, New York (2012)
5. Brown, D.J., Standen, P., Evett, L., Battersby, S., Shopland, N.: Designing serious games for people with dual diagnosis: learning disabilities and sensory impairments. Press in Educational Gaming. IGI Global (2010)
6. Brown, D.J., McHugh, D., Standen, P., Evett, L., Shopland, N., Battersby, S.: Designing location-based learning experiences for people with intellectual disabilities and additional sensory impairments. Computers and Education 56(1), 11–20 (2011)
7. Cook, J., Pachler, N., Bachmair, B.: Ubiquitous Mobility with Mobile Phones: A Cultural Ecology for Mobile Learning. E-Learning and Digital Media. Special Issue on Media: Digital, Ecological and Epistemological 8(3), 181–195 (2011)
8. Davis, P., Florian, L.: Teaching strategies and approaches for pupils with special educational needs: A scoping study. Department for Education and Skills, Nottingham (2004)
9. Freedman, J.L.: Role Playing: Psychology by Consensus. Journal of Personality and Social Psychology 13(2), 107–114 (1969)
10. Kalaitzaki, M., Saridiki, M., Tagoulis, A., Floratos, N.: Case study, based on GBL learning theories, on adult learners with general learning and cognitive disablilities. In: Proceedings of Interactive Technologies and Games Conference, Education vol. 56(1), pp. 81–93 (2009)
11. Kalz, M., Klerx, J., Parra, G., Haberstroh, M., Elsner, J., Ternier, S., Schilberg, D., Jeschke, S., Duval, E., Specht, M.: EMuRgency: Addressing cardiac arrest with socio-technical innovation in a smart learning region. Interaction Design and Architectures Journal (17), 77–91 (2013)
12. Kukulska-Hulme, A.: Mobile usability in educational contexts: what have we learnt? The International Review of Research in Open and Distance Learning 8(2) (2007)
13. Lewis, J.R., Sauro, J.: The Factor Structure of the System Usability Scale. In: Kurosu, M. (ed.) HCD 2009. LNCS, vol. 5619, pp. 94–103. Springer, Heidelberg (2009)
14. Low, L., O'Connell, M.: Learner-centric design of digital mobile learning. In: Proceedings of the OLT Conference, pp. 71–82 (2006)
15. Marchiori, E.J., Ferrer, G., Fernandez-Manjon, B., Povar-Marco, J., Suberviola, J.F., Gimenez-Valverde, A.: Video - game instruction in basic life support maneuvers. Emergencias 24, 433–437 (2012)
16. McCombs, B.L.: Assessing the Role of Educational Technology in the Teaching and Learning Process: A Learner-Centered Perspective (2000), `http://files.eric.ed.gov/fulltext/ED452830.pdf` (retrieved April 23, 2014)
17. Plant, N., Taylor, K.: How best to teach CPR to schoolchildren: A systematic review. Resuscitation 84(4), 415–421 (2013)
18. Ponder, M., Herbelin, B., Molet, T., Schertenlieb, S., Ulicny, B., Papagiannakis, G., Thalmann, D.: Interactive scenario immersion: Health emergency decision training in JUST project. In: VRMHR Conference Proceedings (2002)
19. Reiser, B.J.: Scaffolding complex learning: The mechanisms of structuring and problematizing student work. The Journal of the Learning Sciences 13(3), 273–304 (2004)

20. Rogers, Y., Connelly, K., Hazlewood, W., Tedesco, L.: Enhancing learning: A study of how mobile devices can facilitate sensemaking. Personal and Ubiquitous Computing 14(2), 111–124 (2010)
21. Sauro, J.: Measuring usability with the system usability scale (SUS) (2011), http://www.measuringusability.com/sus.php (last accessed January 23, 2014)
22. Saridaki, M., Mourlas, C.: Integrating Serious Games in the Educational Experience of Students with Intellectual Disabilities: Towards a Playful and Integrative Model. International Journal of Game-Based Learning (IJGBL) 3(3), 10–20 (2013)
23. Semeraro, F., Frisoli, A., Ristagno, G., Loconsole, C., Marchetti, L., Scapigliati, A., Pellis, T., Grieco, N., Cerchiari, E.L.: Relive: A serious game to learn how to save lives. Resuscitation (2014), http://dx.doi.org/10.1016/j.resuscitation.2014.03.306
24. Semeraro, F., Taggi, F., Tammaro, G., Imbriaco, G., Marchetti, L., Cerchiari, E.L.: icpr: A new application of high-quality cardiopulmonary resuscitation training. Resuscitation 82(4), 436–441 (2011)
25. Shatte, A., Holdsworth, J., Lee, I.: Hand-Held Mobile Augmented Reality for Collaborative Problem Solving: A Case Study with Sorting. In: Proceedings of the 47th Hawaii International Conference on System Sciences (HICSS), pp. 91–99 (2014)
26. Smith, C.C.: Using Mobile Technology to Improve Autonomy in Students with Intellectual Disabilities in Postsecondary Education Programs. PhD diss., University of Tennessee (2013), http://trace.tennessee.edu/utk_graddiss/2485 (last accessed April 22, 2014)
27. Sokoloff, D.R., Thornton, R.K.: Using interactive lecture demonstrations to create an active learning environment. The Physics Teacher 35, 340–347 (1997)
28. Soloway, E., Guzdial, M., Hay, K.E.: Learner-centered design: The challenge for HCI in the 21st century. Interactions 1(2), 36–48 (1994)
29. Ternier, S., Tabuenca, B., Specht, M.: ARLearn: Learning activities and interaction in augmented reality. In: Specht, M., Multisilta, J., Sharples, M. (eds.) Proceedings of the Mobile Augmented Reality for Education Workshop, Helsinki, Finland, October 16-17, pp. 10–13 (2012)
30. Tüzün, H.: Blending video games with learning: Issues and challenges with classroom implementations in the Turkish context. British Journal of Educational Technology 38(3), 465–477 (2007)
31. Vygotsky, L.S.: The collected works of L. S. Vygotsky. The history of the development of higher mental functions, vol. 4. Plenum Press, New York (1997), Hall, M. (trans.), Rieber, R.W. (ed.)
32. Wattanasoontorn, V.: Serious games for health and medicine. A cardiopulmonary resuscitation (CPR) case study (Doctoral dissertation) (2013), http://www.tdx.cat/bitstream/handle/10803/128270/tvw.pdf?sequence=2 (retrieved)
33. Wattanasoontorn, V., Magdics, M., Boada, I., Sbert, M.: A Kinect-Based System for Cardiopulmonary Resuscitation Simulation: A Pilot Study. In: Ma, M., Oliveira, M.F., Petersen, S., Hauge, J.B. (eds.) SGDA 2013. LNCS, vol. 8101, pp. 51–63. Springer, Heidelberg (2013)
34. Williams, P., Nicholas, D.: Testing the usability of information technology applications with learners with special educational needs (SEN). Journal of Research in Special Educational Needs 6(1), 31–41 (2006)
35. Williams, P., Jamali, H.R., Nicholas, D.: Using ICT with people with special education needs: what the literature tells us. Aslib Proceedings 58(4), 330–345 (2006)
36. Zhu, E., Hadadgar, A., Masiello, I., Zary, N.: Augmented reality in healthcare education: An integrative review. PeerJ PrePrints 2 (2014)

Building an Mlearning Research Framework through Design Science Research

Merryl Ford, Marlien Herselman, and Adele Botha

CSIR Meraka Institute, Building 43, CSIR, Meiring Naude St, Pretoria, South Africa
{mford,mherselman,abotha}@csir.co.za

Abstract. The purpose of this paper is to provide an explanation of how Design Science research has been applied in order to develop a mobile learning framework for the ICT4RED project which is currently in progress in Cofimvaba in the Eastern Cape Province of South Africa. This is regarded as a complex project and the development of a research framework is essential in order to replicate elsewhere in developing environments with similarities and to scale at a national level. The unique process of developing this ICT4RED mobile learning framework indicates the importance of 12 components which are either regarded as enablers or as drivers for implementation. It is evident that monitoring and evaluation as well as teacher professional development are key to the success of the project and this is also evident from the development of the framework thus far.

Keywords: Mobile learning research framework, Design Science research process, tablets in schools, ICT for Development.

1 Introduction and Background

The Information, Communication Technology for Rural Educational Development (ICT4RED) project of the CSIR Meraka Institute, in South Africa, is funded by the Department of Science and Technology (DST) and the Department of Rural Development and Land Reform (DRDLR). For the DST, the primary purpose of the project is to create an opportunity to examine whether and how the introduction of new technologies, as well as technologies that have been tested in other contexts, will contribute to improvements to the quality of teaching and learning in a rural context. Both teachers (first) and learners are being equipped with mobile tablets and the school environment with infrastructure to use the tablets in that context [1].

The ICT4RED component is a part of the bigger TECH4RED project that will facilitate a range of technology interventions in rural Cofimvaba, including amongst others, in ICTs, water and sanitation, health, nutrition and energy. This list is not exhaustive of the possibilities but is a considered selection of key interventions. To this end, a number of multi-stakeholder project Working Groups (WGs) have been established to examine opportunities and to jointly make recommendations on various themed interventions in support of the wider project objectives. CSIR Meraka will

M. Kalz et al. (Eds.): mLearn 2014, CCIS 479, pp. 107–122, 2014.

only focus on ICT integration in teaching and learning and refer to this project inside the bigger project (TECH4RED) as ICT4RED [2].

In terms of participation outcomes, the project has already completed the first two of the envisaged three phases where 12 schools have already been supported and up skilled to integrate mobile tablets into their practice. Phase three that expands to include an additional 14 schools will commence from February 2014. The project will continue till March 2015.

The project has, amongst others, a research component with a well-defined research agenda. Masters and Doctorate students have been recruited from different universities to do collaborative research on the project. One of the Master's students is a teacher at one of the schools, which adds an additional research depth and opportunity to illustrate impact. Facilitators are assigned to a particular school (where training takes place) and thus establish long term relationships with the individual teachers as they support them to integrate tablets into their teaching and learning. This supports a potentially sustainable formation of a community of practice for the future. At the schools, technical committees have been formed where the teachers reflect on the training and add their own interpretations and voice to the learning. District and circuit officials have also been trained to use tablets.

The sustainability and long term impact of this initiative depends on the long term ownership, leadership and on-going learning of the community formed by the District, Circuit and schools. The project has adopted a stance wherein it is explicitly stated that the teachers are the experts in teaching, the project teams know about technology and are there to learn with and innovate with the teachers in how to use technology to enable teaching and learning. Teaching strategies are demonstrated to teachers during a 10 module course, but teachers have to figure out how to apply these strategies in their classrooms and provide feedback on how this is working in practice [3].

This paper will explain how design science research as a methodology, was applied to advance a *mobile learning research framework* for this project in the Nciba district in Cofimvaba, Eastern Cape Province of South Africa.

2 Use of Mobile Technology and Mobile Learning

The use of mobile technology to support teaching and learning is found to be cheaper and more usable than for instance a PC, which helps people to get access to the learning materials, the facilitators and the fellow learners [4]. Researchers have stressed the special significance of collaborative mobile solutions, that is, solutions enabling peer-to-peer support for mobile learning [5]; [6]; [7]. In brief, mobile solutions are valuable in easing up access (i.e. bridging the digital divide), making learning more flexible and personalized, and providing collaborative (community) support for learning [8]. Collaborative or community support can also function as a bridge from more formal and institutionalized forms of education to informal, work-related and life-long learning.

2.1 Defining Mobile Learning

Tétard et al. [8] define mobile learning as "situated, collaborative and guided teaching, studying and learning, supported by mobile devices that utilise symmetric mobile communications channels by which the learners and the facilitator may use and mould specially designed learning objects for work, hobby or citizenship-related purposes or as an aid to traditional education" (p. 191). This definition is viewed as appropriate for the projects engagement with tablet technology.

3 Objectives

ICT4RED project aims to investigate how the application and deployment of new and existing technologies (which include e-textbooks and other electronic resources) at schools in the Cofimvaba district of the Eastern Cape Province can assist to develop a mobile learning framework. This framework will need to be replicated and scaled to other provinces and across the rural education system. The framework, in addition would be well suited to influence policy on how technology enhanced teaching and learning can be introduced to schools in resource-constrained contexts [3]. Contemplating the aim, the following main research question will guide and frame this investigation:

What elements constitute an ***ICT4RED Framework*** within the Tech4RED project?

In order to address the aim the following objectives are envisaged [2]:

- Design systemic and sustainable approaches to providing access to digital content by learners at resource constraint rural schools in South Africa;
- Design, develop, test and improve new and evolving educational technologies, devices, platforms and processes that support the access to digital content for rural school environments;
- Measure the effect of this project on the 21st century skills of teachers and learners; and
- Use the evidence from the research within this context to inform policy in an integrated and coherent manner.

In order to develop a framework it is necessary to define what it is. According to Lethbridge and Lananiere [9], a framework can be defined as a generic solution to a generalized problem that provides common services to situations that are applicable and consist of a set of ideas or principles and the processes needed to control the implementation of functionalities. 'A framework is also defined as a fundamental construct that defines assumptions, concepts, values, and practices, and which includes guidance for implementing the specific framework'[10]. The conceptual framework of the ICT4RED research framework mainly involves the following knowledge domains, namely models, frameworks, methodology, theories, policy and procedures and practices on the development of an ICT4RED framework. These will be investigated in the fields of Information Systems, Computer Science, Business Information Systems, Information Technology, Human Computer Interaction and the cross-cutting field of ICT4D.

The ICT4RED framework will consist of all these constructs and functionalities in the end and will also involve various tested models.

4 Methodology

The ICT4RED engagement is underpinned by Ford [2] ICT4RED 12 Component Implementation Model that identifies and outlines the different components that are to be addressed within the initiative.

Ford [2] identified the following components after the first phase (pilot at one school):

1. Project Management,
2. School ICT Architecture,
3. Network,
4. Change Management,
5. Teacher Professional Development,
6. Content,
7. Operations Management,
8. Communication,
9. Monitoring and evaluation,
10. Research,
11. Community Engagement, and
12. Stakeholder Management.

In the light of the research question and the ICT4RED implementation model above, the methodology which will be applied to develop this replicable ICT4RED Framework will be Design Science research. Design science research focuses on creation and the purpose of design is "to change existing situations into preferred ones" [11]. Design science addresses 'wicked problems' in Information Systems or IS [12] and is fundamentally a problem-solving paradigm. Wicked problems as explained by Hevner & Chatterjee [13] relate to the ill-defined environmental contexts, creativity and teamwork to produce effective solutions. There are compelling arguments to accept the educational exploitation of ICT within resource constrained environments such as the Cofimvaba school district as a wicked problem.

The research, grounded in the philosophy of pragmatism will apply the deductive reasoning approach, which will be operationalized through a Design Research engagement informed by the ICT4RED 12 Component Implementation Model of which each component will be viewed as a unit of analysis towards the development of the ICT4RED Framework. Each of the 12 components (units of analysis) will be investigated through the use of case study research and these results will inform the development of the framework which will be designed by Design science research.

Table 1. Design-Science Research Guidelines [14]

Guideline	Description	Application in this project
Guideline 1: Design as an Artifact	Design-science research must produce a viable artifact in the form of a construct, a model, a method, or an instantiation	A framework will be developed as an artifact based on knowledge gained from the application of technologies and resources in the Cofimvaba schools.
Guideline 2: Problem Relevance	The objective of design-science research is to develop technology-based solutions to important and relevant business problems	Technology-based solutions will be created to support the teaching and learning at Cofimvaba schools.
Guideline 3: Design Evaluation	The utility, quality, and efficacy of a design artifact must be rigorously demonstrated via well-executed evaluation methods	The framework will be evaluated by various expert reviewers before replication and M&E processes will also be applied to evaluate the different components of the framework.
Guideline 4: Research Contribution	Effective design-science research must provide clear and verifiable contributions in the areas of the design artifact, design foundations, and/or design methodologies	The elements of the framework will be able to contribute to new processes and artifacts which can assist schools to use technology enhanced learning resources to support teaching and learning. There is a theoretical, methodological and practical contribution.
Guideline 5: Research Rigor	Design-science research relies upon the application of rigorous methods in both the construction and evaluation of the design artefact	Rigor will be achieved through evaluation of the framework by expert reviewers from industry and academia. Additions to the knowledge base as well as determining through monitoring and evaluation what worked and what not, with reasons, will assist in strengthening the elements of the framework.

Table 1. *(Continued)*

Guideline 6: Design as a Search Process	The search for an effective artifact requires utilizing available means to reach desired ends while satisfying laws in the problem environment	Each component of the existing 12 model will have their own research question, deliverables and methods which they apply to add to the knowledge base of the elements of the framework.
Guideline 7: Communication of Research	Design-science research must be presented effectively both to technology-oriented as well as management-oriented audiences	Communication will happen through a communication strategy in consultation with the stakeholders of Department of Science and Technology, Department of Basic Education and Department of Rural Development and Land Reform.

4.1 Design Science Research Guidelines

The seven guidelines to perform design science research in Information Systems discipline described by Hevner et al. [14] will be followed and include: design as an artifact, problem relevance, design evaluation, research contributions, research rigour, design as a search process and communication of research [15].

4.2 Design Science Research Process (DSRP)

Peffers et al. [16] developed a conceptual process and mental model for carrying out design science research and presenting it. The proposed Design Science Research

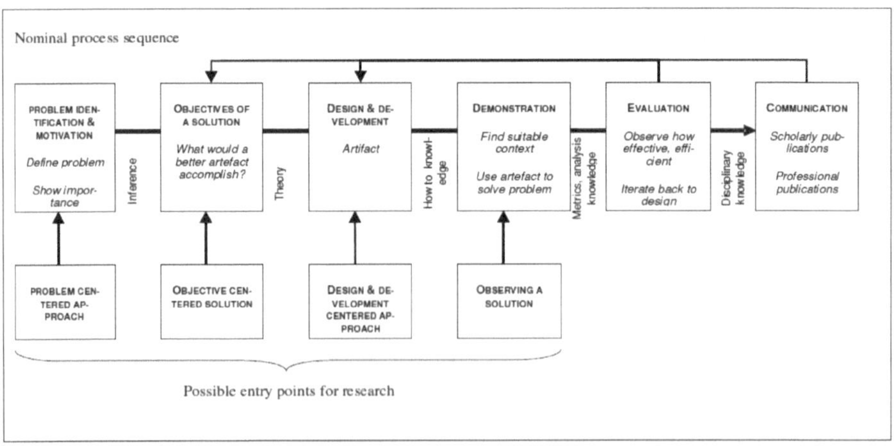

Fig. 1. Design Science Research Process (DSRP) Model [16]

Process (DSRP) is consistent with prior literature [14; 17-20] and includes six steps: problem identification and motivation, objectives for a solution, design and development, evaluation, and communication. The iterative nature of the Design Science Research Process is represented by the arrows between the various steps.

The Design Science Research Process introduced by Peffers et al. [16] will be adapted for the design of the framework for this project as depicted in Figure 2 below:

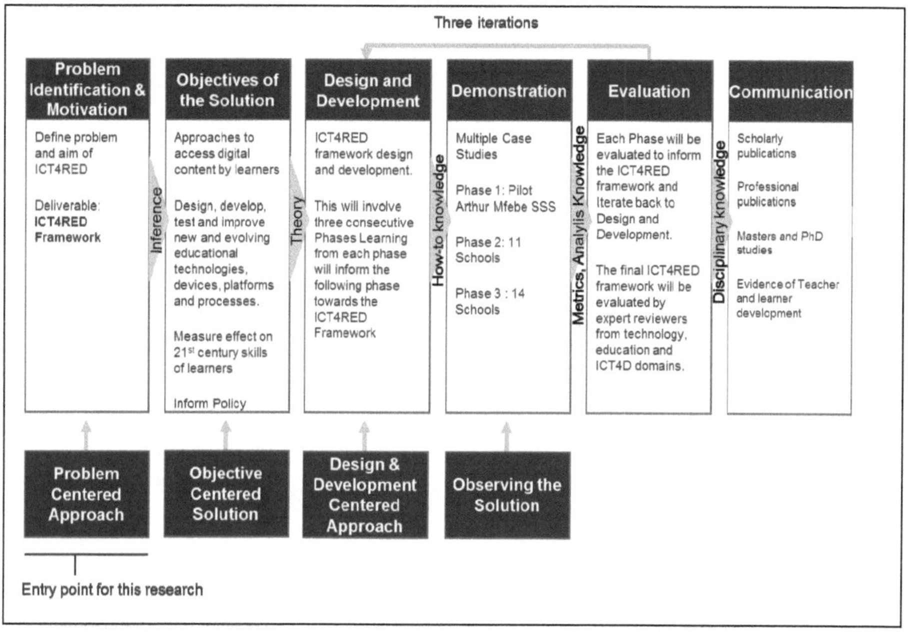

Fig. 2. Design Science Research Process for this project adapted from Peffers et al.[16]

The research process will be guided by Saunders (2007), research onion and therefore will comprise the following aspects in summary:

- **Philosophy:** The philosophy that will be chosen for the study is mainly pragmatism, but interpretivism will also be applied on the results from the multiple case studies (of each of the 26 schools) which will be part of the iteratives of the design science cycle above.
- **Methodology:** The methodology used will be Design science research which will be informed by qualitative multiple case study methodology because it will enable the researcher to answer the research questions.
- **Approach:** The data will be analysed through both inductive and deductive means
- **Research Strategy:** The strategy that will be undertaken to be used in the study is the multiple case study approach.
- **Data Collection Techniques Used:** The data collection techniques that will be used included primary data in form of the validations from the experts and secondary data that will be sought from the literature reviews.

- **Data Analysis:** Will employ hermeneutics, descriptive statistics techniques to make meaningful examination of the collected data as well as within-, cross and holistic case analysis. Triangulation of results will therefore be applied.

In order to develop the final ICT4RED framework where the different phases as iterations will be applied, the following figure explains the final Design science process where the multiple case studies are included:

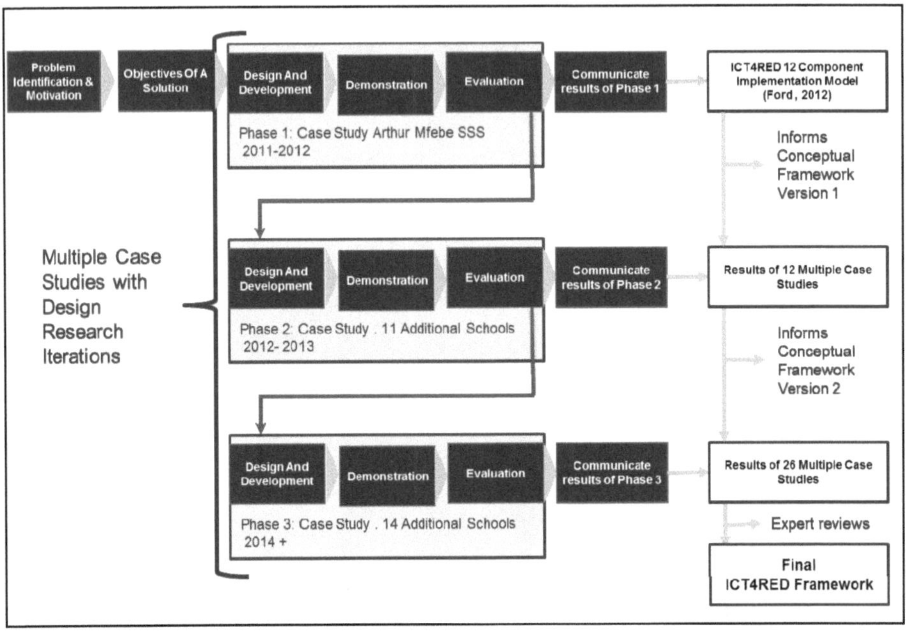

Fig. 3. Design science research process with the multiple case studies and the deliverable from each

4.3 Theory

Theories are constructed in order to explain, predict and master phenomena (e.g. relationships, events, or the behaviour). In many instances we are constructing models of reality. A theory makes generalizations about observations and consists of an interrelated, coherent set of ideas and models. The following theory is applicable to this project:

Design Theory

Gregor [21] identifies five types of theories in IS, one of which is a theory for design and action. The aim of this type of theory is to say how to do something and to give explicit instructions in terms of methods, techniques, principles of form and function for the creation of an artifact. Design theory does not express an opinion about the merits of one course of action compared to another based on moral or ethical considerations. It is concerned with the utility and novelty of the artifact [21; 19].

Gregor and Jones [22] identify eight components of an IS design theory; six of these are regarded as core components while two are additional components. The core components are the purpose and scope, constructs, principles of form and function, mutability of the artifact, testable propositions and justificatory knowledge. The two additional components are the principles of implementation and expository instantiation.

5 Progress so Far with Developing the Conceptual Framework

The development of the conceptual framework, at this point in time, is a result of phase 1 (one pilot school) and phase 2 (11 other schools) and involves the ICT4RED 12 component implementation model of Ford [2].

5.1 Version 1 after Phases 1 and 2

The 12 components, as units of analysis, are viewed as either drivers or enablers and dependencies are still being investigated. Each of these components have to indicate what their theoretical or literature underpinning is, whether they will deliver a process or an artefact, what their respective investigative questions are in support of the main research question. Data collection instruments and specific detail on deliverables also has to inform the development of the conceptual ICT4RED framework in the end (see table 2 below).

The following figure outlines the placement of these components as either enablers or drivers. Identification as an enabler or a driver is pivotal in realising the level of contribution as well as dependencies when one has to build a replicable framework to use when technology is introduced or is enhancing the current rural education system. When a component is identified as an enabler it means that it supports and facilitates the drivers when the implementation of technology in rural schools takes place. A driver creates motion and momentum that act as pillars to support the enablers. Drivers also drive the process of deploying and using technology and produce processes as artifacts within Design Science research. Enablers can produce processes and new artifacts (model, constructs) as deliverables within design science.

After phase 1 and 2 it is evident that the 12 components can be arranged as either drivers or enablers. Table 2 provides a detailed outline of each components of the conceptual framework.

Table 2. Conceptual framework components and deliverables

Component	Research Question	DSR Method (Artefact or process)	Data collection instruments	Deliverable	Theory or literature
Project Management	How can PM assist to deliver a successful project	Process (enabler)	None	Successful project execution Procurement process	PFMA Prince 2 BBBEE
Teacher development	How can we guide the development of relevant teacher knowledge and proficiency to enable classroom practise to portray a 21st century teaching and learning engagement?	Artefact & process (driver)	Observation Questionnaires Focus group interviews Photos Artifacts developed by teachers	Badge micro accreditation system Mobile application Curriculum modules for training teachers (EduPack)	TPAC TEL (Technology enhanced learning) Mobile learning ICT4D Curriculum design Gamification literature
Content	Co-creation with teachers?	Process & Primary driver artefact (driver)			TPAC CAPS documents Development millennium goals
Monitoring and Evaluation	1. How did Design Science, Developmental Evaluation and Utilization focused evaluation influence the artifacts (evaluation framework, evaluation instruments) and processes (data collection, analysis, interpretation and feedback) that were developed in the evaluation of the Cofimvaba e-textbook project? **(M&E as a component)**	Process & artefact (driver)	Case Study data collection instruments	M&E Framework for this initiative	**Developmental Evaluation** (Patton, 2010) **Utilization Focused Evaluation** (Patton, 2008) **Design Science** (Hevner, 2007)

Component	Research Question	DSR Method (Artefact or process)	Data collection instruments	Deliverable	Theory or literature
	2. Was the project designed well, implemented successfully and which results did it deliver? **(M&E reflecting on the project)**		**Implementation Evaluation** Textual analysis of Twitter feed, and Whatsapp group Post training trainer feedback form Attendance registers Post-training teacher feedback forms Quarterly telephonic teacher survey Quarterly telephonic school survey Device usage statistics School functionality checklist	Implementation Evaluation Reports & Continuous feedback Outcome Evaluation Reports & Continuous Evaluation	(OECD Development Assistance Committee, 2010) (DPME, 2011) (Zvoch, Letourneau, & Parker, 2007) (Kirkpatrick & Kirkpatrick, 2006) (Shadish, Cook, & Campbell, 2002)
	3. Which contextual factors and aspects of the Cofimvaba model is necessary for achieving implementation success and outcomes? **(M&E reflecting on Model)**		Re-analysis of school functionality checklist in relation to implementation success and outcomes achieved	Documentation of Sustainability model Documentation of Theory of Action & Theory of Change	(Funnell & Rogers, 2011) (Twinomugisha & Bassi, 2009)

Component	Research Question	DSR Method (Artefact or process)	Data collection instruments	Deliverable	Theory or literature
			Cost Evaluation GeSCI TCO tool Process modelling Scenario building		
Research	How can postgraduate studies contribute to the model and literature & understanding depth? How can the findings Influencing future policy?	Process & Artefact (driver)	Both design science and case studies	M&D studies	Design theory Critical realism Design science methodology Creswell (2011) Mixed methods Saunders, Lewis and Thornhill (2011) research onion
School ICT infrastructure	What process needs to be followed to select appropriate elements of ICT infrastructure for different schools? Which decision models can be used to support infrastructure selection?	Meso Artefact (decision model and process) (driver)	Group decision making – for decision frameworks. Desktop research on technology trends. Consultations with current users and technology providers. User feedback (via call centre; response to technology during training)	Infrastructure selection process Functional specifications Multi-criteria Decision support models (e.g. tablet selection).	Which architecture frameworks? End-to-end ecosystem. Decision science
Network	What elements of the network influence the progress of the ICT4RED project? Levels of dependencics	Process & macro artefact (driver)	Field observations (i.e. school visits to determine appropriate technology in specific location)	Definition of critical elements (bottlenecks to progress, single points of failure)	Scalable, flexible and open network design

Component	Research Question	DSR Method (Artefact or process)	Data collection instruments	Deliverable	Theory or literature
	on the network components.		Information from network vendors and current practices of service providers.	Network architecture	
Operations management	What elements of operations management are critical to the successful implementation of the project? Understanding the complexity of operations in rural areas (high costs, low density).	Process (Operations processes) (enabler)	Alignment with best practices and use CSIR models for technical support (VO model)	Elements of operations management process Process register Financial sustainability of operations in different categories of schools (cost model)	ITIL service management framework. Resource allocation models
Stakeholder Management	What is the role of managing Stakeholders in the implementation of an ICT4D project in rural schools?	Process (enabler)	Qualitative Questionnaire and field visit notes	Rapport built with all stakeholders including the Departments involved, community leadership and existing projects	Theory
Community engagement	What role does Community Engagement play in the successful implementation of the ICT4RED project?	Process (enabler)	Qualitative Questionnaire and field visit notes	A buy-in by the SGB and community surrounding the school. Ownership of project by the school.	Theory
Change Management	What are the key elements of Change Management in the introduction of ICT in teaching and learning at rural	Process & secondary driver artefact (driver)	Focus group interviews Key informant interviews	Increase probability of the ICT4E Project success in line with project	Literature

Component	Research Question	DSR Method (Artefact or process)	Data collection instruments	Deliverable	Theory or literature
	schools?			aims and objectives Provide support to schools and Cofimvaba District office through the change journey and transition Manage resistance to the change Identify risks associated with the change initiative and eliminate or reduce the impact List of documents to be compiled by the school ICT committee. Train District officers to take over the components.	

It is evident from the above table and figure that in this project Monitoring and evaluation plays a pivotal role to ensure that impact can be traced and measured over the three period of the project.

6 Conclusions

This development of the ICT4RED framework was provided and also the role that the application of the Design Science methodology has played to realise the components of the framework. The specific Design Science process [16] which was followed also indicated the role of case studies inside this methodology for validation and evaluation of the framework which is described as the artefact which will be developed. The progress of developing this framework after phase 1 and 2 of the project indicated what each component will deliver and the type of deliverable which is expected from each component. Developing such a framework takes time and constant improvement. Once the whole framework is developed after the completion of phase 3 it will be useful to see how this ICT4RED framework has changed since phase 1 and 2. The significance of the ICT4RED framework lies in the fact that it will be used to replicate the project in other provinces or context with a similar nature and to identify the changes and adaptions based on the contexts.

References

1. Department of Science and Technology: The Cofimvaba Schools District Technology Project: Status Report 2, p. 12. (2013)
2. Ford, M., Botha, A.: ICT4RED 12-Component Implementation Framework: A conceptual framework for integrating mobile technology into resource-constrained rural schools. IST Africa 2014, Mauritius (2014)
3. Herselman, M.E., Botha, A.: ICT4RED Research Framework. Meraka. CSIR, Pretoria (2013) (unpublished)
4. Fozdar, B.I., Kumar, L.S.: Mobile Learning and Student Retention. International Review of Research in Open and Distance Learning 8(2) (June 2007)
5. Tétard, F., Patokorpi, E.: A Constructivist Approach to Information Systems Teaching. Journal of Information Systems Education 16(2), 167–176 (2005)
6. Sharples, M.: The design of mobile technologies for lifelong learning. Computers & Education 34, 177–193 (2006)
7. Dibella, T., Kelly, J.: Developing and applying a learning effectiveness index. Managing Learning for Value Creation (2000), http://www.capworks.com/resnotes/pf000623.pdf (retrieved October 5, 2013)
8. Tétard, F., Patokorpi, A., Carlsson, J.: A conceptual framework for mobile learning. Institute for advanced Management Systems research, Finland (2008), http://ideas.repec.org/p/amr/wpaper/464.html (accessed December 02, 2013)
9. Lethbridge, T.C., Lananiere, R.: Object-orientated software engineering: Practical software development using UML and JAVA, 2nd edn. McGraw-Hill-Education, Berkshire (2005)
10. Tomhave, B.L.: Alphabet Soup: Making Sense of Models, Frameworks, and Methodologies (2005), http://www.secureconsulting.net/Papers/Alphabet_Soup.pdf (accessed July 12, 2013)
11. Simon, H.: The Sciences of the Artificial. MIT Press, Cambridge (1996)
12. Rittel, H.J., Webber, M.M.: Planning problems are wicked problems. In: Cross, N. (ed.) Developments of Design Methodology. John Wiley & Sons, New York (1984)
13. Hevner, A., Chatterjee, S.: Design Science Research in Information Systems: Theory and Practice. Springer, New York (2010)

14. Hevner, A., March, S., Park, J., Ram, S.: Design science in information systems research. Manage Information Systems Quarterly (MIS) 28(1), 75–105 (2004)
15. Wang, S., Wang, H.: Towards innovative design research information systems. Journal of Computer Information Systems (Fall 2010)
16. Peffers, K., et al.: A design science research methodology for information systems research. Journal of Management Information Systems 24(3), 45–77 (2007)
17. Hashim, L., Hasan, H., Sinnapan, S.: Australian Online Newspapers: A Website Content Analysis Approach to Measuring Interactivity. In: 18th Australasian Conference on Information Systems 2007, Toowoomba (2007)
18. Hevner, A.R.: A Three Cycle View of Design Science Research. SJIS 19 (2007)
19. March, S.T., Storey, V.C.: Design science in the information systems discipline: An introduction to the special issue on design science research. MIS Quarterly 32(4), 725–730 (2008)
20. Ellis, T.J., Levy, Y.: A Guide for Novice Researchers: Design and Development Research Methods. In: Proceedings of Informing Science & IT Education Conference, InSITE 2010 (2010)
21. Gregor, S.: The Nature of Theory in Information Systems. MIS Quarterly 30(3), 611–642 (2006)
22. Gregor, S., Jones, D.: The anatomy of a design theory. Journal of the Association for Information Systems 8(5), 312–335 (2007)

A Proposed Evaluation Framework from Mobile Inquiry-Based Learning Activities in Science Education

Sohaib Ahmed[*]

Bahria University, 13-National Stadium Road,
Karachi, Pakistan
dr.sohaibahmed@gmail.com

Abstract. With the advancement in the use of mobile technology has attracted many educators and researchers to develop mobile science inquiry-based learning activities for educational purposes. In school science education literature, many applications are designed and evaluated using various evaluating frameworks. These evaluations are based on either one or more of the following attributes; technological usability, inquiry skills, learners' performance, learners' perceptions about the applications, learners' cognitive load and the longer impact of an application. However, none of these frameworks cover all of these attributes in a single framework. Thus, this paper addresses this issue by conducting a comprehensive literature review about the current state of the use of mobile science inquiry evaluation frameworks. Consequently, a proposed framework, MSI (Mobile Science Inquiry) evaluating framework is presented in this paper.

Keywords: Mobile Learning, Inquiry-based Learning, Evaluation Framework, Science Education, Mobile Science Inquiry.

1 Introduction

Inquiry-based learning (IBL) or Inquiry learning is a pedagogical approach that allows learners to gain knowledge and develop their reasoning skills through exploration and investigation of a given scientific phenomena [1-2]. This learning activity can provide ample opportunities to acquire inter-related objectives including the acquisition of general inquiry abilities, the acquisition of particular investigation skills, and the understanding of the underlying domain in a given scientific inquiry [3].

With the tremendous advancement in learning technologies, educators are tend to use these technologies, particularly mobile devices for promoting science education in different ways such as technology-based instructions in their classes or preparing materials via technology [4]. Further, the affordance of these devices provide powerful multimedia, geo-location capabilities, social networking communication and mobile learning

[*] Corresponding author.

M. Kalz et al. (Eds.): mLearn 2014, CCIS 479, pp. 123–130, 2014.

opportunities in order to perform scientific inquiries in natural environments (e.g. parks, woodlands, museums) and inside the schools [5].

There are numerous studies found in the literature related to mobile science education, state that students can perform better if they equipped with mobile devices while doing scientific inquiries [2], [6]. In addition, there are few other studies highlighted that the use of mobile devices as tools are significant when their instructional scaffolding improve the quality of learners' explanation during scientific investigations [7-8]. Hence, these devices facilitate students in order to acquire underlying domain knowledge and may enhance their reasoning skills compared with those who do not use them during science inquiry activities.

In any application, evaluation can play a vital role for investigating rigorously the utility, quality and efficacy using some well-executed methods [9]. There are many applications found in the literature, which are evaluated either technological usability of the application, learners' performance, learners' perceptions, cognitive load, learners' reasoning skills or combination of these few attributes. However, there is no single framework found in the literature that covers all these aspects comprehensively. Therefore, this paper addresses this challenge and presents an evaluation framework for such mobile science inquiry learning activities.

The remainder of this paper is structured as follows. In section 2, previous literature about the evaluation framework for mobile inquiry-based learning is discussed. Section 3 presents our proposed framework while section 4 contains concluding remarks and discusses some future directions of this research.

2 Evaluation Frameworks for Mobile Science IBL Applications

There are many evaluation frameworks found in the literature, which are used for evaluating different aspects of mobile science inquiry investigations including technological usability, learners' performance, learners' perceptions, cognitive load, learners' inquiry and reasoning skills, and longer impact of an application at organizational context. Here, 15 mobile science inquiry applications are described in Table 1 in which 9 applications were evaluated by using a particular framework while others were evaluated without using any framework.

In these applications, learners are involved in various activities; ranges from data collection to scientific explanations at different locations including indoor (e.g. classroom, laboratory or any place within a school), outdoor (e.g. Museum, Park or any place outside a school) and mixed (both indoor and outdoor). Further, most of the applications were designed for primary and high schools students.

Amongst these evaluation frameworks (see Table 1), the M3 (Micro, Meso, Macro) evaluation framework [10] was used. The M3 consists of three levels; a Micro level deals with usability, a Meso level concerned with the learning experience about the application, and a Macro level related to the longer impact of an application. This framework has been used in couple of applications namely MyArtSpace [10] and ThinknLearn [11]. The first application was evaluated at all three levels while the latter was involved in only two levels; Micro and Meso [12].

Table 1. Mobile Science Inquiry Applications And Their Evaluations

Application	Evaluation Framework	Activity	Evaluation Type	Evaluators (Age/Class)	Location
Ambient Wood Project	Digital Augmentation Framework	Data Collection & Hypothesis Generation	Frequencies of Probing Activities & Perception	11-12 Years Students	Outdoor
BioKid Sequence	----	Data Collection	Usability & Accuracy	5th-8th Grade Students	Mixed
MyArtSpace	M3 Evaluation Framework	Data Collection	Usability, Performance, Perceptions & Long Term Impact	Students & Teachers	Mixed
ThinknLearn	M3 Evaluation Framework	Hypothesis Generation	Usability, Performance & Perception	High School Students	Indoor
MPLS (Mobile Plant Learning System)	Mobile Learning Framework	Data Collection & Collaboration	Usability, Performance & Perception	Avg. 11 Years Students & Teacher	Outdoor
LET'S GO!	----	Data Collection & Collaboration	Performance & Perception	16-18 Years Students	Outdoor
nQuire	----	Data Collection & Collaboration	Usability, Performance & Perception	Avg. 14 Years Students & Teachers	Mixed
WHIRL	----	Classroom Assessments	Frequencies of Use & Perception	Primary School Teachers	Indoor
Zydeco	McNeil's CER* Framework	Data Collection & Scientific Explanation	Performance & Perception	6th Grade Students & Teachers	Outdoor
Mobile G-Portal	TAM/TTF**	Data Collection	Usability, Performance & Perception	Secondary School Students	Outdoor
SDG (Shared Display Groupware)	----	Data Collection & Collaboration	Performance	4th Grade Students	Outdoor
Chu et al.	Two-Tier Test Guiding (T^3G)	Guidance	Performance & Perception	Primary School Students	Indoor
Shih et al.	----	Guidance	Performance & Cognitive Load	Primary School Students	Mixed
Liu et al.	5E Learning Cycle	Data Collection & Guidance	Performance & Perception	4th Grade Students	Mixed
Hwang & Chang	FAML***	Data Collection	Performance, Perception & Cognitive Load	5th Grade Students	Outdoor

* Claim, Evidence and Reasoning Framework, ** Technology Acceptance Model/ Task-Technology Fit Model, *** Formative Assessment Mobile Learning

The other frameworks were used in a single application such as MPLS [6], a mobile plant learning system was evaluated by using a framework defined by Motiwalla [13]. This framework has two objectives: monitor the usage of mobile applications and get feedback about the use of the application and determine the students' opinions about the significance of the application. Similarly, in Mobile G-Portal [14], TAM [15] along with TTF [16] was used for the evaluations. TAM aims to determine learners' perceptions and the technological usefulness of an application [15] while TTF emphasizes on the learners' performance in a particular given task [16].

McNeil's CER framework [17] and 5E Learning cycle [18] were used in Zydeco [19] and Liu et al. [20] respectively for the evaluation of learners' performance and perceptions about the applications. Both these frameworks emphasize on evaluating the inquiry process and its relevant skills. McNeil's CER [17] assists students in developing scientific explanations by dividing a task into these parts; making a claim, evidence to support that claim and the reasoning that connects the evidence with a claim. On the other hand, 5E Learning cycle [18] consists of five phases including engagement, exploration, explanation, elaboration and evaluation that can help students to perform scientific inquiries in a natural environment.

The other two frameworks FAML [21] and T^3G [22] were designed to investigate by focusing on the learning assessments of learners while using mobile devices for conducting scientific inquiries. Another framework namely digital augmentation framework was designed to evaluate Ambient Wood Project [5] in order to identify the number of frequencies of the use of probing and learners' perceptions about the application.

However there are few other applications in the literature such as BioKid Sequence [23], LETS'GO! (Learning Ecology through Technologies from Science for Global Outcome) [24], nQuire [25], SDG (Shared Display Groupware) [26], WHIRL [27] and Shih et al. [2], which do not apply any particular framework but these applications were evaluated by one or two of these aspects; technological usability of an application, cognitive load while using applications, learners' performance and their perceptions about the applications. Therefore, this paper addresses a new evaluation framework that includes all these aspects used in such mobile science inquiry investigations.

3 MSI: A Proposed Evaluation Framework

A MSI (Mobile Science Inquiry) evaluation framework is based on six aspects; three are application-related, two are inquiry-based learning skills and the one is related to the longer impact of both application-related and inquiry-based learning and reasoning skills at organizational context as shown in Fig 1.

3.1 Application-Related

This application-related is further divided into three aspects; technological usability, learners' perceptions and cognitive load for using application.

3.1.1 Technological Usability

Usability can play a vital role in the success of any application. If an application is not usable enough then it obstructs the way for facilitating students' reasoning skills [28]. In addition, quality aspects need to be evaluated for developing mobile learning applications [29]. Thus, this technological usability consists of both [12]; usability aspects (Learnability, Operability and Understandability) [30] and mobile quality aspects (Interactivity, Metaphor and Learning Content) [29].

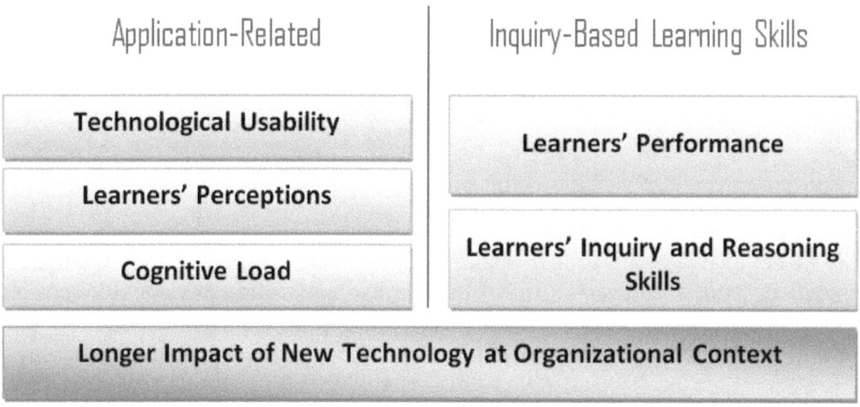

Fig. 1. MSI (Mobile Science Inquiry) Evaluation Framework

3.1.2 Learners' Perceptions

Learners' perceptions or interests are important ingredient to measure while evaluating any application. It is confirmed that almost all of the above applications cover this aspect in evaluations. This aspect is used to examine the learning experience about the given task or activity while using an application [6].

3.1.3 Cognitive Load

Cognitive load is another important aspect in any IBL application that may optimize learners' intellectual performance during learning activities [31]. There are two types of cognitive load [2]; intrinsic and extraneous, the former deals with the nature of the learning materials, the learners' cognitive levels and their inter-related activities, also referred as mental load [21]. The latter, known as mental effort, tells how learning content can be organized and presented to learners [2].

3.2 Inquiry-Based Learning Skills

3.2.1 Learners' Performance

Many studies reported that the measurement of learning performance or achievement is significant in the evaluations of mobile science IBL applications [2], [6], [10-12], [14], [19- 22], [24-26]. In most of these applications, a test was conducted at the end of any learning activity in order to examine the learners' understanding about the underlying domain.

3.2.2 Learners' Inquiry and Reasoning Skills

This aspect is not much evaluated in many applications in the literature. However, McNeil's CER Framework [17] and 5E Learning Cycle [18] are the frameworks that describe how learners' inquiry and reasoning skills can be evaluated. These frameworks target scientific reasoning processes to be evaluated. The scientific reasoning processes are [32]: *orientation*, which focuses on the identification of variables and their relations; *hypothesis generation*, which formulates a model of a particular domain for considerations; *experimentation*, which validates the model or hypothesis while in *conclusion*, a new knowledge of a particular domain is constructed. Learners need to be evaluated at each process so that inquiry-based learning skills can be evaluated properly.

3.3 Longer Impact of New Technology

This aspect evaluates the longer impact of the new technology at the organizational contexts described as macro level of the M3 evaluation framework [10]. In science education, the use of mobile technology in such inquiry learning activities should be examined appropriately by educators and students for a longer period of time.

4 Conclusion and Future Work

In this paper, a mobile science inquiry (MSI) evaluation framework is proposed for conducting science inquiry investigations in school science education. This proposed framework is based on the evaluation frameworks found in the literature discussing science inquiry-based learning applications. This framework consists of six evaluation aspects including technological Usability, learners' performance, learners' perceptions or attitudes, learners' inquiry and reasoning skills, cognitive load and technological longer impact on learning and inquiry skills.

This is an early stage of this proposed framework for the evaluation of any science inquiry learning activity. In the future, this framework will be used with some previous or new applications in order to identify its actual potential so that it can become as a standard evaluation practice for performing mobile science inquiry investigation in schools.

References

1. Lim, B.-R.: Challenges and Issues in Designing Inquiry on the Web. British J. of Educ. Tech. 35(5), 627–643 (2004)
2. Shih, J.-L., Chu, H.-C., Hwang, G.-J., Kinshuk: An Investigation of Attitudes of Students and Teachers About Participating in a Context-aware Ubiquitous Learning Activity. British J. of Edu. Tech. 42(3), 373–394 (2011)
3. Edelson, D.C., Gordin, D.N., Pea, R.D.: Addressing the Challenge of Inquiry-based Learning Through Technology and Curriculum Design. J. of the Learn. Sci. 8(3-4), 391–450 (1999)
4. Lin, M.-F., Fulford, C.P., Ho, C.P., Iyoda, R., Ackerman, L.K.: Possibilities and Challenges in Mobile Learning for K-12 Teachers: A Pilot Retrospetive Survey Study. In: 7th IEEE International Conference on Wireless, Mobile and Ubiquitous Technology in Education (WMUTE 2012), pp. 132–136 (2012)
5. Rogers, Y., Price, S., Randell, C., Fraser, D.S., Weal, M., Fitzpatrick, G.: Ubi-learning Integrates Indoor and Outdoor Experiences. Comm. of the ACM 48(1), 55–59 (2005)

6. Huang, Y.-M., Lin, Y.-T., Cheng, S.-C.: Effectiveness of a Mobile Plant Learning System in a Science Curriculum in Taiwanese Elementary Education. Comp. & Edu. 54, 47–58 (2010)
7. Mulder, Y.G., Lazonder, A.W., De Jong, T.: Comparing Two Types of Model Progression in an Inquiry Learning Enviornment with Modelling Facilities. Learn. and Instr. 21(5), 614–624 (2011)
8. Peker, D., Wallace, C.S.: Characterizing High School Students' Written Explanations in Biology Laboratories. Res. in Sci. Edu. 41(2), 169–191 (2011)
9. Hevner, A.R., March, S.T., Park, J., Ram, S.: Design Science in Information Systems Research. MIS Quarterly 28(1), 75–105 (2004)
10. Vavoula, G., Sharples, M., Rudman, P., Meek, J., Lonsdale, P.: Myartspace: Design and Evaluation of Support for Learning with Multimedia Phones Between Classrooms and Museums. Comp. & Edu. 53, 286–299 (2009)
11. Ahmed, S., Parsons, D.: Evaluating 'ThinknLearn': A Mobile Science Enquiry Based Larning Application in Practice. In: 11th International Conference on Mobile and Contextual Learning (MLearn 2012), pp. 17–24 (2012)
12. Ahmed, S., Parsons, D.: Abductive Science Inquiry Using Mobile Devices in the Classroom. Comp. & Educ. 63, 62–72 (2013)
13. Motiwalla, L.F.: Mobile learning: A Framework and Evaluation. Comp. & Educ. 49, 581–596 (2007)
14. Theng, Y.-L., Tan, K.-L., Lim, E.-P., Zhang, J., Goh, D.H., Chatterjea, K.,... Vo, M.C.: Mobile G-Portal Supporting Collaborative Sharing and Learning in Geography Fieldwork: An Empirical Study. In: Proceedings of the Joint Conference on Digital Libraries (JCDL), pp. 462–471. ACM, Vancouver (2007)
15. Davis, F.D., Bagozzi, R., Warshaw, P.R.: User Acceptance of Computer Technology: A Comparison of Two Theoretical Models. Mangmt. Sci. 35(8), 982–1003 (1989)
16. Dishaw, T.D., Strong, D.M.: Extending the Technology Acceptance Model With Task-Technology Fit Constructs. Info. & Mangmt. 36, 9–21 (1999)
17. McNeil, K.L., Krajcik, J.: Supporting Grade 5-8 Students in Constructing Explanations in Science: The Claim, Evidence and Reasoning Framework for Talk and Writing. Pearson & Allyn & Bacon, New York (2011)
18. Baybee, R.W., Taylor, J.A., Gardner, A., Van Scotter, P., Powell, J.C., Westbrook, A., Landes, N.: The BSCS 5E Instructional Model: Origins, Effectiveness and Applications. BSCS, Colardo Springs (2006)
19. Kuhn, A.C.: Zydeco: Exploring How Mobile Technology Can Scaffold Student Multimedia Data Collection and Use of Science Inquiry. PhD Thesis. University of Michigan, USA (2013)
20. Liu, T.-C., Peng, H., Wu, W.H., Lin, M.-S.: The Effects of Mobile Natural-science Learning Based on the 5E Learning Cycle: A Case Study. Educ. Tech. & Soc. 12(4), 344–358 (2009)
21. Hwang, G.-J., Chang, H.-F.: A Formative Assessment-Based Mobile Learning Approach to Improving the Learning Attitudes and Achievements of Students. Comp. & Educ. 56, 1023–1031 (2011)
22. Chu, H., Hwang, G.-J., Tsai, C.-C., Tseng, J.C.: A Two-Tier Test Approach to Developing Location-aware Mobile Learning Systems for Natural Science Courses. Comp. & Educ. 55, 1618–1627 (2010)
23. Parr, C.S., Jones, T., Songer, N.B.: Evaluation of a Handheld Data Collection Interface for Science Learning. J. of Sci. Edu. and Tech., 233–242 (2004)

24. Maldonado, H., Pea, R.D.: LET'S GO! To The Creek: Co-Design of Water Quality Inquiry Using Mobile Science Collaboraties. In: 6th IEEE International Conference on Wireless, Mobile and Ubiquitous Technology in Education (WMUTE 2010), pp. 81–87 (2010)
25. Mulholland, P., Anastopoulou, S., Collins, T., Feisst, M., Gaved, M., Kerawalla, L., Wright, M.: nQuire: Technological Support for Personal Inquiry Learning. IEEE Trans. on Learn. Tech. 5(2), 157–169 (2012)
26. Yang, J.C., Lin, Y.L.: Development and Evaluation of an Interactive Mobile Learning Environment With Shared Display Groupware. Educ. Tech. & Soc. 13(1), 195–207 (2010)
27. Yarnall, L., Shechtman, N., Pennel, W.R.: Using Handheld Computers to Support Improved Classroom Assessment in Science: Result From a Field Trial. J. of Sci. Edu. and Tech. 15(2), 142–158 (2006)
28. Wong, S.K(B.), Nguyen, T.T., Chang, E., Jayaratna, N.: Usability Metrics for E-learning. In: Meersman, R., Tari, Z. (eds.) OTM Workshops 2003. LNCS, vol. 2889, pp. 235–252. Springer, Heidelberg (2003)
29. Parsons, D., Ryu, H.: A Framework for Assessing the Quality of Mobile Learning. In: 6th IEEE International Conference on Advanced Learning Technologies. IEEE Computer Society, Kerkrade (2006)
30. Chua, B.B., Dyson, L.E.: Applying the ISO 9126 Model to the Evaluation of an E-Learning System. In: Atkinson, R., McBeath, C., Jonas-Dwyer, D., Phillips, R. (eds.) Beyond the Comfort Zone: Proceedings of the 21st ASCILITE Conference, pp. 184–190 (2004)
31. Sweller, J., Van Merrienboer, J.J., Paas, F.G.: Cognitive Architecture and Instructional Design. Educ. Psy. Rev. 10(3), 251–297 (1998)
32. de Jong, T.: Computer Simulations - Technological Advances in Inquiry Learning. Science 312(5773), 532–533 (2006)

On Mobile Learning with Learning Content Management Systems: A Contemporary Literature Review

Asiimwe Edgar Napoleon* and Grönlund Åke

Department of Informatics, Business School, Örebro University
Örebro 701 82, Sweden
{edgar.asiimwe,ake.gronlun}@oru.se

Abstract. Learning Content management systems (LCMS) are important tools for organizing learning material and communication. Increasingly mobile technologies are used for internet access; particularly important in developing countries where broadband is scarce. Mobile LCMS introduce specific challenges, which are yet not fully addressed. This paper reviews the literature on mobile LCMS for the purpose of identifying current research focus, research gaps, and future research directions regarding how to bridge the gaps and leverage CMS technology to support "mobile learning". The concept matrix method is used to collect and analyze literature. Five prominent research areas are found; Use, access, design and infrastructure; communication and collaboration; engagement and knowledge development; content and service delivery; and implementation experiences and evaluation. A major gap identified is that research does neither clearly nor thoroughly address the intersection between learning and technology. Adjusting technologies to learning contexts and environments is a key area for future research.

Keywords: mobile learning, MLCMS, LCMS, LMS, CMS.

1 Introduction

Technology mediated learning, "e-learning", is increasingly important in higher education. E-learning platforms and tools such as Learning Management Systems (LMS) are developed to allow teachers and learners to share education materials, work on assignments, communicate, and collaborate online [19, 20, 21]. Benefits such as simplicity in data mining [36] and good content and social interactive environments [4] have created demand for use. In [20], it is indicated that more than 90% of universities and colleges in developed countries for example the US and UK have implemented e-learning systems (LMSs) for students and faculty use. The authors explored uses and perceived benefits of LMS in supporting traditional classroom teaching and found that teachers and students highly value the teaching and learning tools within the LMSs most especially tools and activities for efficient communication.

* Corresponding author.

M. Kalz et al. (Eds.): mLearn 2014, CCIS 479, pp. 131–145, 2014.

According to Digitalbuzz [7], there are 6 billion mobile phones in the world. As today's world population is 7 billion, this suggests that even though some people have more than one phone some 80 % of the world population have access to, and use, mobile phones. One driving factor for mobile demand is mobile Internet, which is predicted to surpass desktop Internet in 2014 [6]. This steady growing market demand serves as a fertile ground for the integration and implementation of m(obile)-learning systems. M-learning implementation is already a stated need in many education institutions [1].

Research has been carried out on different aspects of learning using LMS and investigating how such systems contribute to learning processes. However, in [23] authors report that most studies rather focus more on technology adoption and use and less on learning processes. There are also claims that mobile learning has not been as successful as other forms of e-learning [12]. Challenges include many kinds of issues like cost, isolation during learning, security, privacy, etc. [32]. Another commonly mentioned challenge is the inability to process multimedia content in a manner suitable for learners' small screens and unsupported data formats of supplied content. In this study therefore we focus on Content Management Systems (CMS) technology with a specific focus on use in learning contexts because of the content management challenge – a challenge that is considered a key factor in web-based learning systems [13]. Even more specifically we focus on the use of mobile applications with such systems, MLCMS.

Content presentation is one aspect that can affect the motivation of learners negatively or positively [29]. From time to time efforts have been made to improve and facilitate access to and management of web content. Some of these efforts involve creation of standards and platforms for content presentation, access and management across different computing devices such as phones and computers.

This study presents a comprehensive review on learning content management systems (LCMS) with a focus on mobile phone use in learning. Our objective is to identify research gaps and future research directions regarding how to leverage CMS technology to support m-learning. The study seeks to answer the questions:

- What are the key focus areas in MCLMS research and practice?
- What are the key future research directions in the MLCMS field, and what gaps are there?

Before going into the literature study we present an understanding of LCMS by distinguishing different CMS technologies in educational contexts.

2 CMS Technology

There are various definitions of CMS. Rivera, Candela, Garcia [35] defines CMS as tools based on web technology that provide an environment for creating, accessing, editing, and managing information in a friendly, reliable and secure way. Many definitions, considering the education perspective, are formulated based on the generic classes of technologies derived from CMS technology, for example Learning Content Management Systems (LCMS), Mobile Learning Content Management Systems (MLCMS), and Learning Management Systems (LMS), etc. Our exploration takes the education strand and focuses primarily on MLCMS, but as the definitions are not used in a strict way, and as many system features cut across different systems, we need to include all of them.

Ibid [35]'s definition of CMS is the most comprehensive one, covering many aspects of other definitions; hence we draw on it to define LCMS/MLCMS as learning CMS that can store and deliver learning content and services to mobile computing devices. In this paper, following the mainstream literature, "mobile" excludes laptops. For the purpose of content separation depending on access device choice, LCMS and MLCMS may be discrete/independent systems or application modules of any CMS technology that uses multi-channel content delivery method, where content is intelligently and automatically separated and sent to a receiving device based on the kind of the device for example a desktop computer or some mobile device. It should be noted that the terminology in the literature is not crystal clear. Often LMS share the same definition with LCMS. While the two kinds of system are complementary, they are often used to address different and unique education challenges [31, 34].

Rengarajan [34] and Qwaider and Hattab [31] further give an overview of differences between LCMS and LMS. The differences are summarized according to applications' core functions in Table 1. The functions listed in Table 1 are some of the key requirements of an e-learning system [31]. The table has been extended to include comparisons on CMS and MLCMS.

Table 1. Comparison of CMS, LMS, LCMS and MLCMS

#	Features (learning perspective)	Core Functions			
		CMS	LMS	LCMS	MLCMS
1	Content creation and management	R	-	R	L
2	Course Catalogue	L	R	L	R
3	Management of Learners and e-learning activities (portfolio management)	-	L	R	L
4	Instructor-led sessions management	-	R	-	-
5	Searchable library of re-usable content	L	-	R	L
6	Registration system	L	L	R	R
7	HRM application integration	-	R	-	-
8	Competence management	L	R	L	L
9	Assessments management	L	L	R	R
10	Location-based aware delivery of content	-	-	R	R

Table 1. *(continued)*

11	Collaboration and communication	R	L	R	R
12	Planning tools	L	R	L	L
13	Content creation and management	R	-	R	L

R = Robust functionality
L = Limited functionality

Table 1 summarizes how various system functions are related to the different kinds of systems in the literature [31, 34]. While there are differences in how various authors define these systems, they have similar functions or characteristics: for example most functions of LMS are also found in LCMS and MLCMS but may not be key functions in those contexts.

The evolution of systems supporting learning, which started as early as 1960s, has undergone four stages according to [31, 37]. These include (1) generic content libraries, (2) learning management systems, (3) outsourced e-learning platforms, (4) learning content management systems. The first stage involved creation and delivery of e-learning material on CD-ROMs until organizations spearheaded the development of content libraries for e-learning courses (stage two; basic LMSs). However in stage two, most events such as enrolment were classroom based. Companies started to offer outsourcing services by providing web space for education institutions to upload their teaching material (stage three). This stage, according to [31], had a drawback, namely inefficiency. Organizations using outsourced services would not manage learning content and events quickly. Finally LCMS (stage four) were introduced as an all-tasks solver platform.

Many studies on CMS and learning have covered various aspects but there are still considerably under-researched areas such as MLCMS. For example, the description of e-learning evolution by [31] does not include MLCMS. Figure 1, based on the Solemon & Sulaiman model [2006], is used to show/add yet another stage in e-learning evolution (stage 5).

According to Wagner (2005), the success of mobile learning will, among many other factors, ultimately depend on content management. While CMS technology has been applied in building various learning platforms, so far these platforms provide fewer services to mobile users compared to desktop users. Examples of such learning platforms include Drupal, Moodle, Blackboard, and more.

To make advancements in this regard, research has been done on numerous aspects of e- and m-learning using CMS, for example infrastructure, user access and interaction [38], learners' engagement [29], quality service delivery and better communication and collaboration [17, 22], and evaluation tools and methods [10].

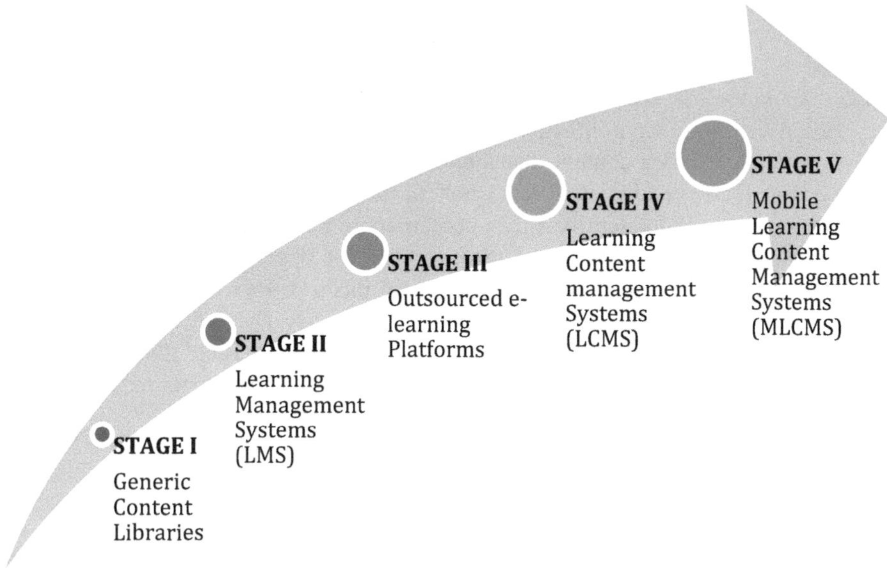

Fig. 1. E-learning Systems Evolution extended based on a model in [37]

Still, less research is done on MLCMS use in learning, more specifically mobile phone use by CMS users. Trifonova and Ronchetti [38] claim this could be because m-learning is a new field. Other studies highlight the challenges of phone use and the obstacle these provide for users. The present research therefore reviews such contemporary discussions on m-learning and CMS technology with the objective of identifying future research direction on how CMS technology can be leveraged for m-learning.

3 Method

This study used the concept matrix method [14, 15, 18, 27, 39]. The matrix helped in identifying the most discussed concepts and made it possible to synthesize literature by categorizing research papers based on concept centricity. Snowballing i.e., further inspection of literature from the bibliographies of reviewed papers was used to acquire more related literature.

There are many suggestions from different researchers on how to apply the concept matrix method. For the purpose of making this review rigorous, we modified Webster and Watson's model [39] to incorporate other researcher's review indicators. For example in Table 2, empirical and non-empirical classifications [2] are added and Table 3 is a merger of views from [15] and [39]. Non-empirical articles are articles that focus on objects such as systems types, products and installations while empirical articles are articles that focus on events and processes such as laboratory experiments, field experiments, field studies, case studies and surveys [15].

Research papers were searched and selected from online journals based on keywords. The keywords, used in combination, were: mobile learning, electronic learning, virtual learning environment (VLE), learning content management systems (LCMS) mobile learning content management systems (MLCMS), learning management systems (LMS) and content management systems (CMS). The papers retrieved (n=87) were reviewed to identify if they contain more than three of the used keywords so as to make sure there was a real discussion in the paper, not just the peripheral reference to a keyword. Forty-four papers passed this test, four of which were found using the snowball method, making the total number 44. The content of the papers was then analyzed and grouped in five themes as shown in Table 2. These themes were identified based on the similarity in discussions of the concepts in the selected papers.

4 Findings

The results show that researchers discuss numerous aspects of learning using e-learning systems. We grouped these into five concepts according to their interrelations, research methods used, for example empirical or non-empirical methods, and their classifications based on intended audience (unit of analysis).

The unit of analysis indicates that researchers mostly address individual issues that affect individual learners, followed by groups, and organizations last. However, some researchers address all units in general. Most research methods used are empirical. The articles were searched using specific keywords and tagged by publication type source and content themes. The individual publication sources containing the highest number of relevant articles were Journal of computers and education (n=4) followed by European Journal of Open, Distance and E-Learning (2 articles) and journal of Educational Technology & Society (2 articles). The rest of the sources had one article each. In the articles, mobile learning was the most used keyword followed by electronic learning, LMS, CMS, LCMS and lastly MLCMS. The keyword MLCMS yielded much fewer research papers compared to the other keywords, which may not be surprising considering that the "M" makes it a subtype of system adding features to a basic set present also in the other types of system. The number of retrieved papers based on keywords above is shown in Tables 2 and 3.

Table 2 shows the number of papers in which different concepts appear. As can be seen, "use, access, design and infrastructure" is the most common theme. Papers are categorized according to the main theme; some papers also discuss some of the other themes. A majority of the papers use empirical methods. On the other hand, Table 3 shows the papers according to publication outlets and year of publication.

Table 2. Total number of papers according to concept and method

#	Concepts	No. of papers addressing 1 or more concepts	Papers that use empirical methods	Papers that use non-empirical methods	Papers that fall in different concepts
1	Use, access/ design and infrastructure	14	14	10	7
2	Communication and collaboration	6	5	3	3
3	Engagement & knowledge development	8	6	7	5
4	Content and service delivery	4	2	2	2
5	Implementation experiences & Evaluation	7	12	3	3

Table 3. Number of papers per publication category and year

	Years											
Publication Category	'12	'11	'10	'09	'08	'07	'06	'05	'04	'03	'02	**Total by category**
Journal Articles	1	5	4	3	4	6		1	2			26
Conference papers		2	3	1		1	2	2	1	1	1	14
Others, e.g. reports				1			1		2			4
Total by year	**1**	**7**	**7**	**5**	**4**	**7**	**3**	**3**	**5**	**1**	**1**	**44**

Table 3 shows that most of the retrieved papers were journal papers (26). Most articles are published in the years from 2007 onwards.

5 Discussion

Below we present and discuss the contents of the categories introduced above, i.e. Use, access/ design and infrastructure; communication and collaboration; engagement, personalized learning and knowledge development; content and service delivery; and implementation experiences and evaluation.

5.1 Use, Access/ Design and Infrastructure

Use, access, design and infrastructure of learning systems are the most discussed aspects in CMS literature. Design approaches can generally take two strands, user or technology centered. LCMS "provides trainers and subject matter experts with an intuitive method to create, review, analyze, and reuse learning content" [12], which is a complex process that requires a user's goals centered approach. When a system is not centered on users' need, technical aspects such as "version control" and "data management" affect the intended use. As a result two problems are common mainly in LCMS, i.e., interface designs focused on data structures rather than user goals, and absence of learning focus due to focus on data management. The next generation LMS designs however are user-centered [12].

From a pedagogical point of view Jesshope and Zhang [13] note that content creation and delivery are considered to be two key factors in e-learning systems specifically in web based learning systems. They show that e-learning system main components are infrastructure (software that allows learning to be created, managed, delivered and measured), services (which involve the planning, customization, integration and management of the e-learning application) and content (information shared or about users). Content is the core of LMS, LCMS, and other systems based on CMS technology.

The general requirements for an e-learning system therefore include accessibility (access to knowledge anytime and anywhere, flexibility (a flexible learning environment), extensibility and interoperability (an interoperable system that allows integration of system components easily and allows data to be shared across other platforms, reusability (ability to reuse content by creators or users), scalability (ability to provide access to many users large volumes of content repositories), security (ability to protect data and information) and be standards compliant [13]. The most adopted framework for learning platforms is SCORM (Sharable Content Object Reference Model). SCORM is a set of standards proposed by ADL (Advanced Distributed Learning) for development of online teaching and web-based learning platforms [5]. Adoption of this framework can be a solution to issues raised above.

In [25], it is indicated that students who enjoy using many of the course management tools do not consider the tools to be effective at enhancing the learning experience. Ibid considers this to be an adoption problem as borrowed from Celsi & Wolfinbarger's framework [3], that, three stages have to be undergone for students to be satisfied; (1) merely support functions, (2) replication of traditional teaching methods, (3) developing innovative learning situations and make learning more active, engaging, and ultimately better for students. In this paper, we link this argument to the usability and design aspects highlighted by Guralnick [12] although use of CMS should reflect Celsi & Wolfinbarger's framework most especially the third stage of the framework which is crucial to learners.

There are other issues researchers are concerned about regarding use and access rather than those highlighted above. Westera [40], Chu, Lin and Lin [5] discuss contextual learning and context sensitivity of content representation. Contextual learning can be defined as learning that is influenced by the operational settings (learning environment and its agents) of a learner. Modern digital media technologies induce the learning context and some of the factors that affect an individual's learning context are human culture and knowledge domain [4]. Design of learning systems should consider such factors. The fact that learning platforms are mainly for desktop computer environments rather than mobile devices, context sensitivity is also required to enable systems to sense and analyze context from various device sources during access [5]. Intersections of context, technology and learners should therefore be prioritized [17].

5.2 Communication and Collaboration

E-learning systems such as learning management systems are believed to provide greater support to tutors and students in areas of collaboration and communication [1, 28, 33] Collaboration tools such as online forums and discussion boards are invaluable resources in learning management systems because they enable learners to communicate with their fellow students as well as teachers hence acting as empowerment tools that allow socialization and group learning and teaching [28]. Alvarez, Alarcon and Nussabaum [1] holds that, collaborative activities can only lead to positive learning outcomes to the extent learners engage in productive interactions such as explanation and argumentation. Learning platforms are therefore said to be of great use in engineering collaborative activities.

Based on observations such as the above, many researchers emphasize collaboration and communication as crucial aspects in learning platforms and that they should be harnessed; with development of new technologies like virtual reality [28], use of various access tools [11] and development of synchronous collaborative models that accommodate both technical and pedagogical aspects [33]. Some communication and collaboration systems such as Skype and Adobe Connect, as well as technically more advanced virtual reality systems enable students to visually be aware of each other, get in contact in real-time, and gain a sense of presence while in remote physical locations. This, according to Monahan, McArdle and Bertolotto [28] facilitates simultaneous access to learning materials and has been used extensively in simulations and visualization of data.

Worth to mention regarding collaboration and communication in learning platforms is that learners can develop culturally meaningful interactions through asynchronous online communications [30]. Learners effectively encounter and explore different learning opportunities through virtual interactions and communications. This promotes cultural tolerance especially in distance learning and also provides opportunities to learn various social skills.

5.3 Engagement, Personalized Learning and Knowledge Development

Many researchers criticize conventional classroom teaching and learning approaches for being limited in terms of providing personalized learning content that would meet the individual needs of a learner. However, that criticism falls back on technology as well. Even though an opportunity for learning content management systems is to provide personalized learning environments [24], many e-learning platforms are not student/user driven but deliver the same content for all the students.

Mbendera [24] suggest three perspectives of learning; associative (learning by or through association), cognitive (learning by exploring the world around, receiving feedback and making reflection on actions) and situative (learning by belonging to a particular community with own social and cultural identity). Like blended learning, mobile learning encompasses the aforementioned learning perspectives [24]. However, learners must be involved in designing learning platforms to make them student-driven. Non-student driven platforms in terms of personalization and engagement can affect design inclusions for such perspectives. On the other hand, learning object standards serve to account for such perspectives and to enforce requirements like simplicity and flexibility, as the two factors are crucial for facilitating use [24].

Kukulska-Hulme notes that many emerging contemporary activities consolidate engagement and personalized learning and knowledge development [16]. These include: (1) social interaction tools for example Facebook and Twitter, (2) mobile phone internet access, (3) multimedia use such as use of audio books and podcasts, (4) location tracking for example use of GPS to locate places and services, etc. While these developments are not very common in learning platforms, they promote a mobile learning culture and allow learners to view themselves as innovators, creators and producers.

High levels of engagement and personalized learning have a positive effect on knowledge development. Macdonald and Chiu [22] show that regular users of mobile phones find "mobile learning materials to be user-friendly, offering increased convenience and flexibility". Learners become flexible in learning processes and acquire knowledge at a pace suiting themselves, which makes the process enjoyable. Given the content and collaboration management tools in learning platforms, the engagement of students during learning would be strengthened.

5.4 Content and Service Delivery

Research on most e-learning systems that use CMS technology indicates that there has been tremendous improvement in content management by some of learning management systems [8, 22]. Learners can manage and access multimedia content using various tools for example desktop and smart phones. Podcasting which involves publishing of audio media files has for example been one successful way of delivering content in learning management systems [22].

In the health industry, handheld computers have become standard equipment for medical students and clinicians to get a handle on the increasingly complex and large amount of information and text messages can be sent through SMS, audio, video and images through Multimedia Messaging Service (MMS) [8]. According to Franziska, Emil, Roland, Guo and Schwaninger [9], e-learning systems are a strong management instrument since they can be used to monitor and manage learners and that, they have a very critical advantage compared to traditional classroom, i.e. ability for learners to take the assessments, repeat them if they fail and get instant feedback at anytime from anywhere.

Although e-learning platforms offer numerous opportunities to deal with data and improve service quality for example managing schedules, students' assessment, transcript and scores display, course evaluations, grading of coursework and collaborative training (Ibid), mobile learning is still hindered by factors such as usability and content navigation due to the nature (size) of mobile devices like phones, costs associated with data access and unsupported file formats for documents [22, 32]. The fact that there are endeavors to overcome such challenges, Ibid notes that, currently m-learning would be "a means of providing an optional delivery mechanism for course content" [22].

5.5 Implementation and Evaluation

Implementation experiences and evaluation are among the most researched areas of e-learning platforms. Many studies evaluate and compare different learning management systems and tools ([10, 31] while others seek to develop and implement new platforms and models [13, 29] and others try to develop evaluation tools [10]. Most studies on implementation focus on identifying challenges and assessing the impact of the systems on learning. This common focus directed towards better ways of knowledge building is urged for in [26]: "content management systems should solve the problem of turning content into information and information into knowledge" (p. ii). Michelinakis [26] continues to state that the most important characteristic of content is the human aspect i.e., issues to do with access and usability. Thus, these should be key indicators in evaluations.

Most of the implementation experiences highlight some major services that mobile e-learning platforms share as listed in [32] i.e., content management, all-the time access and ease in communication. There has not been an evaluation framework for mobile e-learning platforms that facilitate such services [10]. Georgieva [10] therefore proposes criteria for m-learning systems evaluation that include: type of supported mobile devices, type of supported m-learning modes (on-line or off-line), type of information supported by the system (educational or administrative, in multimedia form or other forms), supported LMS, supported e-Learning standards, supported programming, and content adaptation technology used.

6 Conclusions

This paper summarizes discussions on a number of aspects addressed in literature on CMS technology in a context of learning and education. Most of the issues addressed focus on learning management systems in general rather than mobile learning management systems or mobile learning content management systems in particular. However, theoretical and practical strategies that prevail in MLCMS implementation are not completely different from those of other CMS e-learning platforms, but rather extend them. Mobile technology adoption affects the processes of learning positively by complementing e-learning. In summary, our investigations yielded the following response to our research questions (RQs).

RQ 1. What are the key focus areas in M(LCMS) research and practice?
We found five themes in the literature;

1. Use, access, design, and infrastructure
2. Communication and collaboration
3. Engagement and knowledge development
4. Content and service delivery
5. Implementation experiences and evaluation

While the first theme was the most frequently discussed one, there is clearly reason to believe that others pertaining to communication, collaboration, knowledge and engagement are also very important and likely to become more common in the future. One reason is that, so far, research does neither clearly nor thoroughly address the intersection between learning and technology. Technological issues and issues of learning and knowledge are rather treated separately. Broadly speaking, technical issues dominate, with issues of use – e.g. usability, access etc. – come second. Social issues, such as adjusting technologies to learning contexts and environments are very rare.

Massive mobile phone adoption creates more demand for mobile, flexible and ubiquitous systems and affects the process of MLCMS implementation by unveiling the need to provide education services using such systems. However, m-learning is regarded as a part of an e-learning infrastructure. All learning tasks must not necessarily be done with a MLCMS or a mobile device such as a phone. Achieving system interoperability therefore is one infrastructural strategy used to eliminate MLCMS as discrete educational service delivery channel to make it part of the general e-learning systems environment.

Another strategy is responsive web design. This strategy intends to meet the needs more than the wants of users and enforces web standards including mobile web. With the introduction of Web 2.0, responsive web design is more focused on designing for social reality and individual actions of users. Adhering to standards helps solve security and technical problems such as display and content incompatibilities.

In relation to responsive design, effective and efficient usable applications are developed that can minimize cost in terms of access simplicity and connectivity. Applications that require installations on the phone consume memory and power, which may cause inefficiencies. Such shortcomings are overcome by accessing the web directly rather than through installing mobile applications. MLCMS templates and themes should be designed to meet the phone requirements and where required, device detection techniques need to be used to detect devices and redirect content accordingly.

RQ 2. What are the key future research directions in the MLCMS field, and what gaps are there?

There is no indication that the above-mentioned five themes will become obsolete, rather to the contrary. Mobile devices and network infrastructures have limitations which need to be overcome, and they are developing with regards to technical issues. Regarding use in learning contexts, the lack of integration between learning issues and processes and technical ones is apparent. Use of mobile technologies is already widespread in many daily activities of a majority of the world's population, but not yet so in learning processes. In the literature we found some examples of views on learning but the overall focus is more on technology and use. There is hence a need to find ways of incorporating established or emerging popular ways of using technology into the field of organized learning.

We suggest undertaking further experimental studies on assessing how mobile learning content management systems use can contribute positively towards learning outcomes. User needs and experiences should be investigated so as to understand the effects in areas of content management, collaboration, and communication. As mobile content access problems are evident, instructional design for mobiles is another area of concern.

References

1. Alvarez, C., Alarcon, R., Nussbaum, M.: Implementing collaborative learning activities in the classroom supported by one-to-one mobile computing: A design based. The Journal of Systems and Software 84, 1961–1976 (2011)
2. Alavi, M., Carlson, P., Brooke, G.: The Ecology of MIS Research: A Twenty Year Status Review. In: DeGross, J.I., Henderson, J.C., Konsynski, B.R. (eds.) The Proceedings of the Tenth International Conference on Information Systems, Massachusetts, Boston, pp. 363–375 (1989)
3. Celsi, R.L., Wolfinbarger, M.: Discontinuous classroom innovation: Waves of change for marketing education. Journal of Marketing Education 24(1), 64–72 (2002)
4. Chou, C., Peng, H., Chang, C.: The technical framework of interactive functions for course-management systems: Students' perceptions, uses, and evaluations. Computers & Education 55, 1004–1017 (2010)
5. Chu, C.W., Lin, H., Lin, X.: Context-Sensitive Content Representation for Mobile Learning. In: Proceedings of the 4th International Conference on Advances in Web-Based Learning, Berlin, pp. 349–354 (2005)
6. Digitalbuzz.: Infographic: Mobile Statistics, Stats & Facts 2011 (2011), http://www.digitalbuzzblog.com/2011-mobile-statistics-stats-facts-marketing-infographic/
7. Digitalbuz.: Infographic: Infographic: 2012 Mobile Growth Statistics (2012), http://www.digitalbuzzblog.com/infographic-2012-mobile-growth-statistics/
8. Ducut, E., Fontelo, P.: Mobile devices in health education: current use and practice. Journal of Computing in High Education 20(2), 59–68 (2008)
9. Franziska, H., Emil, O.W., Roland, G., Guo, Y., Schwaninger, A.: Using learning content management systems as highly adaptive and efficient supporting tool for (recurrent) training – an applied perspective. In: 43rd Annual International Carnahan Conference on Security Technology, Zurich, pp. 52–58 (2009)
10. Georgieva, E.: A Comparison Analysis of Mobile Learning Systems. In: International Conference on Computer Systems and Technologies, CompSysTech 2006 (2006)
11. Grönlund, Å., Islam, M.Y.: A mobile e-learning environment for developing countries: The Bangladesh Virtual Interactive Classroom. Information Technology for Development 16(4), 244–259 (2010)

12. Guralnick, D.: Next-generation LCMS design: A user-centered approach. IADIS International Journal on WWW/Internet 2(2), 69–78 (2004)
13. Jesshope, C., Zhang, Z.: A Content Management System for the TILE Managed Learning Environment. In: The 3rd International Conference on Networked Learning (NLC), pp. 136–143. Sheffield University, UK (2002)
14. Järvinen, P.: On developing and evaluating of the literature review. In: Workshop on Literature Review in the IRIS31 Conference, Åre, Sweden (2008)
15. Klopper, R., Lubbe, S., Rugbeer, H.: The Matrix Method of Literature Review. Alternation 14(1), 262–276 (2007)
16. Kukulska-Hulme, A.: Learning Cultures on the Move: Where are we heading? Educational Technology & Society 13(4), 4–14 (2010)
17. Kurubacak, G.: Identifying Research Priorities and Needs in Mobile Learning Technologies for Distance Education: A Delphi Study. International Journal of Teaching and Learning in Higher Education 19(3), 216–227 (2007)
18. Levy, Y., Ellis, T.J.: A Systems Approach to Conduct an Effective Literature Review in Support of Information Systems Research. Informing Science Journal 9, 181–212 (2006)
19. Liaw, S., Chen, G., Huang, H.: Users' attitudes toward Web-based collaborative learning systems for knowledge management. Computers & Education 50, 950–961 (2008)
20. Lonn, S., Teasley, D.S.: Saving time or innovating practice: Investigating perceptions and uses of Learning Management Systems. Computers & Education 53, 686–694 (2009)
21. Lonn, S., Teasley, D.S., Krumm, E.A.: Who needs to do what where?: Using learning management systems on residential vs. commuter campuses. Computers & Education 56, 642–649 (2011)
22. Macdonald, I., Chiu, J.: Evaluating the Viability of Mobile Learning to Enhance Management Training. Canadian Journal of Learning and Technology 37(1), 1–12 (2011)
23. McGill, J.T., Klobas, E.J.: A task–technology fit view of learning management system impact. Computers & Education 52, 496–508 (2009)
24. Mbendera, J.A., Kanjo, C., Sun, L.: Towards Development of Personalised Knowledge Construction Model for e-Learning. In: Second International Conference on Mobile, Hybrid, and On-line Learning, St. Maarten, Netherlands, pp. 29–35 (2010)
25. McCabe, D.B., Meuter, L.M.: A Student View of Technology in the Classroom: Does It Enhance the Seven Principles of Good Practice in Undergraduate Education? Journal of Marketing Education 33(2), 149–159 (2011)
26. Michelinakis, D.: Open Source Content Management Systems: An Argumentative Approach. MSc Electronic Business Management thesis report, The University of Warwick (2004)
27. Mike, M.: Author(ity): The Literature Review as Expert Witnesses. Forum Qualitative Sozialforschung/Forum: Qualitative Social Research 4(1), Art. 8 (2003), http://nbn-resolving.de/urn:nbn:de:0114-fqs0301187
28. Monahan, T., McArdle, G., Bertolotto, M.: Virtual reality for collaborative e-learning. Computers & Education 50, 1339–1353 (2008)
29. Omar, Y., Ashaari, N.S.: Futuristic model for school's content management systems: A beginning. In: International Symposium on Information Technology, vol. 3, pp. 1387–1392 (2010)
30. O'Neill, J.E.: Implementing International Virtual Elementary Classroom Activities for Public School Students in the U.S. and Korea. The Electronic Journal of e-Learning 5(3), 207–218 (2007)

31. Qwaider, Q.W., Hattab, E.: Learning Content Management System (LCMS) Versus Learning management System. In: 3rd International Conference on Excellent in Education, Research Strategic Planning-Development, Athens, Greece (2010)
32. Rakkedal, T., Dye, A., Ally, M.: Mobile Distance Learning with PDAs: Development and Testing of Pedagogical and System Solutions Supporting Mobile Distance Learners. ssues in Distance Education series, pp. 51–74. AU press (2009)
33. Raymond, D., Kanenishi, K., Matsuura, K., Baudin, V., Gayraud, T., Yano, Y., Diaz, M.: A Model for Content and Communication Management in Synchronous Learning. Educational Technology & Society 8(3), 187–205 (2005)
34. Rengarajan, R.: LCMS and LMS: Taking Advantage of Tight Integration (2001), http://www.e-learn.cz/soubory/lcms_and_lms.pdf
35. Rivera, A., Candela, S., Garcia, R.C.: WikiDIS: A case of collaborative content management system for educative community. In: IEEE EDUCON 2010 Conference, pp. 601–610 (2010)
36. Romero, C., Ventura, S., Garcia, E.: Data mining in course management systems: Moodle case study and tutorial. Computers & Education 51, 368–384 (2008)
37. Solemon, B., Sulaiman, R.: Rapid ELearning Content Management System (RECOMS). International Journal of Computing & Information Sciences 4(1), 1–8 (2006)
38. Trifonova, A., Ronchetti, M.: Where is Mobile Learning Going? In: Rossett, A. (ed.) Proceedings of World Conference on E-Learning in Corporate, Government, Healthcare, and Higher Education, pp. 1794–1801. AACE, Chesapeake (2003)
39. Webster, J., Watson, R.T.: Analyzing the past to prepare for the future: Writing a literature review. MIS Quarterly 26(2), xiii-xxiii (2002)
40. Westera, W.: On the Changing Nature of Learning Context: Anticipating the Virtual Extensions of the World. Educational Technology & Society 14(2), 201–212 (2011)

The Relationship between Students, Mobile Phones and Their Homework

Peter Ilic

Toyo University, Department of English Communications,
5-28-20 Hakusan, Bunkyo, Tokyo 112-8606, Japan
peter.ilic@gmail.com

Abstract. When considering the freedom of communication in terms of time and space that mobile technology provides, educators need to understand how this ever present communications platform can be exploited to enhance collaborative learning. This research is designed to explore the areas of collaborative learning and the use of mobile phones as a support for collaborative learning through a year-long exploratory multiple case study approach integrating both qualitative and quantitative data analysis. Qualitative exploratory interviews are combined with Multidimensional Scaling Analysis to provide a detailed image of students' mobile use during collaborative activities. The results are triangulated across data sources, and key issues are interpreted and discussed.

Keywords: m-learning, space, MDS, collaborative, case study, EFL, Japan.

1 Introduction

This study will add to the understanding of how mobile phones impact on collaborative learning activities. As instructors determine the most-effective ways to encourage collaborative learning, it is important to consider the context of the students and the tools that are familiar to them. The affordances offered by mobile phones for learning such as anywhere, anytime communication and data gathering contribute to these considerations. The research question is "Does the intervention affect the relationship between students, their mobile phones and their homework?" This study is set at a university in Tokyo, Japan and the research students are native Japanese speaking members of four translation classes who have studied English as a foreign language (EFL) for eight years.

Modern telecommunications is increasingly leading to a description of time as being compressed (Harvey, 1999), global (Adam, 2013), and instantaneous (Urry, 2002). Time is seen less as continuously linear and more as a multitude of individual moments (Adam, 2013) such as home, school, and work. This view could mean very different patterns of interacting between students when collaboration through a mobile phone. The mobile phone supports this idea of time because it allows communication during previously unproductive periods of time (BenMoussa, 2003; Perry, O'hara, Sellen, Brown, & Harper, 2001) such as when traveling so possibly increasing the number of activities (Johnsen, 2001) like finishing homework

M. Kalz et al. (Eds.): mLearn 2014, CCIS 479, pp. 146–155, 2014.

(Virvou & Alepis, 2005). Likewise space is less about localized presence as mobile technology separates space from place (Giddens, 1990). This mobility replaces the impression of being at a place for communication with a telephone, to belonging to a network of communication (Geser, 2004). This network membership means that the importance of traditional boundaries in physical space is changing.

Mobile phones have created "simultaneity of place" (Traxler, 2009) a kind of bridging of physical space, such as home, school, and work, through the creation of a mobile social space by filling the space in between(Bull, 2005). Mobile technologies transport communities and discussions into physical public and private spaces forcing people to adjust their behavior to manage a more fluid environment (Traxler, 2009). Private is no longer just what happens when physically alone (Cooper, 2002). A student on a crowded train may have a private moment enjoying a favorite movie or silently texting a close friend. The advance of mobile communications has brought with it a blurring of public and private boundaries; however, it is still unclear what the impact of such fundamental changes will have on collaborative learning and learning in general.

2 Methodology

Qualitative research through an exploratory multiple case study approach was adopted for a period of one academic year. The purpose was to gain a deeper understanding of the processes and outcomes of the completion of collaborative learning activities through mobile devices by Japanese university students. In this study, the units of analysis were four groups of students, and the context was their participation in mobile collaborative activities within an English as a Foreign Language (EFL) course on the topic of translation. The students were all aged between 18 and 20 years of age, and were all living in Japan. Four groups of between five to eight participants were selected from second-year undergraduate students on a voluntary basis. A gender balance of males and females was attempted but was not possible due to the high percentage of female students in the school. Group one included 5 girls and 2 boys, group 2 contained 8 girls, group 3 contained 6 girls and group 4 contained 6 girls. An ethical consent letter, containing terms and conditions for participating in this project, was given to the participants.

The intervention used in this study was in the form of small group collaborative learning activities presented as weekly modules that were accessible on a private class website by either mobile phone or computer. These activities consisted of homework questions related to translation issues that are relevant when moving between the Japanese and English languages. Prior to each of these activities, the particular topic of the activity was discussed in the classroom and included theoretical concepts, terminology, and practical examples. The homework activities were all collaborative in the sense that they required the individual members of each group to find and upload their answers to the class website, to comment on other answers, and then to decide upon a best-single answer from their group answers. The members of each group communicated as a whole by uploading a text message with or without some other media such as photographs to the website forum.

2.1 Data Collection Analysis

A list of usage patterns emerges from each of the student interviews which then inform the production of probing and follow-up questions. The interviews were 30 minutes long and face-to-face involving one interviewer and one subject at a time. The coding for this research was done in NVIVO, and a form of thematic analysis and coding (Ezzy, 2002) was used. Thematic analysis refers to the identification of themes or concepts that are in the data, the building of a systematic account of what has been observed, and the emergence of a theory through the coding process.

In addition to the interviews, at the beginning and the end of the data collection period the participants were given an on-line word association questionnaire to complete and the results were analyzed using a Multidimensional Scaling (MDS) technique. MDS is a method for capturing efficient information from observed dissimilarity data by representing the data structure in lower dimensional spatial space. The purpose was to gain a picture of the participants' relationship to school, homework, and the mobile phone. It is hoped that by analyzing these word association questionnaires the researcher would be able to see any shift in the changing relational position of the words.

The data is numerical and in the form of symmetric similarity data consisting of student perceptions of the similarity among the following 25 English words: Afternoon, Alone, Computer, Dictionary, Discussions, Education, English, Evening, Groups, Home, Homework, Listening, Mobile-Phone, Morning, Night, Outside, Picture, Reading, School, Speaking, Text, Traveling, Video, Voice, Writing.

The word associations formed a 25 x 25 matrix of all possible word pairs. The students were asked to enter a value from 0 to 5 in the box at the intersection between each set of words that represented their perception of the strength of the relationship between the words. The participants were given the following number to meaning relationships as a reference: 0 = NO relationship, 1 = very distant relationship, 2 = distant relationship, 3 = close relationship, 4 = very close relationship, 5 = extremely close relationship.

After the participants completed the data entry form the total marks for each word intersection over all of the participants were added. So each number represents the total of all values given by the students for each row word and column word pair. A higher number indicates a greater number of participants perceived a stronger similarity between the word pairs. This type of data is known as proximity data which consists of measures of similarity or dissimilarity between objects of interest (Everitt & Rabe-Hesketh, 1997). In this case the objects of interest are the row words and the column words. The output from MDS is in the form of a plot of all the objects (words), and the distance between them shows the value of dissimilarity. In other words, the closer the words appear in the plot, the higher the perceived similarity.

3 Results

MDS analysis data from the first questionnaire results of semester one in April and the second questionnaire results of semester two in January are shown as a plot in Fig. 1. Since this result was a comparison of two results, one from semester one and the second from semester two, data from only those participants who completed both questionnaires could be used in the final analysis. So there were a total of 22 out of a possible 28 response sets that were used for the MDS analysis comparing semester one and two. The arrows were added to Fig. 1 to clarify the movement of the words from semester one to semester two. Most of the points seem to be moving towards the center. The largest changes were seen in the words "alone" (2), "homework" (11), "mobile phone" (13), "outside" (16), "text message" (21), and "writing" (25). After

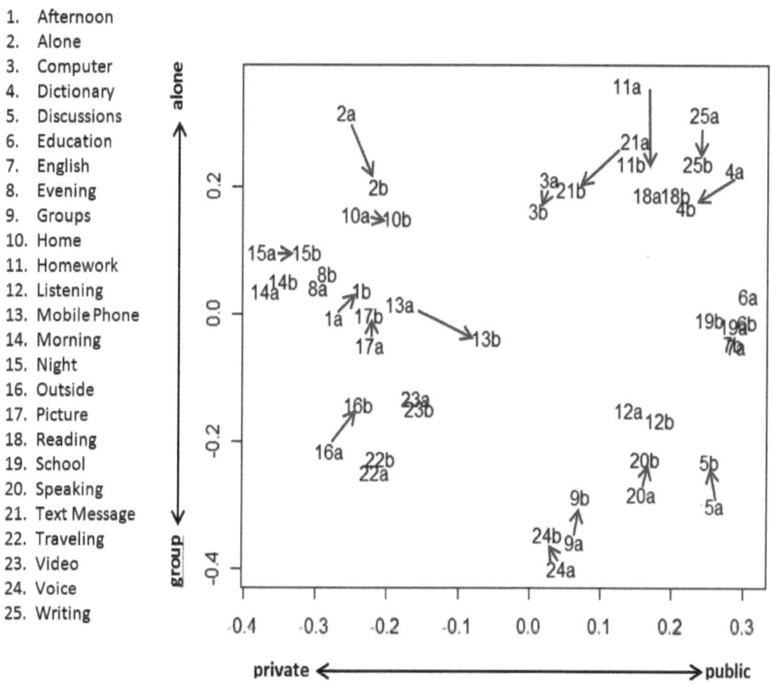

Fig. 1. MDS Plot All Groups

reviewing the placement of the words a possible description of the two dimensions became noticeable. At the bottom of the plot, the word "groups" (9) appears and at the very top is "alone" (2), so the vertical dimension was interpreted as representing the change from words related to a group of people to words related to being alone. On the far left side the words "evening" (8), "morning" (14), and "night" (15) appear which are times that are often spent at home in private. While at the far right side appear "education (6), English (7), "groups" (7), and "school" (19). Since the participants are all students attending school to study English, this all relates to very public activity. Because of these placements the horizontal dimension was interpreted as representing the movement from private life to public life. Following this interpretation, the lower right corner cluster including "outside (16), "traveling (22), and "video" (23) which are all things a person can do privately while surrounded by a group of people. The cluster at the right middle made up of "afternoon" (1), "evening" (8), "mobile phone" (13), "morning" (14), "night" (15), and "picture" (17) could be seen as private times and places when you are surrounded by one or two people. An example might be at home in the morning eating breakfast with their parents. This interpretation is supported by the proximity of this cluster to the word "home" (10). The cluster at the bottom right includes "discussion" (5), "groups" (9), "listening" (12), "speaking" (20), and "voice". These are all related to things a person would do in public with several people. An example might be English language students in a classroom working on an assignment. This interpretation is supported by the proximity of the small cluster made up of the words "education" (6), "English" (7), and "school" (19). Finally, in the top right corner appears "computer" (3), "dictionary" (4), "homework" (11), "reading" (18), "text message" (21), and "writing" (25). These were interpreted as words relating to things done alone that are related to their public life. An example is a student at home alone in their room using a dictionary to write the answers to a homework assignment for school.

As mentioned above, the participants were also asked to participate in two interviews. The first was at the beginning of the Japanese school year in April followed by a second at the end of the same school year in January. At the end of the year, group one had completed 12 individual interviews; group two had completed 15, group three had completed 10, and group four had completed 11. Altogether, this made a total of 48 individual interviews that averaged 30 minutes in length for a total of 1440 minutes. These interviews were transcribed and then coded.

4 Discussion

The coded themes suggested the collaborative intervention did affect the relationship between the students, the mobile phone, and the homework. The themes were examined further in light of the MDS analysis and the intervention did appear to have affected the relationship between the students, the mobile phone, and the homework in several ways. These include a) time of collaboration, b) space in which collaboration takes place, and c) method by which collaboration takes place.

Time

The mobile phone encroached on private time that would otherwise not have been used for homework collaboration. These times included that spent commuting, doing daily activities, and socializing. The mobile phone made it possible for the students to continue the homework activities while commuting to and from school, thereby increasing the available time for the activities. While traveling in trains or walking in the street, the students used the mobile phone to do the homework collaboration and data collection activities. Before this research, the students reported using their train time to sleep or to do some form of entertainment on their mobile phones. There was a very positive response to the opening up of this previously free time to homework activities. They could now use this time to read and reply to their group members. The students also reported that they began recording homework examples with their mobile phones as they walked around the city. The mobile phones' constant presence allowed the students to continue the homework during shopping trips or in restaurants. Since there were many group comments during the day, this mobile access allowed them to keep on top of the group activity.

The students started doing the homework while they were with their friends at lunch or socializing out at night. Many of the students began checking the website during lunch time with their school friends. Some students also began to check the homework site while out at night with their friends. This activity was reportedly quite common when the homework deadline was approaching. However, one student eventually decided to stop using a mobile phone for homework during lunch time because she decided just to spend that time with her friends.

The mobile phone has changed the students' use of time so that now they do not have as great a need to coordinate exact times for communication. Instead, they can use brief moments that are only approximately aligned with their group members' schedules (Ling, 2004; Plant, 2002; Sørensen, Mathiassen, & Kakihara, 2002). The students created moments of temporarily stable contexts while commuting and walking around by searching for and collecting examples. This activity allowed them to enable meaning-making within the flow of everyday activity (Kukulska-Hulme, Sharples, Milrad, Arnedillo-Sánchez, & Vavoula, 2011).

Space

The students began bringing the homework into places in which they had either rarely or never done this type of collaborative homework. This change in the spaces in which collaboration took place was observable in a variety of forms and included mobile as a private device, home as a place for homework, and work as a place for homework.

A mobile phone is considered a private and very personal device to the students. All of the students in this research received their first mobile phone from their parents when they were around ten years old so that they could keep in contact with their family. The personal nature of the mobile phone in the lives of the participants can be seen in Fig. 1. In semester one, the mobile-phone (13) is positioned on the "private" side of the horizontal dimension and between "alone" and "group" on the vertical dimension. In addition, mobile phone (13) is next to the four times of the day, morning (14), afternoon (1), evening (8), and night (15) which suggest it is with them throughout the day. Another commonly reported reason is for entertainment while traveling. The mobile phone seems to allow them to create a private space even when in a crowded train. In this private space, they can comfortably interact with the homework website, check social network sites, play games, or view videos.

Through the mobile phone, homework has gained a stronger foothold in the home. The speed and ease of using a mobile phone were attractive to many students because it allowed them to check the homework without the need to sit at their desk with a computer. In the morning while preparing for school, they would check the website for updates to see if anyone had commented on his or her posts. For some students, the mobile phone was the first thing they reached for after waking up. The device stayed with them as they moved around from the bedroom to the kitchen table and out the door to the train station. At all of these times, the students regularly checked the homework website. Then, after returning home, they would commonly reply to their group while lying in bed before going to sleep, even when a computer is available.

In Fig. 1 there are some indications of this in the movement of home (10) towards public on the horizontal dimension and homework (11) moving towards group on the vertical dimension. This mobility afforded by the mobile phone allowed the homework to move into these small but very personal spaces in the students' homes.

The mobile phone allowed the collaboration to continue into the workplace of the students. Many of the students have part-time jobs which take up a significant amount of their out of school time, often running late into the night. The mobile phone access allows them, for the first time, to immediately connect with their group's webpage and contribute to the discussion, even during a short 15 minute break. The word "work" was not included in the MDS word list, but night (15), when most students worked, did shift towards the public end of the horizontal dimension reducing the distance with homework (11). Without this mobile access option, there would be less time available to the students for homework.

The mobile phone allows the students to separate the idea of learning space from physical space so that it is less about being at a place and more about belonging to a network of learning (Geser, 2004; Giddens, 1990). The mobile phone has blurred the traditionally clear boundaries between school, work and home allowing one to invade another as observed with at home company workers (Schwarz, Nardi, & Whittaker, 2000). Now, even when physically alone at home the students can enter into a virtual public space discussion with their group members (Cooper, 2002; Sheller, 2004). This access allows the learning to be situated in the everyday environment where the students will most likely be using their translation skills (Lave & Wenger, 1991).

Method

The role of the mobile phone as an essential communication tool and its ability to switch effortlessly between the private websites and the public homework website places the device as a bridge between public and private life. These changes in the communication method used for collaboration were observable in a variety of forms and included feelings towards mobile device, non-friends, private websites, and language.

The mobile phone has drawn homework into a more central relationship with many aspects of the students' personal lives. In Fig. 1, homework (11) and mobile phone (13) have both moved towards the center of the plot. The close personal connection that the students have with their mobile phones affects the feelings that they have towards the homework. When using their mobile phone in place of a computer, they reported feeling a greater sense of familiarity with the homework, which they felt brought them closer to the topic under discussion. In addition, they reported a reduction in their nervous feelings that were associated with homework done through other non-mobile methods.

The mobile phone affords the ability to switch quickly and easily between mobile applications whether they are for entertainment or homework. This ability creates a greater sense of closeness between these two separate aspects of their lives – the private entertainment and the public homework- so that they all seem to be in the same space or shared presence on a network. In the interviews, some participants reported that this feeling of shared presence of entertainment and homework reduces both the perceived distance and the physical effort required to switch between them, so reducing the motivational barrier needed to start a homework session.

The boundary between friends and others is blurred by the mobile phone so possibly increasing the development of close relationships. The mobile phone use for homework allowed the students to communicate more with their group members because the phone is always present and reduces the need to meet face-to-face. While the students were happy to discuss the topic face-to-face with friends, they did not like the idea of meeting face-to-face with group members that they did not already know. They mentioned that they would not talk with them at all if the homework were just face-to-face. However, the same students felt that the mobile phone offered a comfortable way to communicate with those students that they would otherwise not speak with, resulting in an increased chance of getting to know them better.

Mobile social network sites that the students previously used only for very private friends began to be used as another channel for communication. Mobile social network sites, such as Twitter and Mixi, were clearly identified by the students as something they would not like to use for homework. Some students strongly expressed a need to keep some part of their lives private from school so that they would have a way to relax. The social network sites were identified by many students as that private place in which homework is not allowed. However, this attitude changed over the year as the students started to incorporate their mobile social networks into the collaboration process. They used these systems as secondary notification channels by which their friends in the group could request an immediate and private comment on their posted homework message. These requests were seen by friends who were not in their group or even in their school so they were able to get feedback and advice. In addition, the students used the mobile social networks as a way to overcome the limitations of the homework site. Some students started to use their social networks because they found the interface to be much simpler than that of the homework website.

The students clearly placed their first language (Japanese), into their private world of communication while their second language (English), was used for public communications. When communicating face-to-face, texting, or on their mobile social network sites, the language of communication was usually Japanese. Alternatively, the homework website, which can be seen by everyone including the course instructor, is an all English language environment. They are allowed short Japanese examples of a few words when explaining a translation, but otherwise everything must be in English. This public forum for their language was difficult for those students who lacked confidence in their English ability. Students used their private mobile communication channels to check their ideas with friends and others with whom they felt more comfortable in order to reduce the chances of potentially embarrassing public mistakes. The students began to fit the mobile learning activities on their mobile phone into their already existing preferred practices(Waycott, 2004). They accessed the homework website while they were playing a game or reading their social network site, switching rapidly between entertainment and education as a way of relieving the boredom of the commute (Lasen, 2002).

The mobile phone allows students to participate in two different interactions at the same time (Rettie, 2005). The students would often enter their private social network sites to share thoughts on the homework in Japanese because it allowed them an alternative channel of communication (Gikas & Grant, 2013). These private websites would allow a deeper level of communication because they were using their first language, and it was not being monitored by the instructor (Gikas & Grant, 2013).

5 Conclusion

The collaborative intervention did affect the relationship between the students, the mobile phone, and the homework, including time, space and method of collaboration. The ability to immediately switch with ease between private mobile activities and more public homework activities created a type of shared space or network where students who would not normally communicate face-to-face found themselves sharing this space so increasing the number of people with whom they could collaborate. Commonly used mobile SNS were drawn into the collaboration acting as a separate channel to communicate in their private language of Japanese which reduced their motivational barriers. This behavior has clear implications for curriculum design such as language learning where students are hesitate to show mistakes.

Mobile phone use for collaboration seems to change the nature of the homework from a very public activity to a more private one by bringing the homework into areas of their lives not possible with traditional technology. The practical implication of this is an improved sense of community (Lave & Wenger, 1991). In addition, when pre-existing negative feelings are present in students the use of the mobile phone appears to erode these as the homework becomes more embedded in the private activities of the students such as mixing education and entertainment activities. This effect is an opportunity for curriculum designers to incorporate what is meaningful to the students into the course material.

The student participants often seemed to blur the lines between the use of the social networking tools and the mobile device, making little distinctions among mobile services, such as homework website, and social media. One implication of this is that the mobile phones may provide a way for instructors to strengthen the network of learning formed by collaboration by allowing students to continue the discussion as they move through physical locations(Sharples, Taylor, & Vavoula, 2005). In addition, instructors could consider incorporating mobile web services already used by the students to leverage the advanced technology and popularity of the sites to further increase motivation and communication.

6 References

1. Adam, B.: Timewatch: The social analysis of time: Polity (2013)
2. BenMoussa, C.: Workers on the move: New opportunities through mobile commerce. Paper presented at the Stockholm Mobility Roundtable (2003)
3. Bull, M.: No Dead Air! The iPod and the Culture of Mobile Listening. Leisure Studies 24(4), 343–355 (2005)
4. Cooper, G.: The mutable mobile: Social theory in the wireless world. In: Brown, B., Green, N., Harper, R. (eds.) Wireless World: Social and Interactional Aspects of the Mobile World, pp. 19–31. Springer, London (2002)
5. Everitt, B., Rabe-Hesketh, S.: The analysis of proximity data. Arnold, J. Wiley, London (1997)
6. Ezzy, D.: Qualitative analysis: Practice and innovation. Routledge, London (2002)
7. Geser, H.: Towards a sociological theory of the mobile phone. Publica (2004)
8. Giddens, A.: The Consequences of Modernity. Polity 53(83), 245–260 (1990)

9. Gikas, J., Grant, M.M.: Mobile Computing Devices in Higher Education: Student Perspectives on Learning with Cellphones, Smartphones & Social Media. The Internet and Higher Education 19, 18–26 (2013)
10. Harvey, D.: Time-space compression and the postmodern condition. Modernity: Critical Concepts 4, 98–118 (1999)
11. Johnsen, T.E.: The Instantaneous Time How being connected affect the notion of time. IT-Users and Producers in an Evolving Sociocultural Context 69 (2001)
12. Kukulska-Hulme, A., Sharples, M., Milrad, M., Arnedillo-Sánchez, I., Vavoula, G.: The genesis and development of mobile learning in Europe. In: Parsons, D. (ed.) Combining E-Learning and M-Learning: New Applications of Blended Educational Resources, pp. 151–177. Information Science Reference (an imprint of IGI Global), Hershey (2011)
13. Lasen, A.: A comparative study of mobile phone use in public places in London, Madrid and Paris. Digital World Research Centre, University of Surrey (2002)
14. Lave, J., Wenger, E.: Situated learning: Legitimate peripheral participation. Cambridge university press (1991)
15. Ling, R.: The mobile connection – The cell phone's impact on society. Morgan Kaufmann Publishers, San Francisco (2004)
16. Perry, M., O'hara, K., Sellen, A., Brown, B., Harper, R.: Dealing with mobility: Understanding access anytime, anywhere. ACM Transactions on Computer-Human Interaction (TOCHI) 8(4), 323–347 (2001)
17. Plant, S.: On the mobile – The effects of mobile telephones on individual and social life (2002), http://web.archive.org/web/20080625013404/http://www.motorola.com/mot/doc/0/234_MotDoc.pdf
18. Rettie, R.: Presence and embodiment in mobile phone communication. PsychNology Journal 3(1), 16–34 (2005)
19. Schwarz, H.J., Nardi, B., Whittaker, S.: The hidden work in virtual work (2000), http://web.mit.edu/sts/pubs/pdfs/MIT_STS_WorkingPaper_30_Schwarz.pdf (retrieved)
20. Sharples, M., Taylor, J., Vavoula, G.: Towards a Theory of Mobile Learning. Paper presented at the mLearn 2005 – 4th World Conference on mLearning, Cape Town, South Africa (2005)
21. Sheller, M.: Mobile publics: beyond the network perspective. Environment and Planning D: Society and Space 22(1), 39–52 (2004)
22. Sørensen, C., Mathiassen, L., Kakihara, M.: Mobile services: Functional diversity and overload. Paper presented at the New Perspectives on 21st-Century Communications, Budapest, Hungary (2002)
23. Traxler, J.: Learning in a mobile age. International Journal of Mobile and Blended Learning (IJMBL) 1(1), 1–12 (2009)
24. Urry, J.: Sociology beyond societies: Mobilities for the twenty-first century. Routledge (2002)
25. Virvou, M., Alepis, E.: Mobile educational features in authoring tools for personalised tutoring. Computers & Education 44(1), 53–68 (2005)
26. Waycott, J.: The appropriation of PDAs as Learning and Workplace Tools: An Activity theory Perspective. (PhD), Open University. Milton Keynes (2004)

Mobile Learning for Development: Ready to Randomise?

Laurenz Langer[1,*], Niall Winters[2], and Ruth Stewart[1,3]

[1] University of Johannesburg, Centre for Anthropological Research, Johannesburg, South Africa
llanger@uj.ac.za

[2] Institute of Education, London Knowledge Lab, London, United Kingdom
n.winters@ioe.ac.uk

[3] Institute of Education, Evidence for Policy and Practice Information and Co-ordinating Centre (EPPI–Centre), London, United Kingdom
r.stewart@ioe.ac.uk

Abstract. Driven by the demand for evidence of development effectiveness, the field of mobile learning for development (ML4D) has recently begun to adopt rigorous evaluation methods. Using the findings of an ongoing systematic review of ML4D interventions, this paper critically assesses the value proposition of rigorous impact evaluations in ML4D. While a drive towards more reliable evidence of mobile learning's effectiveness as a development intervention is welcome, the maturity of the field, which continues to be characterised by pilot programmes rather than well-established and self-sustaining interventions, questions the utility of rigorous evaluation designs. The experiences of conducting rigorous evaluations of ML4D interventions have been mixed, and the paper concludes that in many cases the absence of an explicit programme theory negates the effectiveness of carefully designed impact evaluations. Mixed-methods evaluations are presented as a more relevant evaluation approach in the context of ML4D.

Keywords: mobile learning, development effectiveness, ML4D, developing-country education, impact evaluation.

1 Introduction

Mobile devices, cell phones in particular, are becoming a ubiquitous feature in developing countries. Driven by poor peoples' endogenous uptake, mobile technologies provide an effective platform to deliver development interventions. This assumption is reflected in the success of mobile applications in diverse settings such as finance, health, and agriculture [1] and has, under the umbrella term 'mobile learning for development', inspired the usage of mobile technologies as a tool to support developing-country education [2,3].

---

* Corresponding author.

M. Kalz et al. (Eds.): mLearn 2014, CCIS 479, pp. 156–167, 2014.

ML4D encompasses the process of teaching and learning with mobile technologies in a developing-country context, building on the unique features and applications of mobile devices [4] to provide marginalised learners and teachers with an educational tool that can effectively address some of the specific educational challenges faced by developing countries. This process of learning and teaching with mobile devices can assume diverse forms and is not restricted to formal education. ML4D programmes for example have been implemented in the fields of health care, agriculture, and financial literacy. In the light of the often compromised formal education opportunities in developing countries [5], mobile learning has been framed as being able to extend access to, as well as improve quality of, education. The importance of context to mobile learning [6] further presents a mechanism to increase ownership and relevance of the learning process - a feature of particular significance in resource-poor education systems.

This paper reports findings of an ongoing systematic review of ML4D interventions. It aims to reflect on the experiences of applying rigorous evaluation designs, such as randomised control trials (RCTs), to assess the effectiveness of ML4D interventions. The paper investigates whether the observed trend towards more rigorous, large-scale field experiments has been a beneficial contribution to the practise and design of ML4D. The paper is structured as follows: an introduction to measuring impact in ML4D is presented before assessing the impact evaluation landscape in the domain based on the systematic review findings. The adequate timing of impact evaluations in relation to the maturity of ML4D is discussed with the aim of encouraging further discussions on the role of rigorous impact evaluations in the field.

2 Background

RCTs or random evaluations currently enjoy high popularity as an instrument to evaluate the impact of development interventions. Originating in health science, donors and scholars alike advocate the methodology as arguably the most effective tool to establish 'what works' in international development [7]. RCTs employ the method of random assignment – that is the use of a random process to determine access to a programme, and is distinct from random sampling, which presents a technique to select units to create a group representative of a wider population. Random assignment is the key mechanism that allows RCTs to establish a valid counterfactual and consequently increased power to measure causal relationships.

Based on this, random evaluations lead the wider drive towards more 'careful' evaluations of development effectiveness and their popularity expresses an increasing demand for reliable research evidence to inform programme designs. The UK Department for International Development (DFID) for example regards RCTs as the most reliable study design to evaluate the impact of development interventions [8]. However, the growth of RCTs in international development is not without controversy, with critics coining the term 'randomistas' [9] to refer to the perceived narrow focus on evidence of impact, neglecting e.g. notions of contexts and politics as an important determinant of programme effectiveness. Since development interventions are inevitable embedded in prevailing social structures, programme evaluations need to account how different contexts (e.g. cultural beliefs, historical background, policy environments) influence programme reception and impact. Process evaluations, which investigate in particular whether the applied intervention is implemented as intended and relevant to the context of the target population are thus advocated as an essential exercise prior to the application of a rigorous programme evaluation [10]. Recently, the occasion of celebrating the first decade of evidence-informed development has revived this debate regarding the utility of rigorous impact evaluations in development [11].

ML4D has experienced a similar trend towards an increased focus on evidence of what works. The UNESCO Mobile learning for teachers in Africa and the Middle East report bemoans a 'dearth of evidence-based research' concluding that in order to make a more valid case for mobile learning in the context of development, an increased evidence-base of the intervention's effectiveness is required ([12]: 6, 28). Citing the discrepancy between the convincing rationale of ML4D and the failure of programmes to manifest or to achieve scale, Traxler ([13]: 164) proposes 'the rigorous analysis of evidence' as a guide to address the domain's 'failure to think clearly' and 'failure to thrive'. Evidence defined as proof of impact thus on the one hand is required to improve the design and implementation of mobile learning as a development intervention. On the other hand, evidence arguably determines the scale ML4D programmes are able to achieve. The value proposition of conducting rigorous impact evaluations in ML4D then can be identified as (a) gaining evidence of what, how, and why programmes work and (b) generating evidence to motivate taking programmes to scale.

This paper reflects on the increasing usage of rigorous impact evaluations in ML4D, a phenomenon, which is believed to be influenced by the wider enthusiasm for such evaluation designs in international development. To measure the impact of mobile learning interventions, Sharples and colleagues [14] distinguish between measures of learning gains (effectiveness), measures of motivational changes (satisfaction), and measures related to the mobile device and its features (usability). The measure of impact in this context then is confined to Sharples's first notion of 'effectiveness', i.e. impact defined as learning gains of programme participants. The terms 'rigorous' or 'careful' impact evaluation are defined following Clemens and Demombynes as,

'the measurement of a policy's or programmes' effect with the great attention to scientifically distinguished true causal relationships from correlations that may or may not reflect casual relationships, using well-controlled comparisons and/or natural experiments' ([15]: 2).

3 Methodology

This paper is based on the findings of an ongoing systematic review of ML4D interventions. The review used a systematic approach to identify relevant studies addressing the question of mobile learning's impact in developing countries. Firstly, it applied an exhaustive search effort based on an over-inclusive search strategy in order to identify as many relevant studies as possible. The search strategy included academic as well as grey literature sources, and search hits were screened according to pre-defined inclusion criteria. Included studies were then coded on key indicators such as applied methodology, intervention design, outcomes, etc., and further subject to a rigorous risk of bias assessment. The process of coding and appraisal was facilitated using EPPI-Reviewer 4 software (Version 4.3.5.0).

In order to gain a comprehensive understanding of mobile learning's overall effectiveness in developing countries as well as how and why mobile learning works or does not work, the identified studies were assigned to two different categories depending on whether they aimed to aggregate or configure ML4D impact[1]. This adapted review approach was assumed to better fit the context and demands of a complex development intervention such as mobile learning and allowed the review to establish a more complete picture of the evidence-base of ML4D. This paper reports on the composition of the established evidence-base investigating the diversity and appropriateness of the applied evaluations designs.

4 Brief Overview of the Impact Evaluation Landscape

The systematic review identified 26 studies that aim to aggregate the impact of mobile learning in a developing-country context. These 26 aggregative studies[2] present either programme evaluations or research experiments and date back until 2003. Of these 26 studies, nine studies are designed as field experiments of mobile learning programmes. Each of these nine field experiments has been conducted after 2010 and the studies have considerable larger sample sizes (average n = 1770) compared to the remaining 15 studies (average n = 77). Investigating only the ML4D research since 2010, it is found that researchers undertook only three classroom experiments. This contrasts the past record of ML4D research, as such study designs were applied in all but one of the studies conducted prior to 2010.

Further, this shift towards more large-scale field experiments is accompanied by the ambition to increase the rigour of evaluation designs. This translates in particular into increasing efforts to construct a more careful counterfactual, larger sample sizes, more emphasis on statistical power, and lastly attempts to apply random evaluation designs. The conduction of five RCTs of ML4D interventions, two of which are completed [17,18] reflects these efforts. The remaining seven aggregative studies each used a careful experimental design to evaluate the impact of the mobile learning programme. These evaluations however failed to randomise access to the programme, rather using a purposive sample from which participants were then randomly allocated to either the control or treatment group. This practise compromises the reliability of the counterfactual as it cannot control for unobserved characteristics that potentially could produce systematic differences between intervention and control groups.

In sum, the experiences of conducting large-scale experimental evaluations have been mixed. While it is feasible to conduct rigorous impact evaluations of ML4D and

---

[1] A similar review approach has been applied by Thomas et al [16].

[2] Aggregative studies are defined as experimental research, which design features both, a control/intervention group as well as before/after data, and sets out to measure the impact of an intervention using pre-defined and verifiable indicators.

individual studies[3] have achieved both scale and rigour, most reviewed evaluations were unable to meet the high methodological standard commonly associated with rigorous impact evaluations. Applying a new Cochrane risk of bias tool [19] six of the nine field experiments were judged as displaying a high risk of bias. Most common weaknesses included bias due to departures from intended intervention; bias in selection of results reported; and bias due to missing data (attrition). This leaves doubts regarding the internal validity of these studies.

5 When to Conduct Impact Evaluations?

Even if rigorous, large-scale impact evaluations – including RCTs – of ML4D interventions are feasible, such evaluation designs might not provide the most effective approach to measure mobile learning's impact in developing countries. Similar to experiences made in the education sector, the growing 'evidence movement', i.e. the remit to evaluate programme effectiveness along strict methodological criteria, in international development too has been controversial [20,11]. In education the drive towards more careful programme evaluations caused epistemological concerns, being perceived as 'the unwarranted transfer to the educational domain of a "positivistic" model of knowledge dominated by a "hierarchy of evidence" (…)' ([21]: 26). Evidence-informed development faces similar concerns about the notion and production of evidence. The application of rigorous impact evaluations in development, for example, has been criticised as a political tool to undermine transformational development efforts [22]. In the context of ML4D, Traxler [23] similarly cautions against the unquestioned assumption of evidence as the sole base for policy decisions. Adding to these theoretical challenges, in practise the relevance of rigorous impact evaluations depends on a number of assumptions and contexts to hold – some of which appear questionable in the case of ML4D.

Rigorous impact evaluations are not best suited to answer some types of research questions. They are designed to answer specific cause-effect questions on the difference a programme has made in outcomes. Needs assessments and process evaluations in contrast are better equipped to answer descriptive and normative questions regarding the programmes' performance [25]. Using the case of randomised evaluations, Glennester and Takavarasha [10] moreover show that this resource-intensive evaluation approach can only yield reliable findings of impact if the evaluated intervention has been well-designed taking into account local context and needs (i.e. after a process evaluation has been undertaken); and further, has been implemented strictly following the intervention design. The validity of this assumption faces particular risks when considering programme maturity. The review of ML4D's evidence-base indicates that process evaluations present the minority of applied evaluation designs, appearing secondary in favor of more impact-focused designs such as RCTs. Emerging mobile learning programmes in development countries, eSchools 360[4] for example, increasingly use rigorous designs as the primary and exclusive evaluation tool.

---

[3] For example: [17,18,24].

[4] `http://www/impactnetwork.org/eschool360/`

There are a number of inherent risks when applying rigorous impact evaluations in a new research area in which intervention approaches and designs are fluent and evolving. The programme design is likely to alter significantly as the programme responses to unanticipated challenges in implementation and participant behaviour. This leaves the evaluator with two main challenges: One might find no or few programme impacts, but can neither attribute this finding to weaknesses in programme implementation nor to flaws in the programme approach. The changes made to programme design during the pilot might also be extensive, effectively providing a new version of the programme in need of a new evaluation itself. Having said that, there is also an important reason to conduct a rigorous impact evaluation early on in the programme cycle – the need to provide evidence of programme effectiveness to justify the scale up of a programme to donors or policymakers.

6 Has ML4D Reached the Right Maturity?

The balance between these opposing agendas determines the timing of most evaluations. The findings of our review suggest that many mobile learning interventions in developing countries have not reached sufficient levels of maturity, questioning the recent drive in the field towards more rigorous evaluation designs. This is evident in the composition of ML4D's evidence-base coupled with the results of the risk of bias assessment. Whilst there is an observed increase in the number of rigorous impact evaluations, which have replaced other evaluation designs as a primary evaluation tool, few of these rigorous evaluations have been conducted effectively. The high risk of bias ratings reveal that the majority of these rigorous evaluations encountered methodological challenges and failed to review programmes' impacts adequately.

The evaluation of the Worldreader pilot programme provides an insightful example to underline this case. Worldreader is one of the largest and well-known ML4D programmes. The non-governmental organisation (NGO) is introducing e-readers as an educational tool in nine African countries. Schools are supplied with free e-readers, and the organisation maintains a collection of currently 944,300 free e-books that learners and teachers can access. The programme is targeted at increasing literacy among disadvantaged learners, and Worldreader highlights its early efforts to rigorously examine 'what works and what doesn't when it comes to eradicating illiteracy' [26].

The evaluation of the Worldreader pilot programme, despite a wealth of mainly positive findings, failed to find evidence of Worldreader's impact on pupils' literacy scores ([27]: 38), a measure initially assumed as the programmes' main contribution. Yet, a deeper analysis of the evaluation findings reveals that, due to device breakages and unexpected external events (e.g. teacher strikes and bullying), students spent 40.5 percent ([27:] 19) less time with the e-readers than anticipated by the programme team[5]. The absence of evidence on Worldreaders' effectiveness might thus not be

[5] An evaluation of the TeacherMate programme in Rwanda draws similar conclusions, finding that students had only one-third of the time, which was allocated to learning with the Teacher-Mate device [28].

caused by inherent flaws in the intervention, i.e. using tablets as a learning tool, but rather in the process of implementing the intervention. A more cost-effective evaluation approach, such as a process evaluation would have equally yielded these findings without the risk of drawing premature conclusions on the programmes' impact, which might potentially question the validity of the programme[6].

The evaluation of the Worldreader pilot also highlights another shortcoming of applying careful evaluation designs in immature interventions. Due to the evolving nature of pilot programmes, it is challenging to ex ante identify relevant outcomes, and likewise, to allocate reliable indicators to measure these. While Worldreader initially assumed reading scores to be a critical indicator of programme success, the evaluation team had to concede that the time line of seven months was not accurate in order to establish measurable differences in reading performance that could be attributed to the introduction of e-readers. As a result, the evaluation report primarily presents configurative data on the programme's performance and participants' reception [27].

Lastly, attempts to explain absent findings grounded in methodological flaws might undermine trust in future programme evaluations. In order to explain the lack of the programme's effectiveness on literacy scores, Worldreader for example states that the control group was 'inadvertently influenced to focus on reading and improve its reading performance because they were being monitored' ([27]: 39). An evaluation of the BridgeIT programme in India on the other hand praises its aggregative finding on pupils' math and science scores despite the fact that half of the schools in the control group did not provide endline data [29]. Cherry-picking strengths of careful evaluation designs, just as citing design as an explanation for unsupportive data, does little to reduce evaluation bias – the primary reason why one conducts rigorous evaluations in the first place.

The above shortcomings underline the risks and challenges of applying rigorous evaluation designs in an evolving and nascent field such as ML4D. To assume their value proposition in mobile learning, rigorous evaluations require a detailed understanding of how the intervention is assumed to deliver the desired results. Programme theories (or theories of change) aim to generate such an understanding by mapping the assumed sequence of events triggered by the intervention along a logical pathway [25]. Programmes using mobile learning in the context of development, however, due to the relative immaturity of the field and thus few examples of scale and best practise, often lack such explicit theories of change. Applying careful impact evaluations without prior definition of a programme theory risks placing a rigid focus on evaluating final outcomes at the expense of reviewing practises and intermediate outcomes. Yet, knowledge of practises, e.g. how students in marginalised settings appropriate technologies as a learning tool, is fundamental to any casual understanding of what works, how, and why, in ML4D. Evaluation designs that fail to account for the strengths and weakness in ML4D practises yield an incomplete picture as a narrow concern with impact and final outcomes risks 'cutting out the noise' – that is the complex linkages between the mobile learning intervention, the context it implemented in, and how this influences the reception and practise of ML4D.

---

[6] Worldreader since has moved into the second project phase, taking into account the findings of its initial evaluation. The second phase is evaluated with a smaller sample size and more focused objectives.

7 Discussion

Impact evaluations in ML4D need to make use of different types of knowledge in order to explain if and why mobile learning presents an effective development intervention. Our understanding of ML4D requires reliable information on the aggregative impact of mobile technologies, but we further also need to configure what determines and alters mobiles' potential impact in developing-country education. The need for this type of configurative knowledge is underlined by a RCT of a mobile phone-based adult literacy programme in Niger [17]. The RCT finds that using SMS technology can lead to significant increases in literacy and numeracy acquisition as well as retention. Yet, the programme supplied only one phone per five participants and evaluation data consisted merely of baseline, midterm and final assessments results. Consequently, the evaluation neither provides information on how the learning gains were achieved, nor on what made SMS technology an effective intervention in this setting. The study's findings – as rigorous and reliable as they are – might thus be limited in their contribution to our understanding of how mobile learning takes place in developing countries. This indicates that the proposed value of rigorous evaluations as means to yield a deeper understanding of ML4D is questionable.

The investigative nature of most mobile learning programmes in developing countries presents challenges to the application of rigorous impact evaluations. Aggregative evaluation designs with a narrow focus on establishing impact of what works have encountered important limitations in their effectiveness. As long as the programme theory of ML4D interventions continues to evolve significantly, mixed-methods evaluation approaches using a more configurative manner of investigation might provide a more effective study design. Mixed-methods evaluations, rather than making premature impact/no-impact judgments, can increase our understanding how mobile learning in developing countries takes place, and why programmes might be (or might not be) effective. They combine the strengths of rigorous evaluation designs with a more qualitative inquiry into the processes and contexts shaping the reception and practise of ML4D interventions. Mixed-methods evaluations thereby provide reliable evidence of impact, required to justify a scaling of programmes; but they moreover provide evidence of best practises, required to inform effective and adaptable programme designs. We argue that both types of evidence are of equal importance in improving the effectiveness of ML4D interventions. That said, mixed-methods evaluation approaches can often be resource-intensive due to the need to combine expertise of different research traditions, and can face challenges in interpreting and communicating the research results to different audiences.

ML4D cannot be reduced to the mere provision of mobile devices in resource-poor settings, in particular when it is not clear what pedagogical innovation or value the device features [30]. Likewise, the evaluation of ML4D programmes cannot be reduced to aggregative measures and requires an understanding of the complexity of the social interactions with, and perceptions of, the device that appropriate the technology to serve as an educational tool. To this remit, Velghe [31] uses ethnographic methods investigating the case of a phone-based adult literacy programme in South Africa. One of the key findings is that women develop device-depended literacies and use peer support to reinforce their learning experiences with mobiles. These findings complement the above-cited study by Aker and colleagues [17], showing not merely that women can gain literacies through mobile phone based interventions, but also how and why this process might take place. Albeit anecdotal, this case thus underlines the value of mixed-methods designs to evaluate ML4D programmes.

The drive towards more evidence and more careful evaluations of ML4D programmes is welcome. Yet, the above examples have shown that rigorous evaluation designs need to be applied with caution, as overly stringent evaluations of impact might not be best suited in the context of the infant field of ML4D. Evidence advocates arguably will point out that proof of what works is increasingly required as a precondition for scale up, and while we generally agree with this notion, we would like to refer to the One Laptop Per Child (OLPC) project as an interesting case study in this regard.

OLPC arguably is currently the largest educational technology programme with an explicit developmental mandate[7]. However, the programme achieved this scale despite a lack of rigorous evidence of its effectiveness. A number of RCTs[8] have found that OLPC has at best mixed impacts on learners and that the programme is best implemented alongside other interventions, such as teacher training on how to use the technology. Notwithstanding this ambiguous evidence-base, country governments continued to implement OLPC programmes; and while we acknowledge the complexity of evidence-informed policymaking and the many factors influencing the uptake of research evidence [32], this disregard of the research findings arguably reveals policymakers weighting the appeal of the OLPC idea higher than the absence of convincing evidence.

The role of evidence and its relation to ML4D's evolvement is therefore not necessarily linear. An increased amount and more rigorous knowledge of mobile learning's impact is likely to support its appeal and policy rationale. However, if this focus on evidence of what works sacrifices a deeper understanding of how and why ML4D works, the long-term benefits of this shift are uncertain. Few ML4D programmes have achieved scale and as the OLPC case shows, the relationship between scale and evidence is ambiguous. Achieving scale arguably depends as much on accounting for measurable impacts of programmes' success as it depends on effective programme designs.

Evidence might thus present but one factor determining the uptake of ML4D interventions. We therefore caution against rigidly following trends and best practises applied in more mature and established development interventions. That said, there are also internal drivers that push ML4D to establish more rigorous evidence of what works. Firstly, the domain displays a tendency to market mobile technologies' potential in developing-country education with high claims. The GSM Association (GSMA) [33] estimates for example that 180 million children in developing countries will have the opportunity to stay in school between 2013 and 2017 due to advancements in mobile learning interventions.

---

[7] There is currently some confusion regarding the project's status:
`http://thejournal.com/articles/2014/03/14/whats-new-with-olpc.aspx`

[8] For an overview of studies, see for example:
`http://blogs.worldbank.org/edutech/IDB-research`

In addition, leading institutions in the field often overlook the trend towards evidence-informed policy and practise. For example, the UNESCO mobile learning reports – the flagship publication on the state of mobile learning in developing countries – are informed by desktop reviews, anecdotal case studies, expert opinion as well as expert and policymaker interviews [2,12]. Given the profile of these publications, as well as the scale and expertise of the organisation, a more evidence-informed approach might reasonable be expected. Thus, the habit of overstating mobile learning's potential, paired with the neglect of evidence-informed approaches among leading institutions in the field, might present internal push factors that lead scholars and practitioners to conduct more rigorous impact evaluations in order to highlight the distinctiveness of their programmes.

8 Conclusion

In the context of ML4D the drive towards more careful impact evaluations has thus far rarely been questioned. The mantra of evidence-based practise, and consequent pressure to provide evidence of what works, informs the evaluation design of current mobile learning programmes in developing countries. Yet, given its niche position in international development and the investigative nature of most programmes, mobile learning's faith in and embrace of rigorous impact evaluations might be shortsighted.

Currently, mobile learning programmes in developing countries might not have achieved reasonable maturity in order to be subject to these rigorous evaluation approaches. This does not reject carefully designed impact evaluations altogether but aims to encourage a more balanced application of evaluation designs. Making use of limited resources, evaluators should carefully consider different evaluation options, e.g. process evaluations. Mixed-methods evaluations combining an aggregative and configurative evaluation approach serve as an effective method to assess the impact of ML4D programmes. The advantage of this evaluation approach rests in its ability to aggregate ML4D's overall impact as well as to configure what contexts and practises allow for an effective implementation of mobile learning in developing countries. A more balanced approach to evaluation, coupled with more cautious assumptions regarding its effectiveness might strengthen mobile learning's position as a development intervention and can improve our understanding of mobile learning's role in international development.

References

1. World Bank: Maximizing Mobile. World Bank, Washington D.C. (2012)
2. UNESCO: Turning on mobile learning. Global themes. UNESCO, Paris (2012)
3. Traxler, J., Kukulska-Hulme, A.: Mobile learning in developing countries. Knowledge Series. Commonwealth of learning, Vancouver (2005)
4. Roschelle, J.: Unlocking the learning value of wireless mobile devices. Journal of Computer Assisted Learning 19(3), 260–272 (2003)
5. UNESCO: Teaching and learning: Achieving quality for all. Education for all – Global monitoring report 2014. UNESCO, Paris (2014)

6. Wali, E., Winters, N., Oliver, M.: Maintaining, Changing and Crossing Contexts: an Activity Theoretic Reinterpretation of Mobile Learning. Research in Learning Technology 16(1), 41–57 (2008)
7. Banerjee, A.V., Duflo, E.: Poor economics. A radical way to rethinking of the way to fight global poverty. Public Affairs, New York (2011)
8. UK Department for International Development (DFID): Assessing the strength of evidence. How to note. DFID, London (2014)
9. Ravallion, M.: Should the randomistas rule? Berkeley Electronic Press (2009), http://siteresources.worldbank.org/INTPOVRES/Resources/477227-1142020443961/2311843-1229023430572/Should_the_randomistas_rule.pdf
10. Glennerster, R., Takavarasha, K.: Running randomized evaluations - a practical guide. Princeton University Press, Princeton (2013)
11. Pritchett, L.: An Homage to the Randomistas on the Occasion of the J-PAL 10th Anniversary: Development as a Faith-Based Activity (2014), http://www.cgdev.org/blog/homage-randomistas-occasion-j-pal-10th-anniversary-development-faith-based-activity
12. UNESCO: Mobile learning for teachers in Africa and the Middle East: Exploring the Potential of Mobile Technologies to Support Teachers and Improve Practice. UNESCO, Paris (2012)
13. Traxler, J.: Mobile learning in Africa: the elephants in the room. In: Kinuthia, W., Marshall, S. (eds.) On the Move: Mobile Learning for Development. Information Age Publishing, Charlotte (2013)
14. Sharples, M.: Methods for evaluating mobile learning. In: Vavoula, G., Pachler, N., Kukulska, H. (eds.) Researching Mobile Learning: Frameworks, Tools and Research Designs. Peter Lang Publishing Group, Oxford (2009)
15. Clemens, M.A., Demombynes, G.: When does rigorous impact evaluation make a difference? The case of the Millennium Development Villages. CDG Working paper 174. Center for Global Development, Washington, DC (2010)
16. Thomas, J., Sutcliffe, K., Harden, A., Oakley, A., Oliver, S., Rees, R., Brunton, G., Kavanagh, J.: Children and Healthy Eating: A systematic review of barriers and facilitators. EPPI-Centre, Social Science Research Unit, Institute of Education, University of London, London (2003)
17. Aker, C.J., Ksoll, C., Lybbert, T.J.: ABC, 123: The Impact of a Mobile Phone Literacy Program on Educational Outcomes. CGD Working Paper 223, Centre for Global Development, Washington, DC (2010)
18. Cole, S.A., Fernando, A.N.: The Value of Advice: Evidence from Mobile Phone-Based Agricultural Extension. HBS Working Paper 13-047. Harvard Business School, Boston (2012)
19. Sterne, J., Higgins, J., Reeves, B.: Extending the risk of bias tool to allow for assessment of non-randomised studies, cluster-randomised trials and cross-over trials: a Cochrane methods innovation fund project (Workshop). In: The Cochrane Collaboration. 21st Cochrane Colloquium Abstract Book, pp. 203–204 (2013), http://2013.colloquium.cochrane.org/abstract-book
20. Hammersley, M.: On 'Systematic' Reviews of Research Literatures: A 'narrative' response to Evans & Benefield. British Educational Research Journal 27(5), 543–554 (2001)
21. Oakley, A.: Research Evidence, Knowledge Management and Educational Practice: Early lessons from a systematic approach. London Review of Education 1(1), 21–33 (2003)

22. Eyeben, R., Guijt, I., Roche, C., Shutt, C., Whitty, B.: The Politics of Evidence Conference Report, April 23-24, 2013, at Institute of Development Studies, Sussex, UK (2013)
23. Traxler, J.: mLearning solutions for international development–rethinking the thinking. Digital Culture & Education 5(2), 74–85 (2013)
24. Kaleebu, N., Gee, A., Jones, R., Watson, A.H.A.: SMS story impact assessment report. VSO, Papua New Guinea (2013)
25. Gertler, P.J., Martinez, S., Premand, P., Rawlings, L.B., Vermeersch, C.M.J.: Impact evaluation in practice. World Bank, Washington, DC (2011)
26. Worldreader: Measured Impact: Students Read More & Better (2014), `http://www.worldreader.org/what-we-do/`
27. Worldreader: iRead Ghana study. Final evaluation report. USAID, Washington, DC (2012)
28. Mruz, K.: 2011 Rwanda TeacherMate Report. Open Learning Exchange, Cambridge, US (2011)
29. Wennersten, M., Qureshy, Z.B.: Summative evaluation of the Nokia & Pearson Foundation BridgeIT India Project. EZ Vidya, Delhi (2012)
30. Winters, N.: Mobile learning in the majority world: A critique of the GSMA position. In: Price, S., Jewitt, C., Brown, B. (eds.) The SAGE Handbook of Digital Technology Research. Sage, London (2013)
31. Velghe, F.: 'I wanna go in the phone': literacy acquisition, informal learning processes, 'voice' and mobile phone appropriation in a South African township. Ethnography and Education 9(1), 111–126 (2014)
32. Oliver, K., Innvar, S., Lorenc, T., Woodman, J., Thomas, J.: A systematic review of barriers to and facilitators of the use of evidence by policymakers. BMC Health Services Research 14(2) (2014)
33. GSMA: The Mobile Proposition for Education. GSMA, London (2013)

An M-Learning Solution to Address Critical Healthcare Issues in Ghana: Challenges for Mobile Learning in Developing Countries

Agnieszka Palalas[1], Nicole Berezin[1], Gretchen Kramer[1], Joel Kwesi Appiah[2], and Charlotte N. Gunawardena[1]

[1] University of New Mexico, Albuquerque, USA
[2] Central University College, Accra, Ghana

Abstract. The paper offers practical recommendations on the implementation of a mobile learning project in a cross-cultural setting. This summary of lessons learned includes a brief account of barriers experienced in the first twelve months of the longitudinal PAHELP study in Ghana, Africa, and replicable solutions which emerged from the close collaboration of the stakeholders in Ghana, Canada, and the US. The Design-Based Research (DBR) methodology adopted by the study is presented alongside the discussion of challenges to inform future m-learning studies.

Keywords: Mobile learning strategy, m-learning implementation, limitations, solutions, diverse cultural background, organizational challenges.

1 Introduction

The doctor-patient ratio in Ghana and in many parts of Africa is at an all-time low. The severity of these statistics was confirmed recently by the Director of Health Service for Greater Accra region, Dr. Linda Van Otoo, who stated that "Ghana's doctor-patient ratio is approximately one doctor to 15,259 patients in a year" [1]. She also clarified that similarly "physician assistants also see about 38, 000, patients in a year while midwives and nurses attend to about 6,000 and 1,400 patients respectively in a year." Motivated by this urgency, members of the Physician Assistant (PA) Program at the Central University College (CUC), Ghana, partnered with the University of New Mexico and Athabasca University to address healthcare needs in rural areas by training more PA's using mobile technology. The team, supported by a Grand Challenges Canada grant, launched a Design-Based Research study in May 2013. This project leverages mobile and e-learning technologies to support knowledge building and skill acquisition amongst PA practitioners in remote areas in Ghana, Africa. The educational intervention designed in the study blends innovative distance learning (DE) approaches, including mobile learning content, strategies and tools, e-learning resources, and occasional face-to-face (f2f) sessions to provide a hybrid PA program. Hence, the project has been nicknamed Physician Assistant Hybrid E-Learning Program (PAHELP).

M. Kalz et al. (Eds.): mLearn 2014, CCIS 479, pp. 168–181, 2014.

We are reporting on the key challenges and solutions which emerged during the first twelve months of this four-phase longitudinal DBR study, of which the overarching aim is to educate more medical personnel in Ghana and eventually other similar educational contexts in Africa. While the CUC could only train fifty Physician Assistants at a time in their traditional f2f program, the PAHELP intervention aims to multiply these numbers. Students can participate in the program at a distance at their convenience even while on the job in the clinics. Many of them accessed course resources and other supports on demand and immediately transferred that information to on-the-job solutions. In fact, students stress the benefits of using mobile devices for in-situ application of learning, as reported by one of them: "I currently see forty to fifty patients a day and I never know what is coming next. I use the information from this course, especially the complications of pregnancy lessons, in my clinic". Being able to retrieve the needed information without leaving their community has been identified as a key advantage of the m-learning approach. The PAHELP model leverages medical expertise throughout the country and enables the CUC to train more PA students without requiring them to leave their communities for long periods of time. This format enables students and faculty to remain in their neighborhoods, and meet academic, financial and familial responsibilities, whereas the traditional f2f PA program require students to interrupt their family and community life for up to two years and travel to institutions hundreds of miles away. It helps maintain community support and cultural integrity conducive to learning situated in a real-life context in Ghana and potentially throughout Africa.

The team identified and addressed several barriers in the creation and implementation of the PAHELP system. These barriers included inconsistent telecommunications infrastructure, cultural and technological incompatibility, as well as limited economic and human resources. The team researched Ghana's mobile network, supporting 24.4 million mobile phone users [3] and leveraged this expanding network as a foundation for the development of an m-learning solution that is agile and responsive to the current and future students and faculty needs.

2 Purpose and Outcomes

The key purpose of this DBR study is to design a replicable educational intervention to address a shortage of trained medical staff in remote areas of Ghana. The primary outcome, i.e., a hybrid mobile learning intervention, provides a base for the development of a corresponding instructional design framework, thereby offering design principles (intervention theory) that can be replicated in other similar educational contexts. In addition, the first twelve months of the project have been instrumental in shaping up the project strategy and collaboration between all the stakeholders leading to a modified DBR project framework. This ancillary outcome is vital to the success of the research study and the sustainability of the PAHELP program.

2.1 Educational Intervention: PA Program

The innovative m-learning solution we have been developing provides a redesign of the proven f2f PA curriculum to be delivered at a distance to medical practitioners working in their communities. Based on the feedback collected during the initial stages of the study, Obstetrics & Gynecology and Pediatrics course prototypes have been developed and delivered. This prototypes reflect the teams' focus on the unique educational background and learner characteristics of the target learners as well as their clinical experience. The PAHELP model provides a hybrid learning solution that integrates student control, freedom, and flexibility. In addition, students can access up-to-date content, expert support, and guidance at a distance. These activities are in alignment with our learners' needs and the socio-cultural constructivist theoretical foundation guiding the design. Students can actively maximize their learning experience by connecting to other members of this learning community via their mobile devices. These connections provide peer-support and opportunities to co-create new knowledge through collaborative activities whenever and wherever learners can access the network: "no student can learn in isolation and cooperation is needed in the form of exchange of resources, opinions, feedback, support, and energy" [6].

The course prototypes follow the PAHELP hybrid model that combines elements of mobile and e-learning with occasional f2f interaction. The content of the courses resides in a Moodle platform that is optimized for mobile delivery and accessed by students through tablets provided by the project. The core PA curriculum content has been repurposed and redesigned to integrate "bite-size" Moodle modules with pdfs, audio, video and mobile applications. Quizlet, Study Blue, and Medscape were selected, amongst other mobile apps and incorporated into the courses. In summary, the prototypes have been designed to leverage the affordances of mobile technologies and the existing PA courses at CUC in offering a solution that is sustainable, scalable, and can be replicated in other African countries faced with shortages of doctors and teachers who could, in turn, educate new medical staff.

2.2 Intervention Theory: Design Principles

The current PAHELP prototype was initially based on the design principles gleaned from related mobile and e-learning, as well as DE literature. These guidelines were then modified to accommodate the findings of the first phase, i.e., Informed Exploration, of the reported DBR study. The conceptualization and development of the successive prototypes are driven by the feedback collected through the future cycles of the study. The design guidelines will be refined iteratively, as new feedback emerges from the design tests and pilots, and will target both the pedagogical and technological aspects of the PAHLEP intervention.

2.3 Research Project Strategy: Refined Framework

All aspects of the research strategy and project management have been rigorously planned and documented to give direction to the efforts and focus of the intercultural

PAHELP team. To optimize the resources, time and expertise available to the research study, preliminary needs analysis had been completed prior to the DBR study; a strong research framework was then selected, human and monetary resources were secured, methods and processes were planned and agreed upon by the key stakeholders. However, it was not until that strategy was applied in a real-life context, that the project leaders realized the complexity of conducting in-situ DBR studies in a diverse educational setting. What appeared to the North American researchers to be explicit guidelines for action, were in actuality perceived as optional, and needed to be adapted to the cultural context, country, institution and stakeholders in Ghana. As the international team worked together through challenges that emerged in the process, the original DBR framework adopted by the study had to be modified. More emphasis had to be put in the initial stages of the study on ensuring preparedness and support from all participants, so that they could fully contribute to the design, development, and evaluation activities as well as decisions. The new DBR framework is still being refined and will be presented in detail in future publications; however, the key challenges and recommended solutions, which emerged in the first year of the PAHELP project, are shared in this paper.

3 Methodology

3.1 Design-Based Research Methodology and Research Design

This paper focuses on the first twelve months of the longitudinal DBR study that commenced on May 1, 2013. As mentioned above, the main purpose of this research is to design the hybrid DE model utilizing mobile and e-learning approaches, combined with f2f sessions to train PA students in rural Ghana via mobile devices. One of the key questions explored in the study pertains to the barriers to m-learning faced in this particular educational context, as well as the solutions to those challenges. Our data pertaining to such challenges have been distilled and analyzed for the purpose of this paper and the resultant findings are discussed herein.

3.2 Procedure

A multi-cycle DBR process has been implemented to provide data for design, development, evaluation and refinement of the PAHELP solution. This iterative approach has been adopted to improve understanding of what constitutes a satisfactory learning experience, promotes advancement in medical knowledge and practice, and is attainable within the technical and cultural constraints of the unique context. PAHELP adopted a modified version [6] of the Integrative Learning Design Framework (IDLF) model [2] including three phases: (1) Informed Exploration, (2) Enactment, and (3) Evaluation. Each of the phases would go through multiple iterations allowing for the refinement of the design and collection of up-to-date feedback. The three phases did not succeed each other in a linear fashion, but rather their cycles would overlap and their results inform the following iteration (Fig. 1).

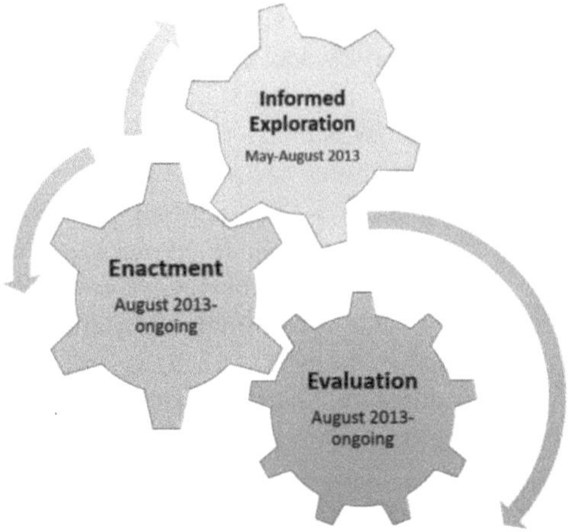

Fig. 1. Iterations of the three DBR phases

The project team has completed one iteration of each phase, which laid the groundwork for the consecutive cycles of Enactment and Evaluation. The three phases are briefly described below.

3.3 Phase 1: Informed Exploration

This phase began with a formulation and analysis of the existing educational problem followed by in-depth examination of the target audience, their needs, background, resources and technology available to them. Results of the earlier exploratory studies of the CUC PA program were also incorporated alongside data and observations collected by researchers from the US and Canada during their visits to Ghana. The technological infrastructure was explored, and consequently a device procurement plan was framed. Feedback from students and practitioners was collected through a student focus group, expert interviews and meetings. Additional data was gathered by the virtual team through regular online meetings, progress reports, and during faculty and instructional designer training sessions conducted at CUC by the North American team. All this data was aggregated and analyzed for common themes. The results were then validated by a comprehensive review of comparable solutions and relevant literature.

The Informed Exploration findings provided the base for the conceptualization of the theoretical construct to guide the design. In addition, Socio-constructivism was selected as the overarching theoretical framework, while Cognitive Apprenticeship as the pedagogic strategy supporting the situated learning approach adopted for the PA learning activities. Furthermore, the analysis of the context brought to the foreground the systemic social, cultural, and organizational characteristics of the educational context and the constraints these put on the DBR process and outcomes.

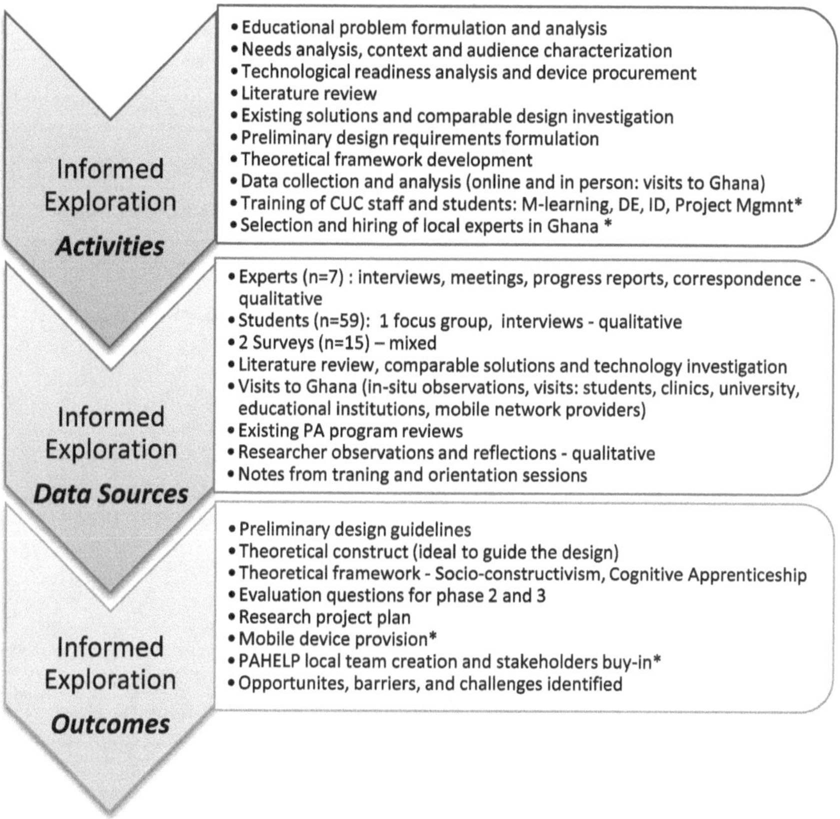

Fig. 2. Informed Exploration activities, data, and outcomes

Organized efforts were also made to support students, faculty and administrators in the transition from the f2f PA courses to the hybrid model. Potential student participants were recruited and offered training in the use of mobile devices. The students and faculty were provided with tablets (supplied by grant funding), which they used to access various learning activities and other components of the coursework. More details on Informed Exploration are presented in the diagram above (Fig. 2).

3.4 Phase 2: Enactment

The main focus of Enactment was the design and development of the first prototype of a hybrid PA program. Two PA courses were developed and made available through the Moodle platform to be accessed by students at a distance. Existing mobile apps were explored, and subsequently integrated into the solution as well. The ongoing discussion and evaluation of the design ideas led to the revision of the conceptual model and the creation of the functional prototype of the first course, Obstetrics and Gynecology, offered to students for evaluation in October 2013 and followed by the

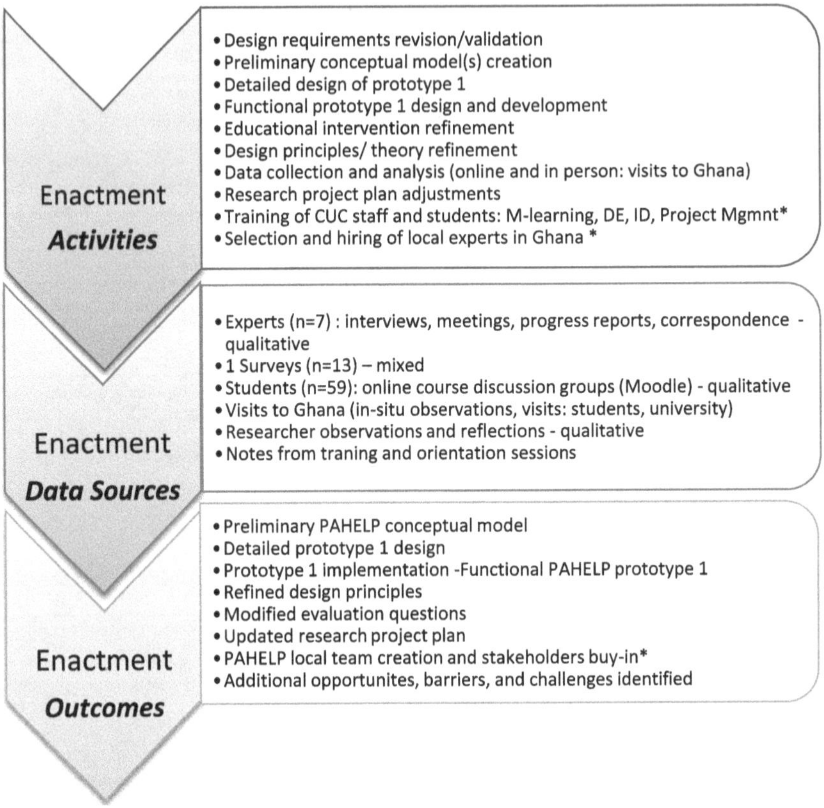

Fig. 3. Enactment activities, data, and outcomes

prototype of the second course, Pediatrics, in January 2014. These prototypes were evaluated through the consecutive cycle of Phase 3 providing feedback for further refinements of the design. The design, development, and evaluation cycles had to correspond with the CUC academic schedules as well as the visits to Ghana by the North American team.

Feedback to support the design and development of the intervention was collected through the methods and instruments summarized in Fig. 3. Additional training sessions for faculty, as well as the ID and IT practitioners were required during this phase to enable the collaboration of all stakeholders on the instructional design activities. Likewise, the project plan and the research strategy necessitated adjustments to accommodate the unanticipated barriers addressed in more detail in *Findings: Challenges*. To ensure adequate human resources and expertise needed for this phase, further professional development initiatives and project team recalibrating were necessary.

3.5 Phase 3: Evaluation

Thus far one cycle of Evaluation has been completed. In this final phase of the DBR process, data necessary for formative evaluation of the design product and process were gathered through the pilots of the two PA course prototypes. This feedback was collected through project meetings, progress reports, correspondence, interviews, surveys, focus groups, as well as researcher observations and memos. As a result, the prototype has been modified and the corresponding design guidelines adjusted accordingly. The design characteristics identified by participants as vital for the mobile educational intervention will be worked into the upcoming prototype 2 before the next evaluation cycle.

Further adjustments to the study framework, the project work plan, and the participant contributions were imperative for the success of the study. The timelines, main activities, outcomes, and data for Phase 3 are illustrated in the diagram below (Fig. 4).

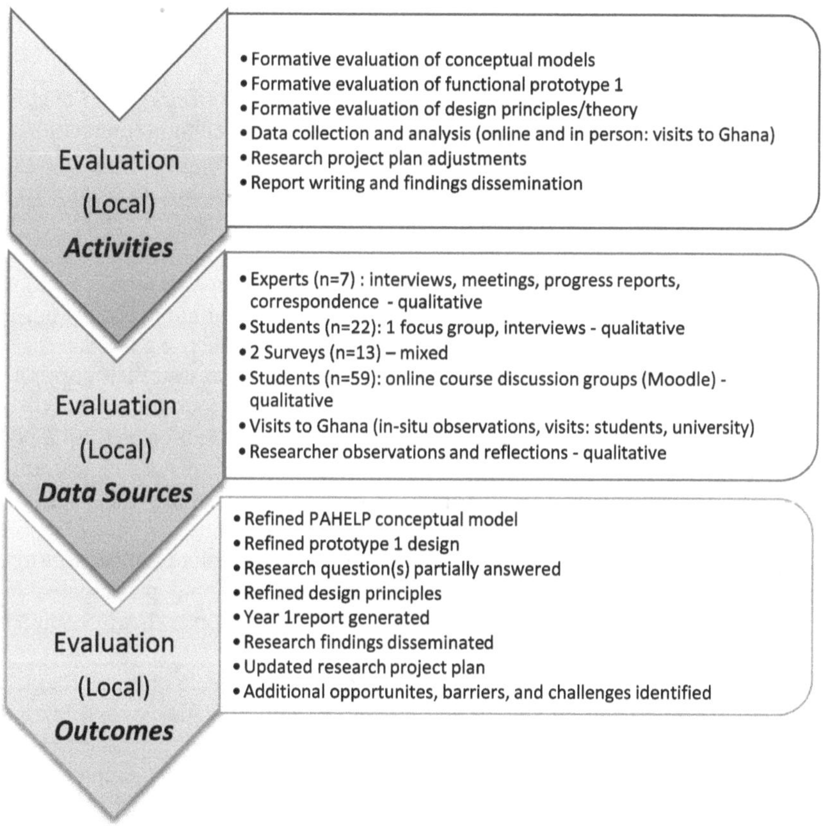

Fig. 4. Evaluation (Local) activities, data, and outcomes

In summary, the findings, reported in the *Findings: Challenges* section, emerged from the qualitative data collected from the student-participants as well as the other stake-holders from Ghana, USA, and Canada, including researchers, instructional designers, teachers, technical support, project managers, and administration.

The feedback was collected through five surveys (four student and one faculty survey) administered at various points of the study and one focus group - at the midpoint of the pilot. The remaining data was gathered on a regular basis through team meetings, interviews, and correspondence, and documented in researcher notes, meeting minutes, progress reports, and written correspondence. The data have been rigorously analyzed and coded by three co-coders. Codes were generated and assigned in a cyclical fashion to phrases and sentences through repetitive thematic analysis focusing on the question of challenges and solutions. The two main codes that emerged were those of organizational and learner-centered issues. These are discussed in more detail after the presentation of participants, their unique educational and cultural background, as well as some of their feedback that highlights their specific needs and attitudes.

4 Participants

The demographics of the participants in this research study reflect a group of students that share similar educational and professional goals and cultural foundations, but vary in age, gender and physical location. There are fifty-nine students enrolled in the PAHELP program with 35 females and 24 males that range in age from 24 to 52. They are all employed as community health workers, nurses, medical assistants (PAs), home health aides, and behavioral technicians.

Each of these students meets the challenge of balancing professional development with career and family: "I may not have children yet, but I still have responsibilities to my mother and father and traveling long distances is expensive and keeps me from home." Students from both urban and rural communities expressed their commitment to learning in this new environment: "I have a new place to look for answers while I am working. I practice alone and don't always have time to call or pick up a book. I can download PDFs and get help right away". Apart from the benefit of learning at a distance, the students appreciate the advantage of exchanging ideas as part of their own learning community: "Some of us talk to each other online in class, I want more of that". The students also communicated that they wanted more connection to each other as resources in class and at work. The PA students that participated in the PAHELP courses consistently practiced support and interaction both while online and in the f2f orientation and focus group sessions.

The current participants of the PAHELP program share socio-cultural characteristics that the research team will continue to incorporate into the design and implementation. According to Samovar, Porter and McDaniel [9] and consistent with our observation, the target learners are highly collective with a great emphasis on the welfare, needs and goals of the group than the individual. Our team observed that that Ghanaians have strong communal values that intertwine traditions, honoring elders and maintaining strong familial bonds. Additionally, personal dignity and proper social conduct are much individual are they are communal. Therefore, learners accept power as part of the culture and that those with power and status consider their subordinates to be different than them. These learners accept hierarchy within a collective and respect the roles of their superiors. The team integrated these cultural components into the design as they impact the dynamics of the teaching-learning relationship.

Furthermore, Samovar, Porter and McDaniel [9] posit that these participants also experience low uncertainty avoidance. They accept uncertainty as a fact of life. This may serve developers of new and innovative approaches to education, as we observed in our study, and explain students' patience and willingness to work closely with the design team to create learning tools for one another. When asked if they wanted to collaborate and share their own stories and real-life cases as foundations for problem solving activities, many of the students were willing to share and learn from each other's experiences in their everyday clinical practice. In summary, these participants form a vital and active group of medical professionals who are highly committed to being part of the solution to increasing access to medical services in their communities. They are willing to work in a highly collaborative environment.

5 Findings: Challenges

For the purpose of this paper, only the findings pertaining to the key barriers to the implementation of the m-learning solution are discussed. The team experienced a variety of challenges in the design, development, and delivery of the PAHELP model. After twelve months of collaboration, including voluntary consultant site visits and training, weekly virtual progress updates and design sessions, the team reassessed the types and levels of support that were being offered to the CUC stakeholders (students, faculty, IT staff and administrators). A systemic review of meeting minutes, including progress updates, established that the following three key areas posed the greatest challenges to the growth and sustainability of the project: (1) high level administrative support (governmental, institutional, and individuals), (2) institutional, collective, and individual readiness, and (3) limited access. This discussion addresses the challenges in combining an emerging organizational infrastructure with the development and adoption of an m-learning and e-learning intervention.

5.1 Securing High Level Administrative Support

The funding organization and CUC administration representatives created steps and processes, identified essential personnel, recruited voluntary consultants from Canada and UNM, and encouraged the CUC partners to create a project plan for the team to review and follow. Despite weekly Skype team discussions and offers of collaboration and support to finalize and implement a work plan, this vital strategic step was notoriously postponed creating a hurdle to the progression of the study. Moreover, once the project plan was drafted, not all stakeholders were equally prepared to adopt it as a binding contractual document. Regardless of the strong leadership from the North American team, the project activities seemed to be stalled by a number of factors that had to be investigated and addressed. Through the analysis of the feedback on these issues it was deduced that these initial challenges occurred due the lack of collective buy-in from CUC and formalized commitment of support from high level personnel at the university as well as government (required for the accreditation of the program). Gaps in understanding in regard to project progress and need occurred due to lack of transparency in the monitoring and communication between the sites. Administrative roles and responsibilities were not clearly defined resulting in missteps that affected outcomes of the project. An organizational and leadership chart along with concise steps was not shared and agreed on by the PA department and university administration to guide as to appropriate procedures and practices. Administration and faculty, new to this arena, required intensive supports to manage bureaucratic process related to receiving a grant, identifying key personnel, and setting up measures for short and long term goals of a cross-institutional research project.

5.2 Readiness for Implementing an M-Learning and E-Learning Intervention

As there is a need to establish solid process and practices at the organizational level and to firm up infrastructure, there is an equal requirement to provide consistent access to course materials, support, and technology for students interacting in a mobile and e-learning environment. A reliable infrastructure is crucial in implementing nascent educational systems driven by technology. The most prominent challenge that came to the fore for students were consistent access to the Internet as well as academic and technical support while engaged in the course. The university's ability to provide anytime access to the course was undermined by a lack of resources that remain in development not just for the hosting institution but for the country at large. Electricity and overall accessibility to the Internet are in high demand yet still in insufficient supply in Ghana. Blackouts are common, thus hindering access to learning materials, which frustrates the mobile learners and ultimately reduces their interest. Additionally, the cost and speed of data bandwidth in tandem with a sporadic electricity supply is a substantial hurdle for students wanting to benefit from mobile learning platforms. In Ghana, wireless hotspots are very rare; therefore most people have to connect to the Internet or to the learning platform through their own telecommunication network. Consequently, students rely on their own providers and, in turn, increasing their cost to access the course and its resources.

In a survey conducted to assess the project's progress, students' responses unvaryingly indicated that lack of consistent access to the Internet and the cost of connection hindered their ability to participate in the course. They reported: "It costs me a lot to use my pad and stay online and study"; "We were promised data packages to use for these courses. Some of us had data the first month but none since then". Students would then incur the costs to access the course: "I have been paying out of my pocket." Additionally, students would travel to the nearest urban center or town known to have reliable connectivity simply to access the material. Access issues combined with limited bandwidth, especially in rural areas where many of the students live and work, work against the success of an online course.

Another challenge experienced by learners relates to support and ease in using technology as a communication and academic tool. Many of the students reported that the PAHELP system was novel to them and was at first both challenging to access and use; therefore, they required more guidance in accessing the courses and interacting within them than originally projected by the North American team. Once the students mastered the steps, they strongly appreciated its flexibility: "I can download PDF's and other information and study at home between cooking meals and doing laundry." As reported by Ntseane [4], learners, including teachers as learners, have learning preferences that differ across the globe and that failure to recognize some of the specific cultural learning differences can defeat the goal of education and e-learning. At this stage of the project, students are in frequent need of support and seek direct communication. They often use the telephone, as they are most comfortable with this first line of contact when struggling with technology or the LMS. In the same survey, students remarked that they preferred to contact the support team most directly by using their telephone, "When I have a problem, I text [IT support] and sometimes call him, and he helps." Students reported that communication through f2f, verbal and non-verbal exchanges, and telephone conversations provided rich information and solutions to technical problems. This observation reflects that maintaining a culturally relevant learning environment ensures that students are being supported within familiar educational contexts. The use of Mobile devices can present challenges in obtaining support that coincides with traditional communication models, impacting comfort levels of students that have limited technological skills.

6 Summary: Solutions

In an effort to document cross-cultural understanding of program goals, priorities and activities, several team members are recommending the creation of a logic model to document and track progress. "A logic model is a systematic and visual way to present your understanding of the relationships among the resources needed to operate your program, the activities you plan and the changes you plan to achieve [10].

Additionally, we have created a collaborative model that extends the onsite f2f training sessions (4 total in the areas of program orientation, instructional design, mobile learning, and technology best practices). The PAHELP team created a dedicated design sub-team comprising subject matter experts, m-learning and IT professionals, instructional designers, faculty and students from both North America and Ghana. This virtual team was responsible for selecting and creating a system of training resources, including pdf documents, audio podcasts, videos and specialized mobile learning tools to be integrated into the Moodle LMS and accessible on students' mobile devices as well as personal computers. Many of the PA students volunteered to assist in the creation of learning resources, which provided an invaluable learner perspective coupled with the understanding of the Ghanaian context. In addition, several of the PA students expressed interest in mentoring and being mentored. As a result, the design team is also collaborating on the development of an integrated f2f and e-Mentoring program to provide mentoring for and between participants at all levels including administration, IT, designers, faculty and students.

As the team continues to collaborate on technology, training and mentoring solutions, they maintain that in developing regions, such as Ghana, difficulties manifest due to the absorption of models that are not entirely applicable to infrastructure that their educational systems stand upon. It would be beneficial to utilize technological tools that are present and successful in the country to construct courses accordingly. In our case, relying on phones, especially those already in students' pockets, rather than tablets would be preferable. Similarly, using SMS and subsidizing SIM cards to augment the existing ICT base has been identified as a more viable alternative. Olaniran suggests that "successful use of IT requires much more than mere installation and application of systematized knowledge. It also requires the application of implied knowledge regarding the organization and management of the technology and its application to the contextual environment in which it is to be used" [5].

Additional recommendations towards troubleshooting the infrastructure issues include a redundant learning platform in the cloud, to be set up and made available towards solving server access problems caused by the irregular supply of electricity. In the event that the institution is without power for an extended period of time, learners would be automatically directed to the alternate platform and be able to access course materials rather than struggle to connect. Furthermore, the institution would best serve PAHELP students by offering either no cost or low cost data packages upon enrollment in the DE course. This would need to be procured by the institution in collaboration with a telecommunication company. Such a partnership could prove lucrative for all parties involved towards providing consistent Internet access to the course, collaboration in mobile and e-learning models, and improvement of infrastructure. It is also recommended for governments to step in as partners towards creating a system conducive to m-learning. Moreover, arrangements should be fostered with telecom companies to reduce data charges for students with specific SIM Cards. Hence, the PAHELP project is engaging a telecommunication company, and a telecommunication device manufacturer, Huawei, to provide Assess Point Network (APN) for our learners. This solution aims to improve access to the learning platform since each student's SIM card is to be logically connected directly CUC's network.

In addressing limited access, the first course of action lies with the institution in evaluating students for preparedness and aptitude for learning in a hybrid, untethered environment – one that many are unaccustomed to. The learners' acquisition of techno-logical skills must mesh with expectations and objectives, as set by faculty and institution. Being that students are the central driving force behind any course, the success of the program is contingent on theirs. Implementation devoid of assessment of participant abilities could negate further advancement of a pilot to something more concrete. As part of the initial phase in assessing participants for preparedness, evaluation of the design should be conducted in a series of beta tests. Target population representatives should be invited to engage with the platform and offer feedback before its launch. This would inform all stakeholders as to what needs reworking in relation to specific learner and facilitator needs. Based on such feedback, the design should be refined and courses created to orient participants into the technology and learning/teaching strategies.

Lastly, the sense of community inherent in the Ghanaian culture has to be leveraged to build a strong community of learners. Apart from regular refresher courses and telephone support via a help line, students require a robust peer support system – both using their mobiles and through regional group meetings where students from the same geographical area meet in person to collaborate on group assignments, help each other and socialize. To support the integration of the work of the individual students and the regional groups into the larger program, facilitators and peer-mentors should monitor students' learning through personal on-site visits and sustained communication. Students, engaged in peer-to-peer teaching and learning within a familiar context provide a vital support system and a sense of community. Not only do they exchange technological and academic knowledge, but they also encourage and motivate each other. "Scaffolding and learning with and from each other is a natural and appropriate mode of learning in many indigenous cultures" [8].

At the same time, the cross-cultural research team continues to communicate openly in order to ensure that all stakeholders share the same paradigm. A shared frame of reference is vital for the project as a foundation from which the virtual team can negotiate, enact and report back on mutual understanding of goals, outcomes as well as roles and responsibilities.

References

1. Ansah, E.M.: Doctor patient ratio alarming, http://www.ghanaweb.com (accessed March 6, 2014)
2. Bannan, B.: The integrative learning design framework: An illustrated ex-ample from the domain of instructional technology. In: Plomp, T., Nieveen, N. (eds.) An Introduction to Educational Design Research. Seminar at East China Normal University, November 2007, pp. 53–69. SLO, Netherlands Institute for Curriculum Development (2009)
3. Eto, D.: Over 24 million mobile phone subscribers in Ghana (2012), http://www.Itnewsafrica.com/2012/over-24-million-mobile-phone-subscribers-in-ghana/
4. Ntseane, G.: African indigenous knowledge: The case of Botswana. In: Merriam, S., Associates (eds.) Non-Western Perspectives on Learning and Knowing Krieger, Malabar, pp. 113–135 (2007)
5. Olaniran, B.A.: Discerning culture in e-learning and in the global work-places. Knowledge Management & E-learning: An International Journal 3(2), 9 (2009)
6. Palalas, A.: Design guidelines for a Mobile-Enabled Language Learning system supporting the development of ESP listening skills. Dissertation, Athabasca University (2012)
7. Rao, K., Eady, M., Edelen-Smith, P.: Creating virtual classrooms for rural and remote communities. Phi Delta Kappan 92(6), 22–27 (2011)
8. Samovar, L., Porter, R., McDaniel, E.: Communication between cultures. Wadsworth, Boston (2007)
9. W.K. Kellogg Foundation: Logic model development guide: Using logic models to bring together planning, evaluation & action. Battle Creek (2004), http://www.wkkf.org/ (accessed July 26, 2014)

Pathways to a Better World: Assessing Mobile Learning Policy against UNESCO Guidelines in a New Zealand Case Study

David Parsons

Massey University, Auckland, New Zealand
D.P.Parsons@massey.ac.nz

Abstract. The UNESCO policy guidelines for mobile learning are intended to guide governments, educational institutions and other relevant organisations in the integration of mobile learning into education policy. Their main goal is to support and enable teaching and learning through the safe, affordable and sustainable use of mobile technologies. Like all policy guidelines, they are open to interpretation in practice, to be adapted to local contexts and conditions. Whilst they are primarily designed to assist developing nations without mature policy implementations, it is useful to benchmark them against possible exemplars in developed nations. This can help to reveal deeper issues in policy formulation that can assist policymakers in other contexts. It can also enable those nations with policies already in place to compare them with global best practice. This paper reports on a study based on a series of in-depth interviews with policy formulators and implementers in New Zealand. Interviewees were asked to reflect on the UNESCO guidelines in the light of the New Zealand experience of pioneering mobile learning policies. The results of this study reinforce many of the assumptions of the UNESCO guidelines, but also provide some insights into detailed aspects of mobile learning policy that are not explicit in the guidelines, yet may be of value to policymakers in other international contexts.

Keywords: UNESCO, Mobile learning policy, New Zealand, case study, interview data.

1 Introduction

Policy formulation works on many levels. It can range from the minutiae of internal organisational regulation, which at its worst reduces flexibility and innovation, and stifles creativity, to the visionary plans that embody our ideals for the future, and provide us with pathways to a better world. This article deals with the latter.

The UNESCO M-Learning Policy Guidelines Project was established to guide governments, educational institutions and other organizations in the integration of mobile learning into education policy. The main goal of the project was to support and enable educational delivery through the safe, affordable and sustainable use of mobile technologies. The "UNESCO policy guidelines for mobile learning" document that emerged from the project was authored by Mark West and Steven Vosloo of

M. Kalz et al. (Eds.): mLearn 2014, CCIS 479, pp. 182–196, 2014.

UNESCO, but contributors to the policy guidelines came from more than twenty countries [1]. The guidelines themselves are intended to be broadly applicable to all levels of educational delivery and all nations, but of course the expectation is that the guidelines will be tailored to local contexts when being used to formulate policy. They are intended in particular to assist the implementation of policy in developing countries. However it may be helpful to examine current policy in developed countries as a way of evaluating how the policy guidelines relate to current practice. The guidelines are still in the process of gaining widespread visibility, but are likely to be increasingly used by policy makers across the world, so empirical investigation into the relationship between the policy guidelines and local policy implementation can provide us with valuable insights into their utility. This paper looks at mobile learning policy in New Zealand as a case study, interprets each of the UNESCO policy guidelines in the light of the local context, and explores to what extent the guidelines are currently embedded in national and local policy.

2 Research Methodology

The case study discussed in this paper addresses how mobile learning policy in New Zealand reflects the UNESCO guidelines. Since these guidelines address policy at both national and local level, and are intended for a range of different institutional types, the study investigates policy in the national government, local government, educational institution and commercial organisational contexts. A purposeful sample of policy makers and other relevant stakeholders (partly selected via snowball sampling) was chosen from a range of stakeholder institutions, and each of these representatives was interviewed using a semi-structured interview. Each of these interviews took approximately one hour, and was conducted face to face at the interviewee's place of work, where possible (6 interviews.) Where the interviewees could not be met face to face due to location, they were interviewed either by video call (2) or by telephone (1). Table 1 shows the interview subjects who kindly contributed to this study. Where direct quotes appear in the text, they are referenced by the name of the sector they represent.

The investigation process took a naturalistic, positivist approach, in that the emphasis of the questions was on the social realities of policy, primarily addressing 'what'-type questions with a fixed set of enquiries. A standard iterative analysis was applied to the data, beginning with data reduction (the interviews were recorded, transcribed and coded in NVivo) then seeking meaning through data display (supported by NVivo tools), and finally drawing some conclusions from the data [2].

In the interviews, the stakeholders were asked to comment on each of the main aspects of the UNESCO mobile learning policy guidelines. In each case, a statement was provided, based on summarising elements of the guidelines, and the interviewees were asked to comment on these statements in terms of (a) whether or not they agreed with the statements and (b) how they felt these statements were, or were not, reflected in their own experiences or professional opinions of mobile learning policy. The statements were:

1. That governments should create or update policies related to mobile learning within existing ICT in education policies. These policies should avoid blanket prohibitions of mobile devices.
2. That teachers are trained to incorporate mobile technologies into pedagogical practice.
3. That curriculum, educational resources and lesson plans are available to teachers via mobile devices.
4. That incentives are created for developers to build learning content specifically for mobile devices, including for local groups and languages
5. That gender equality is ensured for mobile students by encouraging women and girls as well as men and boys to leverage mobile technology for learning.
6. That connectivity options are expanded and improved while ensuring equity.
7. That strategies are developed to provide equal access for all, involving BYOD or central provision.
8. That safe, responsible and healthy use of mobile technologies are promoted, including digital citizenship and management of potential health risks.
9. That mobile technology is used to improve communication and education management, including the collection of educational information following a conflict or disaster.
10. That awareness of mobile learning is raised through advocacy, leadership and dialogue

In cases where the interviewees requested more detail on the statements, relevant extracts from the policy guidelines were referred to.

Table 1. Interviewees who contributed to the study

Sector	Representative
National government (Ministry of Education)	Howard Baldwin (Manager, Sector Engagement)
School (Orewa College)	Kate Shevland (Principal) Mark Quigley (Deputy Principal)
Commercial service provider (Isometric Solutions)	Conrad Stewart (Managing Director)
Educational trust (Manaiakalani Trust)	Dorothy Burt (Professional Learning Programme Leader)
Educational researcher (University of Waikato)	Noeline Wright (Senior Research Officer)
Crown agency (Network 4 Learning)	John Hanna (CEO)
Industry organization (NZTech)	Candace Kinser (CEO)
Local government organization (Auckland Tourism Events and Economic Development)	Brett O'Riley (CEO)

Research Model

Figure 1 shows a research model derived from aspects of policy that appear in the literature. Policy guidelines may come from a number of sources. In this paper we focus specifically on the UNESCO guidelines, but there have been other examples [e.g. 3]. "The impact of mobile learning…has both shaped and been formed by national and [regional] policy" [4 p.14], thus existing practices and policies influence future policy formulation. Variations in context may also shape policy, for example in the United States 'Education leaders, perhaps sensing limited public or policy support, have not yet developed a strategy on how mobile learning should be deployed, or even if it should be used at all.' [5 p.5]. The research model was used to guide the coding process that was used to analyse the interview data, identifying how guidelines, policies/practices and contexts have informed, influenced and shaped policy formulation, and how policies, driven by that process, are enacted in practice.

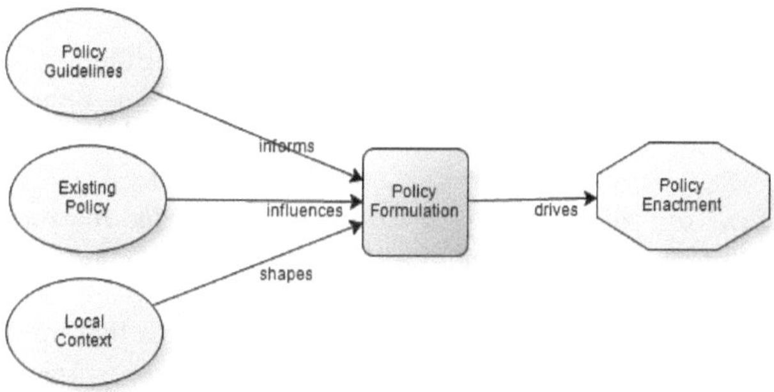

Fig. 1. Research model of mobile learning policy

3 Analysis

In this section, each of the main interview questions are addressed in turn, with related analysis themes explored in detail and illustrated with the voices of the interview subjects. However it is interesting to begin by reflecting on the result of a word frequency analysis of the interview data, which is shown in Figure 2. Whilst this is just an impressionistic view of the data, it nevertheless emphasises that policy is, in the end, all about people, and that thinking about schooling, and what we hope to achieve through it for our future citizens, is how policy is ultimately formed.

Fig. 2. Most frequent words from the interview data

3.1 Creating or Updating Mobile Learning Policy

The first interview question related to whether governments should create or update policies related to mobile learning, and if these policies should avoid blanket prohibitions of mobile devices. The main themes that emerged from this question were differing philosophical viewpoints, with both global and local forces acting on policy.

Although there was general consensus that governments should create or update mobile learning policies, the assertion that they should avoid blanket prohibitions of mobile devices raised some debate from different philosophical viewpoints. There is an argument, based on freedom of choice, that neither prohibition nor compulsion of mobile devices is appropriate.

"Should they mandate the use or mandate the ban? I don't think either position is healthy." – Crown Agency

Some schools in New Zealand have had explicit bans in the past [6], though it is not clear how many (if any) schools still have complete prohibition. On the other hand, there was also recognition that blanket bans are in some contexts a means of repression.

"If the government say you can't have a mobile device for the Internet, because then you'll know what's going on in the country, that's an issue" - School

There was also recognition that there were global changes that might impact local policy, but also an increasing emphasis on local, rather than national, action as the driver of policy change.

"That's all been taken out of the hands of government. People are just using things and so I think it's not a government policy ... a local policy, perhaps" - School

Nevertheless, local policies that have been enacted are often derived from common templates provided by relevant organizations. In New Zealand, NetSafe [7] has developed a set of policy templates that schools can adapt to their own needs. These policy templates cover cyber safety policy and use agreements for staff and students at different school levels (primary, intermediate and secondary.)

"A lot of schools start their policies off a NetSafe one because they can generate one quite quickly ... very generic but it's New Zealand generated." – Commercial Service Provider

3.2 Teacher Training with Mobile Technologies

The analysis of the policy guideline that teachers should be trained to incorporate mobile technologies into pedagogical practice focused mostly on digital skills requirements and their impact on teacher training institutions. This aspect of policy was seen as particularly important by many of the interviewees, though the term 'teacher education' was felt to be more appropriate than 'teacher training' by our educational researcher. A point that was made by more than one interviewee was that common assumptions about young people having good digital skills due to being 'digital natives' are incorrect, and that, in fact, new student teachers are likely to enter (and possibly leave) their professional education with traditional, conservative views of teaching, and poorly developed digital skills, despite their familiarity with social media.

"You get the new grads who come in and, because they're on Facebook all the time, think they know all there is to know" - Educational Trust.

"[the students] seriously speak out now about how frustrating it is to have a teacher come in front of them in February and not know how to run a digital learning environment for them" – Educational Trust

"[new teachers are] good at the social media, games that sort of thing, as are all young people that we employ probably, but actually using it in a more constructive way needs training"- School

A number of solutions to this problem, some of which are already being piloted, were proposed to address this problem.

"One of the programs that we run is for ordinary teachers in our schools...2 day intensive blocks in the holidays, we learn after school, we support them in their classes" – Educational Trust

"We didn't get funding from the ministry but Google gave us a significant amount of money ... we've taken 10 beginning teachers ... they are going through a digital immersion programme" - Educational Trust

"Selecting out master teachers...if you want to be a teacher you actually get apprenticed to one of them...it changes the whole nature of how we train teachers" - School

"The new Masters of Teaching for instance, that this university's just begun as a pilot, one of the underpinning things is about digital literacy" - Educational Researcher

From these concepts a theme of institutional change emerged to balance the rapid evolution of digital skills requirements in the teaching profession. In particular, it can be seen that new concepts in teacher education are being developed.

3.3 Educational Resources Available through Mobile Devices

This aspect of the policy guidelines states that curriculum, educational resources and lesson plans should be made available to teachers via mobile devices. A number of perspectives on this guideline emerged from the interviews. While it was generally acknowledged that resources should be made available through mobile devices, a number of subjects felt that this was in most cases just an extension of materials being available in digital formats, regardless of whether a mobile device was used to access the material.

"The curriculum in New Zealand and educational resources are available already online so that makes them available via mobile devices" - Educational Researcher

The issue of local context was important in this particular question, since it was felt that making specific curricular material available was more relevant in countries with fixed materials for a national curriculum, and not in countries where teaching content is more locally managed.

"In some other countries there's much more central provision of curriculum resources" – School

There was also some scepticism of the value of making lesson plans available.

"A lesson plan is only as good as the teacher teaching it and the class you're teaching it with, so a lesson plan is never a Silver Bullet." – Educational Researcher

The issue of creative commons was also raised. This was seen as important by a number of interviewees to ensure that the correct copyright status was conferred on shared material.

3.4 Incentives to Develop Learning Content

The analysis related to the guideline that incentives are created for developers to build learning content specifically for mobile devices, including for local groups and languages, centred mainly on the conflicting pressures of cost and quality. In New Zealand, the issue of local language is perhaps more important than in many other countries, due to the status of the indigenous Māori language, Te Reo, and the support of Māori language learning is of strategic importance to education in New Zealand [8]. Fishman stresses the role of technology in sustaining indigenous languages [9], and significant efforts have been made to develop mobile learning tools for the Maori language [10].

In their responses, the interviewees highlighted the issues of cost versus quality (including local relevance.) There are clearly a number of forces at play in this area which make for complex decision making. On the one hand, the move towards everyone publishing themselves using social media and Web 2.0 tools has led to a common perception that content should be free.

"People have the mind-set that you can get everything for nothing and why would I pay? ... it's not that people can't afford it but we become accustomed to thinking it's like air, why would you pay for it?" – School

However, a problem with free resources is that they often lack the quality control processes that would be put in place by a professional publishing house.

"If you look back at the current model of producing resources...what comes out the other end is something that's suitable for lots of people. When you've got an individual producing something ... those checks and balances are not in there any more." – School

Perhaps as a result of the competition between free and purchased content, and the pressures on traditional publishers to compete by migrating their content to digital formats, new online forms of content delivery have become relatively more expensive:

"The school has a $10,000 spend a year, say, on textbooks, where we get five years of use. The model the publishers are talking about at the moment is 'we'll give you access to our textbook for one year and then it disappears', which adds up to a lot more than $10,000, much more expensive." – School

Thus there are competing pressures in the area of content generation, and it is clear that market forces alone may not deliver all that is required. This is particularly true where minority culture and languages need to be sustained and promoted.

"We need to ensure that the unique elements of New Zealand's culture are reflected in content." – Ministry of Education

This may mean that targeted incentives will be required to support digital content in minority languages, if market forces will not deliver these resources.

"If the cash is predominantly English or Mandarin the developers and content producers will go where the cash is. From an equity perspective...maybe there needs to be some incentives for content developers and producers to deliver in as many languages as the world thinks is appropriate." – Crown Agency

3.5 Gender Equality

The guidelines recommend that gender equality is ensured for mobile students by ensuring women and girls as well as men and boys can leverage mobile technology for learning. Although all respondents supported the concept of gender equality in principle, there were some views on specific contexts and aspects that were highlighted. A number of the interviewees noted that the introduction of BYOD into schools has addressed some aspects of the digital divide through access to ICTs. This theme was certainly picked up by some of the interviewees, for example regarding the effect in the classroom of each child having their own device, removing any digital access divide that might have previously existed.

"All the girls having digital device...being in charge" – Local Government Organization

Device access is not, of course, the only digital divide. Possession of ICT is "not only a matter of material resources but also of the attractiveness of this technology and the necessary skills to use it among people of different age and gender" [11 p.319]. The potential attractiveness and new affordances of mobile technology, specifically to female students, compared to older industrial images of technology, was noted in some of our interviews.

"It is different with something like mobile technology which is I guess 'cleaner' and maybe lends itself to more visual thinking" - School

A common response from the interviewees was that in New Zealand schools, gender inequality was not a distinct challenge, at least when compared to socio-economic inequalities, which were generally seen as more pressing. Indeed it is recognised from previous research that gender inequality in information technology in education varies from country to country. For example Reinen and Plomp [12] noted that the United States and Bulgaria were, for different reasons, much more gender equal in ICT education than many other nations. New Zealand was not included in their data but would likely follow the United States' model (e.g. in terms of home access to computers and female role models) and be more gender equal than many other nations.

If gender imbalance was seen to be being addressed in a positive way for female students, there was also a potential down side. One interviewee commented that girls seemed to be increasingly doing better than boys in some schools.

"It seems to be the boys who've got more barriers at co-ed schools" – Educational Trust

This idea has certainly gained a lot of traction in recent years in a number of countries. In the UK, for example, there is concern at the underachievement of boys in national assessments [13]. Evidence from the OECD suggests that this is certainly the case in some subjects. However the picture is different when we focus on engineering and science, or higher degrees; "Efforts to increase mathematics and science performance among girls…can promote gender equality even further in education. Meanwhile, initiatives to break down gender stereotypes in fields of study and progressive corporate policies can increase women's employment opportunities." [14 p.4].

In fact the issue of gender inequality once female students leave school and enter the workforce was raised as more significant by some of the interviewees.

"At a senior management level, the fortune 100 technology companies, less than 12 or 13 CEOs are female, 10% roughly, whereas I know at that level there's got the be women in industry who are capable." – Industry Organization

This issue is of course not confined to the boardroom; "The lack of women in leadership positions is only a reflection of the low numbers of women in the sector overall" [15]. Recent international studies in the literature regarding women in the IT workforce are limited but there are many country specific studies that are illuminating. For example a study of ICT training in the UK suggested that education alone will not address the masculine norms of the IT industry, and will thus not remove barriers to participation [16].

"The digital divide between genders tends to be around computer science and programming [we need] to make things more open for women to get involved in coding and application software development." – Ministry of Education

3.6 Connectivity and Equity

The UNESCO guidelines recommend that connectivity options are expanded and improved while ensuring equity. In New Zealand, currently, a large scale government initiative to roll out broadband access to all schools is already well under way, such that it is possible to state that:

"No matter where you are you going to get fibre, you going to get your network upgraded, you are going to have a wireless overlay, and the next bit of the puzzle is Network for Learning" – Ministry of Education

This resonates with a common view that access to ICTs is becoming as much of a basic human right as many other aspects of modern societies.

"Technology should be a right that we are able to extend access to much in the same way as health care and education" – Industry Organization

It was, however, noted by more than one interviewee that a number of initiatives that have taken, or are taking, place rely on some kind of short term investment funding.

"Subsidies are unsustainable over the long run" – Local Government Organization

"Hopefully [the teacher training initiative] will backfill into the teacher training system because certainly what they're doing is not a sustainable, affordable program" – School

The analysis for this theme suggests that equitable access to mobile learning asks serious questions about socio–economic inequalities in wider society, and whether equality of access can not only be provided but be sustained over the longer term.

3.7 Equal Access Strategies

The UNESCO guidelines suggest that strategies are developed to provide equal access for all, which may involve various strategies for putting mobile devices into the hands of learners, whether that is based on a Bring Your Own Device (BYOD) approach, a leasing model, or some kind of centralized provision. The type of provision of mobile devices will have an effect on the nature of the activities that can be undertaken using them [17]. Thus decisions need to be made about the purpose of a BYOD programme when deciding on what types of device should be allowed, recommended or mandated. Having an open policy may potentially lead to difficulties in effectively delivering the curriculum.

"You may paint yourself into a corner if you've not got a good enough device. Does that mean that the teaching has to go to the lowest common denominator?" – Commercial Service Provider

However the main issue of BYOD is an economic one, so schools may feel the need to allow a mix of devices for financial reasons.

"BYOD has got a bad name .... the schools call it 'device for learning' just because it seems to have got negative connotations...parents have to pay for a device" - Commercial Service Provider

While BYOD approaches come with their own challenges, alternative models, where devices are controlled and managed by the school, may result in a lack of flexibility, and excessive management demands on the school's ICT support infrastructure.

"You buy the computer from the school, it's a specific model, it gets attached to the domain. That's not BYOD any more, that's a domain computer and it's controlled by the school" - Commercial Service Provider

One positive development as schools group together into larger clusters to negotiate procurement is that some economies of scale or even marketing power may become evident.

"We purchase more devices per year than [large individual schools] do and we have gone from being the amusement to having Apple come to us saying can we loan you 2 classes of iPads because you're not using our devices" – Educational Trust

The analysis of this theme reveals that there are a number of different access models, and that cost issues may impact on parents, students and teachers in different ways, with no single model providing all the answers.

3.8 Safe, Responsible and Healthy Use of Mobile Technologies

The guidelines recommend that safe, responsible and healthy use of mobile technologies be promoted, including digital citizenship and management of potential health risks. This recommendation therefore covers both social and physical concepts of safety. A number of issues were raised by the interviewees around cyber safety. One of the more interesting comments was that the whole concept of 'safety' did not resonate with school students, and that other approaches had to be taken.

"We don't like to use 'safe' because no child feels unsafe about anything; sex, drugs, the internet...kids just can't conceptualize any reason to be unsafe so we talk about being 'smart'" – Educational Trust

Further, digital citizenship was defined as being much broader than just issues of online safety.

"Digital citizenship is more than just safe and responsible use. It's actually being deeply critical and aware of what's out there and how other people use these things" - Educational Researcher

In addressing potential health risks, there were frequent references to relying on expert advice on the safety or otherwise of technologies such as Wi-Fi networks. The issue of personal choice was also raised in this context, in that any parent who wished to send their child to a school where they were not exposed to electromagnetic radiation should be free to do so. In one New Zealand school, parents have campaigned successfully to have Wi-Fi removed due to fears about possible cancer risks [18]. However, a number of interviewees raised the inescapable issue of background levels of electromagnetic radiation that would be beyond the control of schools even if they did not install their own wireless networks. These may occur both at home and in the general environment, even in schools where there is no on-site wireless network.

"Parents have far more dangerous devices going on in their homes than is happening in the schools" - Educational Trust

"At [...] school there is something which affects their wireless and it's just this massive band of frequency ... so even if that school had no wireless network, students are still getting it. There's not a lot that they can do." – Commercial Service Provider

3.9 Mobile Communication and Management

The guidelines recommend that mobile technology is used to improve communication and education management, including the collection of educational information following a conflict or disaster. Not surprisingly, this particular aspect of the policy guidelines resonated strongly with New Zealand stakeholders as a result of the Christchurch earthquake in 2011. In the aftermath of the disaster, all schools were temporarily closed, and all the students from nine of the city's 163 schools had to be relocated to other local schools because their own schools were too badly damaged to reopen [19]. In addition, over 1,000 school students became 'refugees' in the neighboring region of Otago, with others dispersed to even more distant areas of the country [20].

"The problem of schools having all of their material on site, and those sites then becoming inaccessible for periods of time" – Ministry of Education

"The continuity of education in Christchurch showed good role models on how some schools managed really well because they did have their work out there on the web, in the cloud, and others didn't" - School

ICT can contribute positively in post-earthquake recovery by enabling civic participation [21]. The same effect is observable in educational participation.

"Even if the school itself wasn't functional the teacher was functioning" – Local Government Organization

It was also felt that some broader education was needed around public use of communication resources during a disaster:

"Maybe some policy around citizens' behaviour, mobile device behaviour, in the event of a disaster, to ensure that the network's going to support the communications needs associated with a disaster." - Crown Agency

3.10 Advocacy, Leadership and Dialogue

The need for awareness of mobile learning to be raised through advocacy, leadership and dialogue was acknowledged by all of the interviewees. Indeed, many of them saw a major part of their role as addressing this aspect of policy. Perhaps the most important aspect of dialogue highlighted during the interviews was that which takes place between schools and parents.

"You have to involve parents really early, have to talk to them and meet them and discuss, because all of those questions about policies come up. How are you going to filter the Internet? How are you going to deal with broken devices or stolen devices?" – Commercial Service Provider

One of the most interesting responses that emerged around advocacy and leadership was the way that it is often the students, rather than the staff, who demonstrate this in the classroom:

"They start sharing with each other, they start helping each other, those different experts arise in the classroom" – Educational Researcher

4 Conclusions

This paper has reported on the results of an interview-based study designed to benchmark mobile learning policy in a developed, innovative nation against the UNESCO mobile learning policy guidelines. The intent of this study was to gain new insights into how generic policy guidelines may be implemented in practice, and to reveal subthemes in the guidelines that will inform policymakers wishing to promote mobile learning in different national and cultural contexts. The study reaffirms some core recommendations, such as the need to introduce the use of mobile devices into teacher education and the importance of maintaining equity. However the study also reveals that, in some cases, the promotion of mobile learning may have to be balanced with opposing principles such as user choice. The conflicting demands of cost and quality have also emerged in more than one context, including the provision of mobile devices and the sourcing of digital educational resources. It is therefore one role of policy to ensure that market forces alone do not dictate practice, for example by ensuring that indigenous language learning resources are provided. Further, some positive aims of the policy, for example promoting gender equality, may prove to be highly complex in practice due to specific local conditions. Throughout the analysis, it is clear that in most cases, while the recommendations are seen as positive and worthy in their intents, the detail of policy on the ground will need to deal with conflicting demands, philosophies and constraints. It is hoped that the New Zealand experience will be informative for other policy makers around the world and help prepare them for building pathways to a better world.

Acknowledgments. The author wishes to thank the interviewees who contributed their knowledge and experience to this article.

References

1. UNESCO: Policy guidelines for mobile learning, `http://unesdoc.unesco.org/images/0021/002196/219641e.pdf`
2. Folkestad, B.: Analysing Interview Data: Possibilities and challenges. Eurosphere working paper series 13, `http://eurospheres.org/files/2010/08/Eurosphere_Working_Paper_13_Folkestad.pdf`
3. Naismith, L., Lonsdale, P., Vavoula, G., Sharples, M.: Literature Review in Mobile Technologies and Learning, Futurelab Report 11, `http://telearn.archives-ouvertes.fr/docs/00/19/01/43/PDF/Naismith_2004.pdf`
4. Kukulska-Hulme, A., Sharples, M., Milrad, M., Arnedillo-Sanchez, I., Vavoula, G.: Innovation in Mobile Learning: A European Perspective. International Journal of Mobile and Blended Learning 1(1), 13–35 (2009)
5. Shuler, C.: Pockets of Potential: Using Mobile Technologies to Promote Children's Learning. The Joan Ganz Cooney Center (2009), `http://www.joanganzcooneycenter.org/wp-content/uploads/2010/03/pockets_of_potential_1_.pdf`
6. Wright, N.: e-Learning and implications for New Zealand schools: A literature review. Education Counts (2010), `http://www.educationcounts.govt.nz/publications/ict/e-learning-and-implications-for-new-zealand-schools-a-literature-review/pedagogy-and-e-learning`
7. Netsafe: Policies and Use Agreements (2010), `http://www.netsafe.org.nz/keeping_safe.php?pageID=81&sectionID=education&menuID=81`
8. Māori Language Commission: Promoting Positive Attitudes to the Māori Language in the Classroom (2000), `http://www.tetaurawhiri.govt.nz/english/pub_e/classroom/booklet.shtml`
9. Fishman, J.: What do you lose when you lose your language? In: Cantoni, G. (ed.) Stabilizing Indigenous Languages. Northern Arizona University (2007)
10. McKenzie, T.: The Challenges and Opportunities of Using Mobile Devices to Attain Māori Language Proficiency. Victoria University of Wellington PhD Thesis, `http://researcharchive.vuw.ac.nz/xmlui/bitstream/handle/10063/3291/thesis.pdf`
11. van Dijk, J., Hacker, K.: The Digital Divide as a Complex and Dynamic Phenomenon. The Information Society 19, 315–326 (2003)

12. Reinen, I., Plomp, T.: Information Technology and Gender Equality: A Contradiction in Terminis? Computers & Education 28(2), 65–78 (1997)
13. Paton, G.: GCSE results 2013: Girls stretch to record lead over boys (2013), http://www.telegraph.co.uk/education/educationnews/10260163/GCSE-results-2013-girls-stretch-to-record-lead-over-boys.html
14. OECD. Education Indicators in Focus, http://www.oecd.org/education/skills-beyond-school/49986459.pdf
15. Henderson, J.: Is NZ hi-tech industry starting to address gender imbalance? Techday (2014), http://techday.com/it-brief/news/is-nz-hi-tech-industry-starting-to-address-gender-imbalance/183082/
16. Gillard, H., Mitev, N., Scott, S.: ICT Inclusion and Gender: Tensions in Narratives of Network Engineer Training. The Information Society 23, 19–37 (2007)
17. Wenmouth, D.: Thinking about BYOD, http://blog.core-ed.org/derek/category/mobile-technologies
18. National Business Review: Manawatu school removes wi-fi over cancer fear, http://www.nbr.co.nz/article/manawatu-school-removes-wi-fi-over-cancer-fear-ck-150478
19. Ham, V., Cathro, G., Winter, M., Winter, J.: Evaluative study of co-located schools established following the Christchurch earthquake. Education Counts, http://www.educationcounts.govt.nz/publications/schooling/115174
20. Lewis, J.: 1010 refugee pupils swell schools' rolls. Otago Daily Times, http://www.odt.co.nz/regions/otago/150423/1010-refugee-pupils-swell-schools-rolls
21. Mitomo, H., Otsuka, T., Jeon, S., Cheng, J.: The role of ICT and mass media in post-disaster restoration and recovery progress: A case of the Great East Japan Earthquake. In: 24th European Regional Conference of the International Telecommunication Society (2013)

Mobile Learning Trends among Students in Vietnam

Angela Murphy, Warren Midgley, and Helen Farley

University of Southern Queensland, Toowoomba, Australia
helen.farley@usq.edu.au

Abstract. Mobile learning has the potential to expand access to education in developing countries. Little is known about the preferences of students in some Asian countries such as Vietnam. Some of these countries have restricted internet access and may be subject to internet censorship. A study was conducted with forty-four Masters students in Vietnam to identify informal mobile learning trends. Results indicate that although rates of ownership of mobile technologies are still low in comparison to many other countries, students do use these devices to support their studies. A third of students had access to a tablet computer, smartphone or MP3 player and many students had access to more than one device. Most students used Wi-Fi and considered internet quality to be moderate or fair. Access to high quality internet and the impact of internet censorship needs to be taken into account when developing mobile learning content for students in Vietnam.

Keywords: M-Learning, Mobile Learning, Internet Access, Developing Countries, Higher Education, Vietnam, Internet Censorship, Social Media.

1 Introduction

As a result of the incremental improvements in design, mobile technologies are increasingly perceived as essential to the conduct of people's everyday lives [1]. The ubiquitous connectivity and portable nature of these technologies enables access to contextualised learning experiences which translate into greater ownership of learning processes [2]. Furthermore, these technologies are becoming more accessible and affordable, thereby presenting unique opportunities for facilitating the flexible delivery of contextualised learning for diverse student cohorts.

Proponents of social constructivism maintain that students learn best when undertaking authentic tasks within relevant and meaningful contexts [3] and emphasise the importance of communication and collaboration in the construction of knowledge [4]. A number of research studies have focused on the use of mobile technologies to support learning in both formal and informal settings. Mobile technologies have been found to encourage greater communication and collaboration in classroom settings facilitating social presence as students share information with fellow students [5]. The portability of mobile devices as well as their sophisticated geo-location capabilities (GPS) enable educators to provide students with opportunities to engage in contextualised and interactive learning [6].

M. Kalz et al. (Eds.): mLearn 2014, CCIS 479, pp. 197–204, 2014.

1.1 Internet Access and Censorship in Vietnam

In developing countries, mobile technologies have been adopted at greater rates, as compared to personal computers, probably because tablets and smartphones are more affordable and easier to use [7]. However, to fully leverage the affordances of these devices for learning, there must also be access to high-speed internet. The provision of internet access and other technical infrastructure varies widely across Asia. In Vietnam, the internet has been available since 1997 with the number of subscribers increasing 15-fold since 2000 [8]. According to Vietnam Government statistics, only about a third (34%) of the population are connected to the internet, as compared to 83% in South Korea and 80% in Japan [9].

A report by the Berkman Centre of the Internet and Society revealed that Vietnam has sophisticated and effective filtering systems that resemble those of China [10]. It is important for educators to understand the extent to which internet censorship may impact on mobile learning in Vietnam. Social networks, for example, are often used in mobile learning scenarios to encourage collaboration and sharing of information. In Vietnam, however, local authorities partially or wholly block access to sites such as Facebook [11]. Only a third (33%) of consumers in Vietnam over the age of 15 have a social media profile on a platform called Zing Me and 28% have an active Facebook profile [12].

1.2 Mobile Technologies and Mobile Learning in Vietnam

According to a survey conducted by Nielsen, mobile technologies are gaining in popularity in most countries in Southeast Asia [12]. Devices such as tablet computers, MP3 players, internet-capable games consoles, eBook readers and internet-capable televisions are being used with greater frequency by consumers across all of Southeast Asia, except Vietnam. According to this study only 32% of consumers over the age of 15 have access to an internet-capable mobile phone compared to 64% in the Philippines, Thailand (77%), Malaysia (77%), Indonesia (78%) and Singapore (85%). Smartphone ownership in Vietnam is also much lower than other regions with a penetration of only 11% and less than 1% for tablets. Consumers in Vietnam are still highly reliant on desktop computers (95%) for access to the internet with only 30% accessing the internet through laptops. However, ownership of mobile technologies is growing rapidly [13]. Between 2005 and 2010, mobile phone subscriptions in Vietnam increased by 72.4% with 175.3 mobile phone subscriptions for every 100 inhabitants. This rate of change was much higher than in more developed Southeast Asian countries such as in Singapore (7%) and Thailand (17%) [14].

Research indicates that students have complex relationships with these technologies, and that these relationships evolve as new technologies enter the marketplace [15]. Kennedy and Fox found that current research provides limited insight into the mobile technology ownership and usage trends of students in Asian countries [16]. They conducted research with first-year undergraduate students at the University of Hong Kong, finding that nearly all students had access to mobile phones (98%) and notebook computers (81%). They compared these findings to those from a study conducted by Kennedy, Judd, Churchward, Gray and Krause, and found that rates of ownership of these technologies was on a par or exceeded those of Australian first-year undergraduate students [17]. Unfortunately, there was no distinction provided between feature phone and smartphone ownership and no indication as to rates of ownership of tablets or eBook readers. No previous studies were found that specifically examined mobile learning trends among students in Vietnam.

1.3 Background and Objectives of the Study

In an increasingly connected world, partnerships between universities in different regions are becoming more common. One such partnership exists between the University of Southern Queensland (USQ) in Australia with the Ho Chi Minh City Open University (HCMCOU) in Vietnam. This arrangement involves specialist lecturers from USQ travelling to HCMCOU to teach intensive courses several times a year. These courses constitute the Master of Applied Linguistics program at USQ.

Each course operates with a series of cohorts of between thirty and forty students. All students already hold an undergraduate degree and have demonstrated through previous study a level of English proficiency deemed appropriate for this program (IELTS 6.5 or equivalent). The program is designed for those with careers in the field of 'Teaching English as a Second or Foreign Language.' Usually, there are two or more cohorts taking intensive courses in any given semester. At the time of data collection for this project, there were two cohorts of students with a total enrolment of 72 students.

One of the authors of this paper is involved in this initiative. In each of several visits, the lack of internet access and consequently, use of digital learning and teaching tools, were significant challenges. This was particularly true for the Computer Assisted Language Learning course. In order to investigate the extent and nature of the possible challenges to digital learning and teaching in this specific context, a survey was developed and deployed to students enrolled in the two courses running at the time. The survey was designed to assess not only the challenges, but also the aspirations and possibilities for the use of digital technologies for learning in this particular context.

2 Methodology

2.1 Participants

Fifty students responded to the online survey. After removing incomplete responses, a total of 44 responses were retained, representing a response rate of 61%. The sample consisted of mostly female participants (32, 73%) as compared to males (12, 27%), ranging in age from 24 to 46 with a mean age of 31. The majority of students were employed in addition to studying (35, 80%), working between 6 and 72 hours a week (M=34, SD=17.30). Most students lived with family (24, 55%), or with a partner and/or children (13, 30%). Very few lived with housemates or friends (3, 7%) or in single accommodation off campus (4, 9%). Participants spent between 4 and 96 hours per week studying (M=25.21, SD=18.62).

2.2 Data Collection and Analysis Procedures

A survey was developed with 28 closed and 3 open questions about students' access to and use of mobile technologies and internet for learning. The survey was in four sections: 1) student demographics, including questions about gender, age, current employment, and hours available for study each week; 2) the availability of internet access; 3) ownership of mobile devices; and 4) use of mobile devices for learning activities.

The survey was hosted online using the Qualtrics survey platform and a link to the survey was distributed by the course examiner during the period of face-to-face lectures in March 2013. Completion of the survey took around 15 minutes and was voluntary. The data file was compiled in SPSS for Microsoft Windows version 19.0 and analysed using descriptive methods.

3 Results

3.1 Student Access to the Internet

Students were asked to indicate from where they currently accessed the internet and which internet services they used most often. The research showed that for study purposes, most students accessed the internet at home (91%), at university (57%) or at work (52%). A large proportion of students also accessed the internet in public places such as libraries or cafés (41%). Wireless internet (Wi-Fi) was the most common method of accessing the internet, used most often by 75% of the sample. Nearly half of the sample (43%) also frequently used 3G and 43% frequently used ADSL or ADSL2+.

Students were asked about their perceptions of the quality of their access to the internet for study purposes. While more than half of the sample considered their access to be good (55%), more than a third considered their internet access to be fair (36%). Sample sizes were too small to statistically compare perceptions of the quality for the different types of internet access options.

3.2 Ownership and Access to Mobile Technologies

Students were asked about their ownership of, or access to a range of technologies such as desktop computers, laptops, netbooks, feature phones, smartphone, tablets, MP3 players or eBook readers. Netbook computers and laptops were not considered to be mobile devices for the purposes of this study.

Most students owned a laptop computer (95%) and 5% had access to one. This is possibly because it is required for students to bring a laptop to the intensive courses to participate in structured activities. A large proportion also owned (66%), or had access to (23%) desktop computers. Despite the portability of netbooks as compared to laptops, few owned (9%) or had access to (9%) these devices. Students owned up to 4 mobile devices (M=1.79, SD=1.06), with most owning feature phones (70%) and only 32% owning smartphones. A third of students (15, 34%) owned one device, which was either a feature phone (n=11) or a smartphone (n=3). Further analysis showed that 23% of students owned both a smartphone and a feature phone and 22% owned a feature phone and had access to a smartphone. Students also supplemented their ownership of mobile phones with tablets. Nearly half of the students (43%) who owned a feature phone also owned or had access to a tablet, whereas 25% who owned a smartphone also owned or had access to a tablet. Four students did not own any devices, but still had use of them, and only one student did not own or have access to any devices.

Students who owned or accessed these technologies were asked if they ever used them to support their studies. Unsurprisingly, laptops (93%) and desktops (64%) were the technologies used by most students. The use of tablets among students for study purposes was unexpectedly high with 57% of students who owned or had access to these devices using them for this purpose. Despite the large proportion of students owning feature phones, only 12% used them to support their studies. Smartphones were used by a higher proportion of students for study purposes (23%). This indicates that feature phones and smartphones are less suitable for the majority of learning activities, particularly when alternative technologies are available. This is possibly because of the small screen sizes of these devices as compared to tablets or computers.

3.3 Learning Using Mobile Technologies

Students were asked about the types of learning activities they conducted with their devices. These questions were filtered to appear only to students who indicated that they used each of these technologies to support their studies in the previous question. Laptops were most commonly used to complete most learning activities. However, tablets were used for a variety of activities by the students who used them for creating content including photos or videos (77%), sharing information with other students (77%) and searching online databases (77%). Tablets were also used for communication with course leaders and fellow students via email (77%), audio or video conferencing tools such as Skype (69%), or social media sites such as Facebook (69%). Additional activities using tablets included reading textbooks (62%), accessing course materials (62%), taking notes (62%), or accessing the learning management system (62%).

Of the six students who used smartphones to support their studies, most used them for taking photos or videos (83%), communicating on social media sites (83%) and sharing information with other students (83%). Smartphones were also used to send and receive emails from the course leader or other students (67%), search online databases (67%), search the internet for course-related information (67%), take notes (67%), and use audio or video conference tools (67%). Few students used smartphones for accessing and viewing course materials. This suggests that the small screen size of smartphones is not optimal for these types of activities. Students were also less inclined to complete assignments on their tablet computers (8%) or smartphones (0%). All students who used MP3 players used them to listen to course audio materials such as lectures or podcasts, and eBook readers (n=3) were used to read course materials. Students were also asked if there were any apps (mobile applications) that they used on their mobile devices to support their studies. A third of students (33%) indicated that they used apps and most claimed they used dictionary apps. Additional apps included iBooks, QuickOffice, GoodReader, PDF annotation apps and Skype, suggesting that the apps used most by students were those that enabled them communicate with peers or to read or edit documents.

Those students who used smartphones for study used them on a daily basis. This is probably because these devices are always with the student. Despite the frequent use of tablets for learning purposes, these devices were not always with the student and only 62% used them on a daily basis. Feature phones were used infrequently for study purposes, with only 20% of students using them daily.

To determine the mobility of students while using mobile devices, they were asked about the locations in which they used mobile devices for learning. The majority used their mobile devices while stationary at home or work and there is little evidence of students using their mobile devices while physically moving.

To determine future mobile learning preferences, students were asked whether they would use mobile technologies for learning if they had access to these devices. They were also asked about the types of learning activities they would like to be able to undertake on these devices. The majority would be very likely (68%) or likely (19%) to use tablets to support learning, though fewer were willing to use them for taking photos or videos (59%) or listening to audio course materials such as podcasts (52%). A large proportion also noted that they would be very likely (46%) or likely (24%) to use smartphones for learning activities. Again students considered most learning activities as being suitable for these devices except for completing assignments (44%) and reading prescribed course textbooks (48%).

4 Discussion and Conclusion

The results of the study demonstrate that despite the low penetration of mobile technologies in Vietnam among general consumers [12], mobile technologies do have a clear role in supporting learning. Although few students had access to mobile technologies such as tablets, smartphones and MP3 players, a significant proportion of those who owned or had access to these technologies used them for learning activities. The anticipated increase in adoption of mobile technologies presents an opportunity for the provision of learning activities that leverage the unique affordances of the technologies. However, due to the small sample size, it is difficult to obtain clear insight into the learning patterns of students owning mobile technologies. These findings are therefore an indication of potential trends rather than a representative overview of the characteristics of Vietnamese students' use of mobile technologies for learning.

Without access to high quality internet, the opportunities to provide students with access to mobile learning experiences are more limited. Students in the current sample had acceptable to moderate levels of internet access for their studies, which suggest that the ability to access certain activities from mobile devices would be negatively impacted. Most students accessed the internet via Wi-Fi and it is notable that many students accessed the internet from public locations such as libraries or coffee shops. Students in Vietnam may therefore be hindered by barriers such as available bandwidth and varying levels of access in different locations. A limitation of this study is that no information was obtained on the extent to which internet censorship impacted on students' access to learning content using mobile technologies. There was also no information obtained on students' current usage of social media.

Due to the limited sample size, it is difficult to draw conclusive insights into the types of learning activities that students prefer to conduct on various mobile technologies and the reasons for these preferences. There appears to be a trend to use smaller devices such as smartphones that are always in the possession of the student to conduct activities such as information searches, sharing information with other students and taking photos or videos. The use of dictionary apps on smartphones is a clear example of instant access to a small but vitally important piece of information to assist learning, which is one of the clear benefits of using mobile technologies for learning. Therefore, these devices appear to support learning activities that are immediate and enable students to engage with learning content in a more interactive manner. Tablet devices are popular with students and appear to have the greatest applicability across multiple types of learning activities. Future surveys will need to incorporate questions that identify reasons for preferences of certain mobile technologies for certain activities and more detailed information on the how best to incorporate mobile technologies into enhancing the learning experiences of students in Vietnam.

References

[1] Evans-Cowley, J.: Planning in the Real-Time City: The Future of Mobile Technology. Journal of Planning Literature 25(2), 136–149 (2010)

[2] Wong, L.H.: A learner-centric view of mobile seamless learning. British Journal of Educational Technology 43, 19–23 (2012), doi:10.1111/j.1467-8535.2011.01245.x

[3] Loyens, S.M., Gijbels, D.: Understanding the effects of constructivist learning environments: Introducing a multi-directional approach. Instructional Science 36(5), 351–357 (2008)

[4] Wu, A.: Supporting electronic discourse: Principles of design from a social constructivist perspective. Journal of Interactive Learning Research 14(2), 167–184 (2003)

[5] Engel, G., Green, T.: Are we Dialing up Disaster? TechTrends 55(2) (2011)

[6] Kearney, M., Schuck, S., Burden, K., Aubusson, P.: Viewing mobile learning from a pedagogical perspective. Research in Learning Technology 20 (2012), doi:10.3402/rlt.v20i0/14406

[7] Zambrano, R., Seward, K., Ludwig, S.: Mobile Technologies and Em-powerment: Enhancing human development through participation and innovation. United Nations Development Programme (UNDP) (2012), `http://www.undp.org/content/undp/en/home/librarypage/democratic-governance/access_to_informationande-governance/mobiletechnologiesprimer.html` (retrieved)

[8] Vietnam Internet Network Information Centre. Report on Vietnam Internet Resources 2012. Ministry of Information and Communications, Vietnam (2012), `http://vnnic.vn/sites/default/files/tailieu/ReportOnVietNamInternetResources2012.pdf` (retrieved)

[9] Internetworldstats. com. 2012 Top 20 countries with the highest number of internet users (June 30, 2012), `http://www.internetworldstats.com/top20.htm` (retrieved)

[10] Berkman Cente. OpenNet Initiative Vietnam Report: University Research Team Finds an Increase in Internet Censorship in Vietnam. Berkman Center (August 5, 2006), http://cyber.law.harvard.edu/newsroom/opennet_vietnam (retrieved)
[11] Subramanian, R.: The Growth of Global Internet Censorship and Circumvention: A Survey. Communications of the International Information Management Association (CIIMA). Advance online publication (2012), http://papers.ssrn.com/sol3/papers.cfm?abstract_id=2032098 (retrieved)
[12] Nielsen: The Digital Media Habits and Attitudes of Southeast Asian Consumers (October 2011), http://www.nielsen.com/us/en/newswire/2011/surging-internet-usage-in-southeast-asia-reshaping-the-media-landscape.html (retrieved)
[13] ROA Holdings. Asian Mobile Market Forecast 2012-2015 (2012), http://global.roaholdings.com/press/press_name.html?type=0&num=66 (retrieved)
[14] So, H.J.: Turning on Mobile Learning in Asia: Illustrative Initiatives and Policy Implications. United Nations Educationa, Scientific and Cultural Organization (UNESCO) (2012), http://unesdoc.unesco.org/images/0021/002162/216283E.pdf (retrieved)
[15] Dahlstrom, E.: ECAR Study of Undergraduate Students and Information Technology, 2012 (Research report). EDUCAUSE Centre for Applied Research, Louisville (2012), http://net.educause.edu/ir/library/pdf/ERS1208/ERS1208.pdf (retrieved)
[16] Kennedy, G., Fox, B.: 'Digital natives': An Asian perspective for using learning technologies. International Journal of Education and Development using Information and Communication Technology 9(1), 64–79 (2013)
[17] Kennedy, G., Judd, T.S., Churchward, A., Gray, K., Krause, K.: First year students' experiences with technology: Are they really digital natives? 'Questioning the net generation: A collaborative project in Australian higher education'. Australasian Journal of Educational Technology 24(1), 108–122 (2008)

Crucial Factors of Implementing Mobile Learning in Sector of VET Organisations in Poland

Aleksandra Mykowska, Iwona Kostrzewa, and Magdalena Orda

Dom Szkoleń i Doradztwa, Kraków, Poland
{amykowska,ikostrzewa,orda}@domszkolen.com

Abstract. In this article we tried to present some crucial factors of implementing mobile learning software into the environment of VET trainers and providers and VET organisations in Poland. This paper describes what issues can help or disturb the testing phase, implementation and dissemination process based on our own experience gained from taking part in a European Union project. Our goal was the creation of a new, mobile tool EVAL aimed to increase the learning process results and to help during the evaluation process. In this article, we present the success, failure factors and the solutions how to deal with them during the time of introducing new technology tool into the educational and business market.

Keywords: mobile learning, evaluations, case study, implementation.

1 Introduction

Since October 2012 the company Dom Szkoleń i Doradztwa has been engaged in developing and adjusting mobile tool tailored to the needs of the business training sectors in Poland. We are partner in the UE project: Innovative Peer Learning Assessment System for Evaluation of trainers and Quality VET professional programs. The Leader of this project was Sør-Trøndelag University College (HiST) organisation from Norway, which is also the creator of the software we were testing and implementing in Poland. In the project we have adapted the software for the evaluation purpose. The system is supporting the evaluation process of the trainings and courses based on 4 levels Kircpatrick evaluation methodology. The consortium adjusted the program to the needs of VET trainers and providers to measure the effectiveness of their trainings. The new version of the program is called EVAL.

In the software, trainers can find many useful functionalities like open and closed (multiple choice) questions which may be given different point values and Likert's scale questions. The software is easy to use and very intuitive in the opinions of its users. The questions are fitted to the screens of the smartphones and tablets, which was basically, one of the aims of our work. After conducting the survey, teacher has the possibility to send the correct answers onto the participants' devices where a very clear view of the correct, incorrect or missing answers appears. The use of the immediate feedback makes participants' knowledge increase even though the course is finished. Trainers receive immediate information about the results from the test and

M. Kalz et al. (Eds.): mLearn 2014, CCIS 479, pp. 205–211, 2014.

can present them in different ways in a real life during the course inside the classroom. After all, participants answers and results can be saved as an excel file which make it easier for further analysis. The idea of the software arose from the needs of the target group. It is dedicated to VET instructors and VET providers like training companies, universities, NGO etc.

This paper describes the reflections and conclusions about the adjustments and implementation of the program into Polish market of business and vocational trainings. We would also like to share some hints regarding how to introduce mobile learning tools in a developing country.

2 Our Observation/ Experience

Below, we present a list of factors that we believe have an impact on the implementation of the new mobile technology tool into the training market. These factors have influenced the success and failure of the implementation stage into the VET organisations and trainers.

2.1 Adjustment to the Target Group Expectations and Needs

Offering Benefits to the Trainers. "Mobile technology can offer new opportunities for learning that extend beyond the traditional teacher-led classroom scenario." [2].Trainers, who have tested EVAL, found the EVAL software very useful in the context of saving time and extending possibilities during the evaluation process. The software developed by HiST gave all trainers the opportunity to create numerous tests in a very short time. There was no need to give participants the paper survey thus trainers didn't need to spend much time to print tests, transfer the data from the paper sheets into the computer, do the short analysis of the results etc. All data from the tests and surveys can be easily gather as an excel file and can be analysed more deeply and in more detailed way in the other programs. Taking out a smartphone from the participants' pockets and doing the test on a phone or tablet with one finger took less time in order to get the information form the participants than it usually took with the usual paper surveys.

„The use of mobile learning technology encourages student's active participation in the learning process." [2] Many trainers agreed, that the mobile learning tool is very useful, legible and generally very interesting and what is the most important- fun for them and the participants! It has a variety of questions which can be used in the tests (open, closed, multiple choice, with Likert scale), it is easy to implement during any course and most of all it is mobile. What also pleased EVAL's users was the possibility to use it before the particular training (pre-test) and at the end of it (post-test) or even few days (or weeks) after the particular training took place. It is possible because EVAL allows the testing to take place outside the training room without using pencil and paper. In a very clear way it made it easier to monitor the growth of the participants' knowledge.

Increasing Trainee Motivation. Learning is not compulsory, it is contextual. New technologies are usually innovative programs or systems that increase the effectiveness of the learning process. All of the trainers, who tried EVAL in their work, agreed that the use of mobile technology during the trainings/lectures and courses has a very positive effect on the participants' involvement and motivation. Following, the already mentioned book Pilot Projects and Initiatives "The mobility of digital technologies creates fascinating opportunities for new forms of learning because they change the nature of the physical relations between teachers, learners, and the object of learning. "[2] People like the new idea of learning via their phone, they think it's fun and liked the EVAL a lot, especially its functionalities like the immediate feedback, different type of questions etc. Generally, "the use of mobile learning technology encourages student's active participation in the learning process."[3]

EVAL gives the opportunity to do the tests outside the classroom, courses participants can do the test anywhere and anytime via their mobile phones. EVAL is supporting the interactive learning activities. The tool is increasing the class communication, interaction and collaboration. Users can't just sit at the back of the classroom, they are asked to do a test- survey, which can be created in a very amicable way so that it will force the trainees to stay focused and to pay attention throughout the whole training. This was reflected in the results measured after the course and it indicates the growth of the participants' knowledge, which is higher than during the usual trainings without the use of mobile technologies.

Resistance to Use New Solutions. The resistance of using new technologies can be found in both environments: The trainers and trainees.

In Poland, there are still a lot of people who don't have any access to mobile devices and if they are given such devices they encounter the problem that they don't know how to use them and they seem to be very difficult to convince. During the dissemination process we came across the situation in which several people in the middle age didn't want to use nor tablets or smartphones because using these devices was difficult for them. If we want people to use the new technologies during the learning process, they must get familiar with them before they will use them.

As long the resistance of the users can be explained, the resistance of trainers is difficult to understand. Some trainers appear not believe that mobile technologies can be of a real use and of a great help in the teaching process. They do not see the point in using new technologies or they seem to be too resistive of spending some time in changing their old habits.

Expectations That We Couldn't Answer. Despite the general positive attitude to EVAL, during testing we were facing difficulties connected with the expectations from the people and companies responsible for the software-testing phase. A lot of trainers and organisations wanted the tool to be exactly suited to their individual needs after the testing stage. It is of course impossible to meet the expectations of all testers and what is more, this is impossible without pre-established cooperation details with a regard to the possible future consideration and implementation of the tester's needs.

On the other hand, during the workshops we got some very interesting advices about functions that should be implemented to the program like: the possibility of the anonymous voting (in EVAL you can log in as a guest but it's not possible without login in) etc. Unfortunately, during the project, it wasn't possible to develop and then implement all suggestions but all of them were carefully analysed and hopefully, thanks to them, the program will be still developing.

2.2 Technical Issues

Nowadays, the technology is growing by leaps and bounds. Almost every day we experience the launching of new, better, faster and more intuitive smartphones onto the market. However, developing a program that works on all available devises and systems is a bit tricky and more complicated than it may seems. In this section we described our experience about technical issues, which have influenced the mobile training process.

Adjustments to the Systems Popular among the Target Group. Similar case as with the technical issues we met on the way of adjusting the EVAL software to the systems popular among the target group. To deal with this situation we had to remember that the software must be well suited to the needs of the target group. In order to do this, we needed to explore, examine and verify the target group demands. In our case, there were problems especially with different browsers and devices. The program was developed in Norway where the Apple devices are very common which is the contrary to situation in Poland where the most popular device is Samsung. At the beginning, the software developers from Norway worked hard on adjusting the system to the reality of the Polish VET trainers and providers. This was due to the differences among Polish and Norwegian market. In Poland, the majority of people are using the Firefox Internet browser and in Norway the Google Chrom is the most popular one. That's why, from the very beginning, we were forced to do the job related to the adaptation of the software to various companies environments in different EU countries.

Being User Friendly for All Users. At the beginning of the project a lot of work was done to make the students' interface attractive, intuitive and easy to use. Most of the final users – trainings participants/ students –liked the tool. They informed us that it is simple, user-friendly and intuitive. This was very important, as we wanted to make it easy to use from the trainees' point of view. Unfortunately, the teacher's side of the EVAL still requires some changes, amendments and adjustment to suit it best to the needs of our target group- VET trainers and teachers, otherwise, teachers may encounter many difficulties in the usage of new technologies on a daily basis. In their opinion the navigation system is not user friendly and required a lot of practice in order to create assignments during their everyday work. Unfortunately, we collected opinions that this was rather time-consuming and considered as a big inconvenience for them.

Connection Problems. We also need to bear in mind that there can always be some problematic situations that do not dependent on us. "Mobile learning and its technologies are in their infancy and there are not always easy or technology-based solutions to accessibility problems."[1]

We may come across some factors such as limited network in universities or training rooms/hotels. Sometimes these elements forced the trainer to withdraw from the testing phase and later to make the decision of not using the software in future as such situations may cause many impediments during the planned courses. Of course, these are the elements that are beyond us and fortunately, nowadays more and more hotels and training centres meet the needs of the trainers and introduce better access to the WiFi network.

2.3 Organizational Matters

The main idea of the project was, that we will be testing the tool in Poland trying to tailor it to the needs of Polish trainers. Later, we will test new functions and will introduce amendments made by the Norwegian developers after the feedback given from the target group- VET trainers and providers.

That's why it was substantial to think through, well organise and plan the work in the project.

During our work we came up with a very good model of introducing a new mobile technology tool, which has been contrived and verified by us (and is definitely recommend to be considered). We worked in cycle of:

1. analyse the needs for new functionalities among the target group,
2. develop/introduce new functionalities to the already existing system,
3. test and check the new software,
4. improve it,
5. show the improved software to a wider audience.

In our case this model has proved to work well. However, it is necessary to keep in mind the fact, that all the stages involve a lot of work, time and commitment. Moreover, it requires unceasing involvement and great experience of the developers. That's why the organisational issues must be thought through and must be well planned. To even better tailor the device to the expectations of the target group, we recommend conducting the cycle one more time to make sure that the needs of the testing trainers were taken into account. As of today, from our perspective, we see that we could use one more of the cycle to respond to the needs relating to the teacher interface, navigation for teachers. The key phase in this stage is to verify that adding a new option and functionality is necessary and is required by all testing teachers. Next, try to make all the necessary changes in the software that according to the testers were obligatory, see if it works and then present the new software to a wider audience. Alternatively, repeat the cycle again if needed. Note, that omitting any point of the cycle can lead to a situation where the system won't work properly, which then may lead to a situation where many potential users will resign from using the tool because of the errors occurring in the software.

Sticking to the Planned Actions. The testing phase of a newly launched mobile software dedicated to increase the learning process is a time where the software is deeply tested and checked, so the assumption is that some problems with the tool may occur. The most important thing is that they will be detected and eliminated as soon as possible, before the dissemination or exploitation phase begins. Generally speaking, many system errors can bother people who are engaged in the testing phase. It can discourage trainers to the further work with the software. Thus, the testing phase requires to be done by people who are tolerant and realise that during this stage of the project some errors or problems with the software are possible to appear. It is also very important to assure the testers that in the next stage of the project all bugs will be eliminated. Of course it is worth remembering, that during the next phase, the system must be stable so that it does not cause crucial problems in the future usage because people will be discouraged and won't use it during their trainings. We must remember that we cannot pre-empt or skip certain phases of the project cycle even if the circumstances are in favour of showing the tool to potentially interested people. From our experience, we know that before a certain project stage is finished some bugs and some technical issues can be faced during the testing and implementation phase. That's why we we recommend to complete every stage of the project according to the plan, even if there is an opportunity to show the tool earlier to a wider audience. As, you have to make sure that the system is stable just to minimise the risk that the software will not work as it should which can later discourage its potential users.

License Conditions and What Next? To begin with, it should be clarified from the beginning, what will be the conditions of using the program after the testing process – VET instructors and providers don't want to engage their time just for testing, they need to see the future advantage coming from using such programs in their organisations. All come down to the circumstance that all terms and conditions should be previously (before the testing phase) agreed.

At the beginning of project, VET training organizations were very excited and involved in the testing stage. But in the middle of the phase we've encountered some difficulties regarding the future actions. We tried to inspire potential EVAL clients to use the new software during their courses and many of them were really interested but finally, trainers didn't even want to try to use the system in their work because of the issue of no concrete conditions about the future usage like: how many people/teachers can use this program, for how many training participants it can be used, who will pay for the servers etc. Not clear rules make trainers stop being interested in the mobile tools. It is important to keep in mind that trainers who are interested in implementing the new technologies in their courses must have a clear view of the conditions and rules of how they will be able to use the program. It includes the clear rules of the licence and all costs also regarding the servers, customer service and technical support. It's worth remembering, that the implementation of mobile technologies in a big organisations or universities is always time consuming and it requires a lot of work from the teachers. If the rules of using the mobile technologies are not clear or they cover only few months of the usage the authorities cannot take the risk of implementing them in their institutions. The conclusion is that as well as in business, other organizations also must know what will happen to the software after the testing phase is over.

3 Conclusions

During the time spent on the project and while working on its goals we faced many different opinions about the mobile learning and we came across many difficult situations during the implementation of each stage of the project. Nevertheless, we also experienced several negative attitudes towards the mobile learning tools and programs. We have made every effort to ensure that during the project many crucial options and functionalities will be added to the software and thanks to this a lot of companies liked our software. Three categories of factors must be taken into account while implementing new mobile tool.

First is to tailor the software to the target group needs in such a way that it corresponds to their expectations.

Next, the software should be compatible with the most popular systems and devices among the target group and it must work on the most popular devices in the country of the target groups. The crucial point is that it must be stable and free of any critical bugs which will make it impossible to show the software to the wide audience or that the errors will discourage potential users from working on the new tool.

The last point that must be taken into consideration is the organisational phase that must show the clear rules and conditions on which the exploitation stage will be based and it needs to take into account the time that will be needed to meet all standards and expectations.

In this article we listed the main factors, problems and difficulties that can be faced while introducing new mobile learning systems. We used our own experience and we want to share it with everyone who wants to implement a new system before the implementation phase even starts.

References

1. Ally, M.: Mobile Learning Transforming the Delivery of Education & Training (2009)
2. Guy, R.: Mobile Learning: Pilot Projects and Initiatives. Informing Science Press (2010)
3. Kukulska-Hulme, A., Traxler, J.: Mobile Learning A handbook for educators and trainers. Routledge Taylo

The Impacts of Tablet Use for Eliminating the Time-Space Barriers in University Education: A Turkish Experience

Sarphan Uzunoglu, Ismail Hakki Polat, and Murat Akser

Kadir Has University, Istanbul, Turkey
{sarphanuzunoglu,ismailpolat}@gmail.com,
makser@khas.edu.tr

Abstract. Mobile learning applications are widely used in various levels of education process. In developing and developed countries, educational institutions use tablets and personal computers for supporting learning processes. Mobile learning practices are generally used for overcoming time-space constraints in traditional learning process. This study covers both lecturer's and students' tablet usages and achievements of tablet usage on Introduction to New Media Course in Kadir Has University undergraduate New Media program, including a comparison with traditional and online-blended lectures in previous years Thanks to mobile course tablet application developed, students have been able to watch live broadcasts and video records of lectures, see lecture presentations and read e-materials submitted online, while they were able to submit their assignments, exams and response papers. Interaction between lecturer and students is improved by tablet application, lecture narrations were followed online and archived. Mobile application is integrated to Facebook for improving students' social interactions with the course materials and lecturer which paves the way for social learning concept. A course which has already been complemented by social networks and another online education software was chosen for the study. With almost same syllabus that was used for two years before, comparative data about student and lecturer performances have been obtained. It is found out that average class success increased by %8 compared to previous years, mobilization and online interaction level increased, average time spent for class increased and 3G was used more than Wifi technologies during the semester that enables the mobility and allows time-space independency for the students.

1 Mobile Learning in the Age of Time and Space Constraint

Scholars, tried to run research and development processes together in order to create better learning and teaching practices in contemporary education spheres. Paul Virilio [1] defined time-space constraint as an essential facet of contemporary life stating that "Today we are entering a space which is speed-space ... This new other time is that of electronic transmission, of high-tech machines, and therefore, man is present in this sort of time, not via his physical presence, but via programming." Education

M. Kalz et al. (Eds.): mLearn 2014, CCIS 479, pp. 212–221, 2014.

today, can not be thought free from this time-space compression. Castells [2] extended this definition regarding time and space compression stating that ''individuals are becoming increasingly adept to wireless communication; the idea of being connected anytime, anywhere.'' This meant that, people started not to identify the difference between time and space boundaries that began to blur. Therefore, the concept of 'school' has changed today in relation to the concept of learning.

2 School: History of an Educational Space

The concept of grouping students together in a centralized location for learning has existed since Classical antiquity. Formal schools that have been established by central authorities have existed at least since ancient Greece, ancient Rome ancient India, and ancient China. School in Ancient Greece refers to Plato's school of philosophy, founded approximately 385 BC at Akademia, a sanctuary of Athena, the goddess of wisdom and skill, north of Athens, Greece. It is thought to be earliest form of formal education practices. Renaissance era had been the era that concept of 'academy' had deeply evolved once again. Roman academies included a description of Papal academies in Rome including historical and bibliographical notes concerning the more important of these. Roman Academies[1] were associations of learned men and not institutes for instruction in 15th century. In 16th century it was seen in Rome that literary and aesthetic circles or 'academies', more or less inspired by the Renaissance have increased in number. As a rule these academies, were all very much alike and were merely circles of friends or clients gathered around a learned man or wealthy patron, and were dedicated to literary pastimes rather than methodical study. By 17th century, while the Roman Academy in its older form still survived, there began a new epoch. The Academy was constituted as a public body, i.e. it was no longer confined to a small circle of friends.

This evolution can be understood as formalization of academics and its massive reformation. In that period, academy set itself a fixed and permanent scope in the field of science, letters, and arts, often of a polemic or apologetic character. At same period, French Academy was founded by Richelieu and then that academies became practical and efficacious instruments of culture, with a direct influence on public opinion; in this way, too, they claimed the special attention of the heads of the state. This was the period that opened ways for reformation of education and especially by 19th and 20th century, especially after the establishment of nation states and because of states' general approach towards using academy as an ideological apparatus as Althusser proposed [3], schools, universities or generally the concept of academy had gained a new, formalized and structured meaning. The concept of 'school' gained a 'physical meaning'.

3 School and Potential Rivals of School

In Turkey, all legal educational institutions are supervised by National Education Ministry and Higher Education Council (for undergraduate, graduate and postgraduate education) including private courses designed for preparing students for various exams.

[1] `http://en.wikipedia.org/wiki/Roman_academies`

And most of these institutions use 'school' as a physical space designed for education and where the process of education is exercised in except some private courses and open-university system that enabled remote education practices. In 2012, National Education Minister of Turkey, Ömer Dinçer, declared that there are 607 thousand people who were registered for open-high school and open technical schools[2]. Anadolu University, as the center of open-university education in Turkey has 1.350.000 students who are registered for mutual-learning programs. However, these programs are not completely digitally practiced and still there is a physical conception of school or there are many traditional educational materials used for these processes.[3]

4 Should We Go to School for Attending the Class?

However, there comes the speculative question of the century: Do we still need to go to the school for education. Are classrooms still necessary for educational practices? Do mobile learning practices commute traditional educational practices? This question can be handled in various ways. This paper tries to respond this question regarding a study with a group of undergraduate students.

The study relies on the observations obtained from a course in Kadir Has University, consisting 23 students and one lecturer in 2013 Fall and 2014 Spring. Students' and lecturer's actions on tablet platform are observed through the research and impacts of 3G driven tablet education have been tried to be evaluated. Motives of this study include necessity of distant learning in various levels of education life. Study analyses students' and lecturer's performance within NM 101 and NM 102 courses supported by Vodafone Tablet supports 3G connection regarding the impacts of time-space independence on processes of learning and teaching.

What differentiates this study from previous studies is the use of tablets with consistent 3G connection that provides a stable, instant and continuous educational process between student, lecturer and course. Study's main objective is not evaluation. The lecturer for the courses observed in 2012 Fall and 2013 Spring was also active through the courses in 2013 Fall and 2014 Spring and there was nearly no change in the syllabus and pre-planned outputs of the course. Computer based online education model was available in 2012 Fall and 2013 Spring term while 3G driven tablet education model was used in 2013 Fall and 2014 spring. Previous year's courses were not available on tablet. A Facebook page was used during courses. Students used their personal computers and mobile phones for their online interactions with the class. Lectures were not broadcasted or recorded via Youtube or any similar media. Computers were used for mediating educational process rather than running that process. So we can talk of our study as a study on an education year that causes a shift from ''blended learning'' to ''online learning''. Garrison and Kanuka [4] refer to blended learning as "convergence of text-based asynchronous internet-based learning with face-to-face approaches". The reason that we mention a shift in terms of our course is the fact that, there are still some problems that lecturer and students faced throughout the year. Garrison and Kanuka [4] defined the

2 http://kamudan.com/galeri.php?iID=5707&rrID=12740

3 http://www.anadolu.edu.tr/

categories of e-learning as enhanced, blended and online learning. All those categories are subcategories for e-learning while, they might also be perceived as a route for education specialists. Garrison and Kanuka's debate on the advantages of e-learning relies on possible advantages of having a face-to-face class to meet and build community. Benefiting from their conception of community of inquiry, we'll discuss our educational community composing of three main elements: Cognitive, social and teaching presence [5].

5 Technical Background of the Study and Project

5.1 Technical Features of Our Project

Throughout two semesters, we used Vodafone SmartTabl10, which operates with Android 4.0.3 and has a storage capacity and 10.1' screen. There is a SimCard slot and 3G Bundle available in the device as well. For this project, Vodafone provided teacher and students with an annually free, 8 gb per month membership. The use of quotas and shift from wireless use to 3G use will be clarified in further parts of the paper.

Vodafone's existing technical capacity is known to be 4.3 MB/s average download speed, 2.1 MB/s average upload speed and 479 ms average latency. Network reliability of Vodafone equals 92%. [4]

Fig. 1. 3G Network Coverage of Vodafone in İstanbul

[4] http://opensignal.com/networks/t%C3%BCrkiye/vodafone-tr-kapsama

We used an application available for Android devices through the project. Android was preferred as Android has a dominant market share around the world. Moreover, Turkey's government is also using Android for Fatih Project, wich is the most comprehensive education project in Turkey. Also, the opportunities that Vodafone provided us with were available through a simcard slot and it was available in the Vodafone's tablet that we preferred.

The application that we used have been designed by software developers that we cooperated with and it has been upgraded afterwards regarding feedback of users and teacher. The application had some features that today's social networks and learning applications necessitate involving, weather condition module, location module, semester calendar and live lecture broadcasts which were made available through Google Hangout & Youtube. There was a course material module that involved course syllabus, lecture notes in presentation form, lecture video records, course e-books, relevant documents, images, links and videos.

5.2 Technical Features of the Application

The application was also integrated to a Facebook group that consisted of, lecturer announcements, course material submissions, replies to students, student notifications and submissions of daily life news relevant to the course. Student posts as questions, comments and sharings relevant to the lecture are also available there.

There is an extra note taking module which relies on ecological perspective of our project as well. Its success will be evaluated in part regarding student reactions. No paper has been used through the lecturing and examining progresses. Only 3 students declared that they used paper while studying for exams. Exam & assignment announcements have been made through special collection/submission module created for this ecological perspective.

For evaluation, personalized performance chart for each student has been made available in the application. This provides both students and lecturer a general view of student's performances through the course.

6 Ethical Questions about Mobile Learning

Prior [6] states that:

"It is undeniable that new technologies (mobile communication, digital and social media and a variety of another social tools) have allowed the creation of platforms for interaction and debate that enhance our capabilities and increase, exponentially, the access to participate in the public sphere. But it is precisely what allows the establishment of a communication without spatial barriers and the easy access to knowledge, services and trade, which makes us more vulnerable to electronic surveillance".

His statement should be considered as important for our study as well, because mobile learning project's strategic tools used for the education process brings about the question of privacy of personal information, which is a matter of global ethical debate. From integration to the Facebook to use of Google Hangout, there is a great data that users/students provide for us. Especially, using their mobility, through GPS, it is possible to detect their positions or their involvement in listening through their activities. In traditional sense, surveillance has always been defined as a purposeful, routine, systematic and focused activity [7]. As our project also involves evaluation based aims, students' interactions, submissions and logins have all been recorded with time, location and connection information. This creates an ethical dilemma about obtaining data as mobile learning also means time-space free learning and lecturer or the software gains a huge data about students' personal preferences.

Especially, emerging news and new facts about surveillance and spying around the world would create problem for students. As techniques of espionage develop, not only passwords, whole encryptions are observed. Madsen, [8] states that:

"The National Security Agency of the United States of America (NSA) has maintained for years a secret agreement with Crypto AG, a Swiss company that sells encryption technology to various countries and business organizations, allowing the NSA and intelligence services that collaborate with it full access to allegedly secret communications".

In Turkey, legislations towards surveillance over Internet caused a similar debate regarding the Turkish government and intelligence institutions' monitoring scandals in recent years. Therefore, we need to question and specify the ethical weaknesses of the new media environment we're in. Storage, location and full Internet access is available for the application. All hardware preferences such as taking photograph, recording video, directing vibrations and auto starting are enabled for the application. These technical factors rise a question about ethical future of mobile learning application that we use.

7 Findings

We started our study on September 2013 with 23 students. While 1 of the students was younger than 18, only 1 of them were before 22-24 and 21 of the students were between 19-21 years old.

Any of the students in first semester stated that they were super disciplined in terms of attending classes. While four students had tendency towards spending whole their non-attending rights, five students stated that they had nearly used all their rights of not attending the class. Fourteen of them declared themselves to be ordinary members of class. While only 13% of the students stated that they always submit their homework, 48% of them stated that they rarely skipped doing their homework. 30% of them stated that they did not spent that much time on homework and 2 of them (9% of the population) stated that they never spent time on doing homework. They revealed their regular attendance average as 80%, which is 10% above university regulations.

While average time spent per hour for a student was equal to about 30 minutes for 35% of class, 30% of them declared that average time they spent studying was between half an hour and an hour. Other 30% stated they studied around 1-3 hours per week for a course.

As we look for potential impacts of social interaction through new media on learning, students are asked to define their characteristics. While 48% of them defined themselves as sociable, 39% of them declared themselves to be people with a few but good friends.

39% of the students had stated that they preferred listening to the lecture on the classroom. It was the most popular way of learning for the students in first semester. Homework and projects were the secondary favorite method of learning for the students while student interaction and collective studying were the third choice for them. Students also stated that, they mostly communicate with their friends via instant messengers rather than face-to-face communication.

Another important factor in our study was amount of time the students spent at school and their capability of using new media. All of the students were using computer for more that 3 years while 87% of them have been using computers for more than 8 years. They were also regular Internet users. 96% of them stated that they spent more than one hour per day on the Internet while 65% of them feel themselves connected to the Internet more than 3 hours a day. Only one of the students stated that he/she doesn't have a smartphone. So their capability of using smartphone was high, but 65% of them had not experienced using their own tablets. There are no students who have never been member of a social network. Facebook is the leading social network among the students. It reflects the correctness of our preference of Facebook integration for our application. Whatsapp and iMessage are the leading instant messaging services among the students.

Students have nearly experienced all of similar instant messaging services. They also use use communication platforms like Viber, Tango, Skype and Google Hangout. Skype, Viber and Tango were the leading ones among these similar services while Google Hangout, the service that we used, had the fourth position in their preferences. They are all aware about what 3G is and what it serves for. While 57% of them were intended to attend courses away from the class in starting period, 35% had worries regarding this mobile/distinct attendance. Only 22% of students revealed that they had technology supported courses before. So, that would be a type of new experience for them. Their lack of experience in online education should be considered while analyzing the results.

8 Outcomes of Mobile Learning

In this section, primarily we will make a comparison between 2004, 2013 and 2014's statistics regarding educational performance and the impact of mobile learning. According to statistics recorded in previous years, in 2004, when traditional education progress was available, average time spent per week for a course was more than 3 hours, while 3 hours of lecture time was also involved. By 2012 fall and 2013 spring, while blended learning was available, this average time reserved for course was 4 hours and more. By 2013 Fall and 2014 Spring, this average time spent was equal to 5 hours and more, regarding the statistics taken from our application.

Same positive impact can be found out in students' attitudes towards fulfilling responsibilities. While it was 50% in 2004 and 70% in 2013, by notifications and other similar enhancements by our software and mobility, this has reached 85%. 98 of 115 total homework assignments have been responded by students. Another important finding regarding student's reaction time to teacher notifications is that, in

2013's blended learning model, student's feedback time was around 1 day. However, after 3G support, feedback time reduced to 3 hours in 2014's mobile learning model.

Not only the reactions, but also student interactions and activity have been effected by shift to mobile learning as well. While share per person was 17 in 2013 per semester, it increased to 28 in 2014 per semester. Also, there is a difference among the share of wifi, and 3G connections within two semesters. In first semesters, even if their devices had 3G, 45% of connections were through wifi, and 55% were through 3G band. In second semester, wifi usage reduced to 30% and 3G usage increased to 70%. This demonstrates that, after an orientation process, 3G use becomes more reliable.

There has been no great change in student success average. While 2004's success average was about 65% and 2013's average was 67%, average success of the mobile course reached 71%. Of course there are some other factors than mobile course that impact this success. Moreover, lecture content's suitability and such factors really matter in this sense.

From first semester to second semester, according to surveys on December 2013 and May 2014, average study time spent per course have increased. Integration to the tablet education enhanced students' involvement in course. By second semester, all negative perceptions about role of mobility in student-lecturer relationship have disappeared while 87% of the students revealed that they totally agree that mobility enhanced their relationship with lecturer . There has also been a 25% more positive perception towards the strong characteristics of mobile communication compared to traditional communication.

In first semester, 3 students stated that application had no role in their attendance or participation to the course, while survey in 2014 May reveals that only one of them said it had made no impact. There is a positive impact of tablet application to the learning process. But still, classical (traditional) mode of learning has been preferred by students more in both semesters. On the contrary, students stated that they have a tendency towards learning through social media rather than face to face communication. There is a kind of conflict in their statements. Students are happy about digital course content, because of its ergonomic and easily reachable characteristic. As library module hasn't been that much effective, there has been no change in students attitudes towards using the library for studying. 91% of students also stated that notifications have increased the interactivity and motivation.

In video interviews that we've made with students, it is found out that, most important role of mobile learning in educational process is that, it makes following the course easier and sustainable. Especially on a day that İstanbul had faced a traffic problem regarding weather condition, all of the students have attended the class, while the courses in same level (undergraduate) in same time period had 25% attendance, which is much lower than their regular attendance which is assumed to be 80% by the lecturer. However, of course there are some weaknesses of the tablet application. Students have complained about some problems regarding software development that we'll discuss in future opportunities part.

Our study has revealed that, the concept of mobile learning have been positively experienced by the students and the lecturer. Still, there are some negative points that need to be resolved through software development and a change in general policies of the faculty administration regarding mobile learning. Because change of attitudes towards mobile learning can't be enhanced sufficiently as long as traditional and blended learning methods are available for students in other courses. Lecturer's role in this course shouldn't be underestimated as he is an engineer and has been lecturing on new media for more than a decade. So, expanding study to other course with different lecturer would be academically more benefitable for comparison.

Student's lack of chance to send messages to each other, lack of instant message module within application, also causes some problems of interactivity.

Hangout's user attendance permission limits should also be resolved with strategic partnerships or shift in group video conference model for a better interaction. There are still problems about submission because of keyboard problems. As a result, lecturer has limitations about preparation of exam questions and assignments.

Tablets that we used have been criticized as their ergonomic characteristics do not enhance the experience. Especially their portability is not that easy because of its weight. We categorized some possible enhancements in software and hardware to enhance education process.

In the future, the following software developments are foreseen:

- Gamification: Up-Down scoring, badge (Rookie, Master medals), rewards (extra points given to respondents), leaderboard (a kind of list based on experience points), scoring, recommendation (students can gain points through recommending the course in social networks), challenge, quiz and tournament (a tournament that aims to evaluate existing information degree of students).
- Engagement: Each student and lecturer is sent notifications through the software and that provides a time-space independent progress of education. Students and lecturer are able to schedule free of physical constraints through notifications.
- Broadcast: Eliminating the cameraperson, broadcast setup personnel is planned to be eliminated in near future.
- Interaction: Both in-group and inter-personal messaging is required.

9 Further Academic Opportunities of the Study

In conclusion, our study revealed that, shift from blended learning to mobile learning must be supported as an education policy by faculties and universities as a whole shift rather than a partial shift for better results. Because, today's conditions regarding education should be perceived as a whole rather than a specific condition regarding our course. A new media themed single course can't be an ideal sample for a study like ours to get a concrete conclusion. To develop the study academically, further research and literature, application of the project on different community or samples, and applying project with a different content is essential. Lecturer's role in the project is another fact that impacted the outcomes of the study

References

1. Decron, C.: Speed-Space. In: Armitage, J. (ed.) Virilio Live, p. 71. Sage, London (2001)
2. Castells, M.: Communication power. Oxford University Press, Oxford (2009)
3. Althusser, L.: .Ideoloji ve Devletin İdeolojik Aygıtları İstanbul: İletisim yayınları, İstanbul (1994) (in Turkish)
4. Garrison, D.R., Kanuka, H.: Blended learning: Uncovering its transformative potential in higher education. Internet and Higher Education 7, 95–105 (2004)
5. Garrison, D.R., Anderson, T.: E-learning in the 21st century: A framework for research and practice. Routledge/Falmer, London (2003)
6. Prior, H.R.: Public Sphere and Panopticism: Considerations about surveillance in everyday life in Sphera Publica No: 14, vol. 1(29) (2014)
7. Murakami, D., Ball, K., Lyon, D., Norris, C.: A Report on the Surveillance Society. For the Information Commissioner by the Surveillance Studies Network 5 (2006)
8. Madsen, W.: Crypto AG: The NSA'S Trojan Horse? Covert Action Quarterly, http://mediafilter.org/caq/cryptogate/

Mobile Learning Perspectives from the United States: Superintendents and Researchers Share Experiences and Insights

Michael Corke[1], John Franey[1], Maria Kelley[1], Shawn Gross[1], Timothy Baird[2], and Terry Grier[3]

[1] Mobile Technology Learning Center, University of San Diego, San Diego, California, USA
[2] Encinitas Union School District, Encinitas, California, USA
[3] Houston Independent School District, Houston, Texas, USA

Abstract. The University of San Diego's Mobile Technology Learning Center (MTLC) is honored to propose a session that includes its researchers and two of the top district superintendents in the field of mobile learning from the United States. Together these leaders and their districts represent a multi-dimensional cross-section of mobile learning in the United States. Their perspectives include two different regions of the US, two different size school districts and two different points along the mobile technology integration continuum. Their perspectives as mobile technology innovators will resonate with policy makers, implementers and researchers. Session participants will be able to engage these leaders as they provide an overview of their mobile learning programs covering various themes from building an initial mobile learning plan to managing and monitoring.. These policy leaders will be joined on the panel by researchers from the University of San Diego's Mobile Technology Learning Center (MTLC) who will add their knowledge of emergent trends in mobile technology including the MTLC's emergent theory of action, the Contextual Review.

2 Introduction

The proliferation of mobile technology in learning environments is well underway. Mobile devices are being leveraged as means to make education more accessible, equitable and effective for students all over the world [1, 2]. The process of mobile integration is complicated and takes time and resources. It has not been without its pitfalls [3].

As learning organizations such as districts and schools progress along the continuum of technology adoption they face challenges. There is no one size fits all solution to technology integration in schools as districts operate within their unique contexts recognizing, considering and overcoming various factors. The way each organization goes about this process influences the degree to which its technology vision is carried with fidelity into its classrooms. This session offers participants the opportunity to learn about mobile implementation from two district leaders at various points along the technology integration continuum as well as from the only university based center in the United States focused exclusively on mobile learning.

M. Kalz et al. (Eds.): mLearn 2014, CCIS 479, pp. 222–227, 2014.

3 Brief Literature Review

The rapid evolution of digital technology has moved educators away from a traditional model of instruction where learning passes from teacher to student, to a multilateral model for learners [4]. However, research comparing the effects of digital learning to traditional classroom instruction has yet to show a consistent and significant advantage for digital learning [5]. Some studies report that digital classrooms outperform traditional classrooms [6,7], while others report no difference or the reverse [8, 9,10].

Institutional orientations to technology can take on multiple forms. Orientations that operate as though technology alone will improve learning are known as technology-centered approaches [11]. In these cases the focus is on what technology can do for teachers and students and on taking advantage of the latest technological advances to promote learning. Initiatives focused on specific devices, software, or the latest Internet innovations, reflect such an approach. One shortcoming of a technology-centered approach is the emphasis on technology itself as it is by nature ever changing. This orientation can lead to a constantly shifting focus as organizations seek to implement the newest technology rather than focus on how to use the technology they already have in ways that promote learning. In contrast, a learner-centered approach focuses on how to incorporate technology to support and enhance skillful teaching processes, with an emphasis on how people learn and how technology can play a role in those processes [11].

In their report, *The Technology Factor: Nine Keys to Student Achievement and Cost Effectiveness*, Greaves et al. [12] conclude that implementing technology initiatives is not enough to attain desired learning outcomes. They outline specific conditions for technology to make a difference in learning. The report notes student learning related to the extent to which the nine conditions were implemented, as did other educational outcomes. Research by Shapley, et al., [13] supports the importance of a system-wide approach to technology implementations as well. Their research indicates that the success of digital learning relies on student, family, and community context, school factors (e.g., leadership, technical support, innovative culture), and teacher and student factors. These notions can be extended to a global forum through international organizations such as United Nations Education Scientific and Cultural Organization (UNESCO) and Commonwealth Technology Organization (CTO) who similarly cite the importance of context [14], [2].

In sum, recent research suggests a systems-focused, learner-centered framework is the most appropriate way to articulate the mechanisms through which technology can impact student learning. This thinking serves as the foundation of MTLC's work. MTLC posits in order to reach transformative levels of digital instruction teachers and students must be supported to teach and learn with technology in ways that extend beyond traditional instructional techniques. This level of technology integration can only be achieved when the contextual factors within which the technologically driven instruction is occurring serve as supports of digital integration. The Commonwealth Telecommunications Organization (CTO) noted the importance of considering context and avoiding copying the efforts of others, "Think creatively in your own context. There are no best practices, only a range of good practices from which to choose. Develop solutions that best fit your learning needs, and then get on with implementing them!" [15].

3 Session Description

This session includes researchers from the University of San Diego's Mobile Technology Learning Center and two of the top district leaders in the field of mobile learning from the United States. The leaders include:

Dr. Terry Grier, Superintendent, Houston Independent School District (HISD), Houston, Texas.

- 210,047 PK-12 students (7th largest in US): 80% economically disadvantaged; 92% non-white
- In Fall 2014 HISD is entering year two of a 1:1 technology integration that will provide devices to all 46,000 HISD high school students.
- HISD was winner of 2013 Broad Prize as the top urban school district in the US.

Dr. Tim Baird, Superintendent, Encinitas Unified School District (EUSD), Encinitas, California.

- 5500 K-6 students: 13% economically disadvantaged; 30% non-white
- In Fall 2014 EUSD is in year four of its 1:1 iPad initiative. EUSD has issued devices to all of its students and adopted digital curriculum in three of four core subjects.
- All nine EUSD schools have earned the California Distinguished School Award and four are National Blue Ribbon Schools.

Together these leaders and their districts represent two cross-sections of mobile learning in the United States. Their perspectives encompass different regions of the US, different size school districts, different device types (laptops and tablets), and their districts occupy different points along the mobile technology integration continuum. Their insights as mobile technology innovators will resonate with policy makers, implementers, practitioners and researchers.

Participants will be able to engage these leaders as they provide an overview of their mobile learning programs covering the following themes:

1. Garnering student, parent and administrator support
2. Student and teacher training
3. Device selection versus BYOD
4. IT Infrastructure and processes for managing bandwidth
5. Selecting instructional content
6. Management and monitoring
7. Developing an acceptable use policy

This session will combine the experience laden superintendent perspectives on technology integration with the work of the University of San Diego's Mobile Technology Learning Center (MTLC). MTLC has conducted local, regional and national projects. In this work MTLC focuses on technology readiness and degrees of implementation within various educational organizations. This work has included districts with robust 1:1 technology programs, those that have adopted BYOD policies and those on their way toward 1:1 but at various points along the path to full device implementation. In addition, MTLC recently completed the first technology audit of the 42 San Diego County school districts and is currently engaged in projects in Houston Independent School District (HISD), New Bedford (Massachusetts) and Amman, Jordan.

Through the literature and its work in schools MTLC has learned what integrated classrooms look like. They are places where students and teachers are tapping into the full capability of mobile devices to make learning and teaching experiences formerly impossible, possible. One consistent theme across MTLC's work is although fully integrated classrooms exist they are rare. Schools typically have a classroom that represents the site's "model of technology integration" but these "model integrations" vary dramatically from school to school. One thing however does not vary - there are few classrooms where technology integration is happening to the degree it could be.

At the district and school levels MTLC has identified a number of factors that may in part account for the challenges between introducing technology into classrooms and transforming instruction through technological capacity. According to MTLC and others [2], [12,13,14], the context into which the integration of technology for learning occurs is critical, within each various factors have a strong impact on the use of technology and how to best integrate it for learning purposes.

Contextual factors include:

- students (their demographics and needs)
- teachers (their capacities, priorities, and level of autonomy)
- school leaders (vision for the school or district's communication, capacities, and priorities)
- school resources (funds, supports, school structure, climate, and community resources)
- technology resources (infrastructure, devices, training/coaching for teachers, curriculum integration support, and technical support)

All of these factors greatly influence what occurs when technology is introduced into a school or classroom setting. It is not meaningful to describe the use or effects of technology in the classroom without taking these contextual factors into account. MTLC theorizes that when these contexts function as supports rather than barriers, technology can improve access, enhance communication, free up teacher time, extend the purpose and audience for student work, and shift teacher and student roles.

The MTLC has built a process that takes into account these factors within learning environments. The *Contextual Review* is a method through which individual schools and school districts can better understand the contexts within their learning communities. This process provides an opportunity to examine the factors that either support or act as barriers to the readiness of educational stakeholders for the effective integration of technology. MTLC's *Contextual Review* uses a mixed-methods approach. It takes a broad quantitative inventory of each factor area through online surveys of key stakeholder groups. This is followed by a series of classroom observations, semi-structured interviews and focus groups. The findings that emerge from this research process enable districts and individual school leadership teams to make evidence-based decisions in order to address contextual barriers, build stakeholders capacities, and prepare for the effective integration of technology and the transformation of learning environments.

This session will include a presentation by the panel covering the various aspects of each district's technology implementation. MTLC researchers will add broad perspective to the presentation including elements of the *Contextual Review* and findings from MTLC's work in other districts. The presentation will conclude with a question and answer session between participants and the panel of presenters.

References

1. Farell, G., Isaacs, S., Trucano, M. (eds.): Survey of ICT and Education in Africa: A Summary Report Based on 53 Country Surveys. ICT and Education Series. Infodev (2007)
2. West, M.: Turning on Mobile Learning, Global Themes. United Nations Educational, Scientific and Cultural Organization (UNESCO), Paris, France (2012)
3. Klein, K.: Many of L.A. Unified's iPad project wounds are self-inflicted. Los Angeles Times (November 21, 2013), `http://articles.latimes.com/2013/nov/21/news/la-ol-lausd-ipad-2013112` (retrieved)
4. Clark, R.C., Mayer, R.E.: E-learning and the science of instruction: Proven guidelines for consumers and designers of multimedia learning, 3rd edn. Pfeiffer, San Francisco (2011)
5. Bernard, R.M., Abrami, P.C., Lou, Y., Borokhovski, E., Wade, A., Wozney, L., Wallet, P.A., Fixet, M., Huant, B.: How does distance education compare with classroom instruction? A meta-analysis of the empirical literature. Review of Educational Research 74, 379–439 (2004)
6. Clariana, R.: Ubiquitous wireless laptops in upper elementary mathematics. Journal of Computers in Mathematics and Science Teaching 28(1), 5–21 (2009)
7. Suhr, K.A., Hernandez, D.A., Grimes, D., Warschauer, M.: Laptops and Fourth-Grade Literacy: Assisting the Jump over the Fourth-Grade Slump. The Journal of Technology, Learning, and Assessment 9(5), 42 (2010)
8. Clark, R.E.: Media will never influence learning. Educational Technology Research and Development 42(2), 21–30 (1994)
9. Dillon, A., Gabbard, R.: Hypermedia as an educational technology: A review of the quantitative research literature on learner comprehension, control, and style. Educational Psychology 81, 240–246 (1998)
10. Warschauer, M., Grimes, D.: First Year Evaluation Report Fullerton School District Laptop Program. Fullerton District, Orange County California (2005)
11. Mayer, R.E.: Multimedia learning, 2nd edn. Cambridge University Press, New York (2009)
12. Greaves, T., Hayes, J., Wilson, L., Gielniak, M., Peterson, R.: The Technology Factor: Nine Keys to Student Achievement and Cost-Effectiveness (2010), `http://pearsonfoundation.org` (retrieved May 15, 2014)

13. Shapley, K.S., Sheehan, D., Maloney, C., Caranikas-Walker, F.: Evaluating the Implementation Fidelity of Technology Immersion and its Relationship with Student Achievement. Journal of Technology, Learning, and Assessment 9(4) (2010)
14. Unwin, T., Wong, A.: Global Education Initiative: Retrospective on Partnerships for Education Development 2003-2011. World Economic Forum, Geneva (2012)
15. Commonwealth Telecommunications Organization (CTO), `http://www.cto.int` (retrieved August 7, 2014)

A Comparative Analysis in Evaluating 'ThinknLearn' from Science Educators and High School Students Perspectives

Sohaib Ahmed[1,*] and David Parsons[2]

[1] Bahria University, Karachi, Pakistan
dr.sohaibahmed@gmail.com
[2] Massey University, Auckland, New Zealand
d.p.parsons@massey.ac.nz

Abstract. The current use of mobile technology in the classroom has attracted many educators and researchers to develop mobile web based learning tools for educational purposes. In school science education, deductive or inductive forms of inquiry are used, while the use of abductive form of inquiry has been sparsely explored. 'ThinknLearn', an abductive enquiry tool, was developed and evaluated with students in earlier work. However, this paper addresses the use of such learning practice from the educators' perspective and compares this with previous evaluations in which students were involved. This may help to investigate the longer term impact of this learning practice in classrooms as students and educators are the main stakeholders of this learning practice. Content analyses of data indicated that both educators and students support the use of abductive forms of inquiry-based learning activities with mobile devices in the classroom.

Keywords: Mobile Learning, Abductive Science Inquiry, ThinknLearn, Science Education.

1 Introduction

The rapid development towards mobile device deployment in classroom environments may offer students new opportunities for increasing engagement, motivation and learning [1]. Mobile devices not only revolutionize school education but also transform the traditional classroom into an interactive form of learning for students [2]. These devices provide computing power and wireless capability, which can make learning expedient, immediate, authentic, accessible, efficient and convenient [3]. Hence, these devices may increasingly become a convincing choice of technology in classroom environments [4].

There have been great strides in the affordances of these mobile devices, attracting many educators and researchers to use them in classroom environments for various domains, specifically in science education [5]. Various studies in the literature show

* Corresponding author.

M. Kalz et al. (Eds.): mLearn 2014, CCIS 479, pp. 228–237, 2014.

that the use of these mobile technologies provides enjoyable learning experiences for students, increasing motivation towards science learning and enhances their reasoning skills and learning performance, to some extent [6].

Generally, deductive or inductive forms of inquiry can be supported by classroom learning activities using a number of applications, where students develop meaningful explanation through their pre-defined hypotheses or observational data [7]. Among these, BioKids Sequence [8], WHIRL [9] and nQuire [10] are few prominent science projects in which deductive or inductive forms of inquiry are used. In contrast, abductive science inquiry mainly generates meaningful explanations through the combination of both inquiries: deductive and inductive, in which students can comprehend the given topic by using generated hypotheses from the observed phenomena [6].

Oh's abductive Inquiry model (AIM) [11] consists of the following phases: exploration, examination, selection and explanation. In the exploration phase, students explore the given scientific phenomena by collecting data and find ways to explain it scientifically. The examination phase then follows in which students find out scientific theories or facts to develop scientific hypotheses. During the selection phase, students examine all the previously inferred or possible hypotheses and choose those that provide the most plausible explanation of the observed phenomena. At this phase, if students find any problem then they may revert back to the previous steps for any modifications required. In the last phase of this model, explanation is used for the development of more sophisticated explanations of the observed phenomena where students recommend complete explanations of the given phenomena using the rules and hypotheses chosen in the earlier phases.

The mobile web-based learning tool 'ThinknLearn' [12] was designed on the principles of Oh's AIM model [11] in which students have to perform a real time science experiment in the classroom (laboratory). In one particular example, they have to compare how three tins with different surface colors radiate heat energy. The tool helps students to explain the observed phenomena by involving in all the phases discussed in the AIM's model [11]. This tool was previously evaluated with students in their classrooms with couple of experiments using the M3 evaluation framework [13]. However, the inputs from the educators and their comparison with previous evaluations about this tool were not discussed in earlier articles.

Educators are the main driving force for nourishing high performing and talented students in their society. Hence, their opinions about the established or new forms of learning practice can play a vital role in what may become an established practice for school science education if appropriately implemented [1]. For this purpose, 'ThinknLearn' [12] was evaluated with school science educators in order to comprehend their perspectives in comparison with their students' evaluations. Thus, this paper discusses the semi-structured interview data collected from the educators while evaluating 'ThinknLearn'.

The remainder of this paper is structured as follows. In section 2, related work about the topic is discussed. Section 3 presents the methodology for conducting experiments while section 4 explains the content analyses of the data collected from the educators and students. In addition, the comparisons between educators and students' responses are described in this section. Finally, the last section contains concluding remarks and discusses some advancement in this research.

2 Related Work

Inquiry based learning (IBL) is an educational activity that allows students to learn science by doing science, offers resources to provide hypothesis generation and explanation of the observed phenomena [14]. In the literature related to school sciences, the importance of using mobile devices in the classroom is discussed a number of times such as in Huang et al. [15] and Shih et al. [16]. They stated that students can perform better if they are equipped with mobile devices while doing scientific inquiries. They further elaborated that these devices may improve students' knowledge and reasoning skills compared with those who do not use them in the classroom. Similarly, a few studies [17-18] also highlighted the use of mobile devices for generating correct explanations or hypotheses if appropriate tools are provided to the students.

Most of these studies follow deductive or inductive form of inquiries in which students either use pre-defined hypotheses or observational data for developing meaningful explanations. In contrast, the abductive form of scientific inquiry is based on the theory of abduction that was proposed as a form of reasoning by C.S.Pierce (cited in [19]). In this type of reasoning, students are required to develop meaningful explanations using both hypotheses and observational data [19]. The example can be seen in our earlier research paper [13] that shows the differences between these forms of scientific inquiries; deductive, inductive and abductive.

According to that given example, it can be observed that in abduction, the Rule (Condition) and the Result (Observation) are involved together to identify a Case (Hypothesis). On the other hand, deduction and induction can be processed with either a Rule or a Result to generate the other component. Thus, abduction is well-suited to scientific inquiries in which students are challenged to formulate scientific hypotheses and explain the observed phenomena [11]. It also provides us with an opportunity to identify the educators' perspectives on this abductive form of scientific inquiry.

3 Methodology

In mobile learning tools, evaluation plays a vital role in examining the effectiveness of the tool to enable learning opportunities with the support of new technology [6]. For evaluation purposes, the M3 evaluation framework [6] was applied to verify the effectiveness of the tool 'ThinknLearn' [12]. The M3 evaluation framework consists of three levels; Micro, Meso and Macro [6]. However, only the Micro level evaluation comparison between students and educators is discussed in this paper.

3.1 Experimental Design

The underlying principle of this experimental design is to evaluate 'ThinknLearn' in educators' perspectives, which may assess whether such abductive science inquiry tool can be used for high school science students [12].

Usability serves as a means to provide a successful learning tool. If in case, a learning tool is not usable enough then it may obstruct learning and students may spend more time in understanding how to use such tool instead of learning the provided contents [20]. In addition, usability also enables a platform to identify opinions of the users about such given tools [21]. For that reason, qualitative data were gathered from the science educators in order to evaluate their opinions about 'ThinknLearn'.

3.2 Participants

A meeting was conducted with a group of eight science educators of diverse background (Chemistry, Physics, Biology) from a local high school in Auckland, New Zealand. In the beginning, information was disseminated about how abductive science inquiry can work in general and how 'ThinknLearn' works in particular. After describing this information to educators, mobile devices were provided to each of the participants to evaluate the given tool in terms of its usability [21] and mobile quality aspects [22]. Further, qualitative data were also collected in a group discussion held during the meeting. Previously, 86 students in 25 groups from three classes had participated in a group discussion [13], which is used for comparison in this paper.

4 Content Analysis and Discussion

Qualitative data were collected in semi-structured group discussions between eight school educators and 86 students from three classes. Five questions in the discussions were asked of the educators for evaluating the given tool while only three questions were posed to students as the last two questions were related to educators (see Table 1). For analysing responses from the participants, qualitative content analysis was used, which is a research method that can be used to define the characteristics of language as communication describing the contextual meaning of textual data [23]. In that method, a large amount of data can be divided into a small amount of content categories or codes [24]. There are three forms for developing codes in a content analysis [25]; conventional, direct and summative. In this experiment, direct content analysis was used in which categories are derived during the group discussion and few key categories were used beforehand. In other forms of content analysis, categories are identified either prior to the experiment or after conducting the experiment.

4.1 Content Analysis

In group discussion with science educators, the researchers used some prior knowledge about the topic to identify some key categories such as 'Easy to understand', 'Enjoyable experience' and 'Helpful'. As analysis proceeds, additional categories emerged during the group discussion. For instance, 'Difficulty in use', 'Confusion in hypothesis' and 'Uneasiness' highlight the negative aspects of the tool while 'New or innovative' and 'Interactive' are the positive aspects regarding the given tool, as depicted in Table 1.

Table 1. Questions asked to educators during semi-structured group discussion

Q.No.	Questions	Categories	Frequencies	Code No.
1	What type of difficulty do you find in using this tool?	Easy to use	3	I
		Easy to understand	3	II
		Difficulty in use	2	III
2	How do you feel after using this tool?	Enjoyable experience	6	IV
		Different experience	2	V
3	What do you think about the adaptive suggestions given in the tool?	Relevant suggestions	6	VI
4	What do you think about this abductive form of inquiry used in the classroom?	New or Innovative	4	VII
		Confusion in hypothesis	3	VIII
5	How do you feel the use of mobile devices in the classroom?	Uneasiness	2	IX
		Interactive	3	X
		Helpful	3	XI

After the identification of key categories from the group discussion and their frequencies, some codes were defined. These categories can be further used to organize and group into themes [26]. For this paper, the identified categories of the Table 1 were combined into themes according to the nature of the responses as shown in Table 2. For instance, 'Helpful' and 'Interactive' categories were formed into 'Useful' (see Table 2). In addition, the total frequencies of each of these themes were also calculated by adding the frequencies of the combined codes as presented in Table 2.

During the analysis of the group discussions conducted with the students, the identified key categories are added in response of question 1 such as 'Help in understanding a topic', 'Difficulty in understanding questions' and 'Confusion in hypothesis', as depicted in Table 3. Similarly in question 3, students group discussions also found some more categories including 'Provide guidance in generating hypothesis, 'Correct answers straightaway' and 'Less explanations as suggestions' (see Table 3). In addition, only those frequencies are considered in this analysis, which are at least greater than 10 in number (11.6 % of total frequencies).

Table 2. Categories, themes and total frequencies during educators' group discussions

Q.No.	Combined Codes	Categories	Themes	Total Frequencies (Out of 8)
1	I, II	Easy to use Easy to understand	Usable	6 (75%)
	III	Difficulty in use	Difficult	2 (25%)
2	IV	Enjoyable experience	Engaging	6 (75%)
	V	Different experience	Unpleasant	2 (25%)
3	VI	Relevant suggestions	Relevance	6 (75%)
4	VII	New or Innovative	Innovative	4 (50%)
	VIII	Confusion in hypothesis	Confusing	3(37.5%)
5	IX	Uneasiness	Uneasiness	2 (25%)
	X, XI	Interactive Helpful	Useful	6 (75%)

Table 3. Categories, themes and total frequencies during students' group discussion

Q.No.	Combined Codes	Categories	Themes	Total Frequencies (Out of 86)
1	I, II, III	Easy to use Easy to understand Help in understanding a topic	Usable	53 (61.6%)
	IV, V	Difficulty in under-standing questions Confusion in hypothe-sis generation	Difficult	16 (18.6%)
2	VI	Enjoyable experience	Engaging	64 (74.4%)
	VII	Different experience	Unpleas-ant	10 (11.6%)
3	VIII, IX	Relevant suggestions Provide guidance in hypothesis generation	Relevance	46 (53.4%)
	X, XI	Correct answers straightaway Less explanations as suggestions	Irrelevant	17 (19.7%)

4.2 Discussion

According to the responses, question 1 revealed that most of the science educators consider this tool as usable enough to guide students and encourage them to think about the topic as shown in Table 2. One of the educators said "*...this application is very easy to use and the navigation is very straightforward*". However, there were two educators in discussion who found this tool a bit difficult to understand. One educator responded how "*...It may confuse students because they may not understand the purpose behind the use of the given adaptive suggestions in the application*".

As far as the students' responses are concerned (see Table 3), they responded well in saying that 'ThinknLearn' is a usable application there but not enough understanding of what was expected by their educators. One of the students' groups highlighted that "*... questions were difficult and the given suggestions were not easy to understand*". It could be argued that they did not relate suggestions to understand the given topic. However, the concepts covered in 'ThinknLearn' had already been discussed in their theory classes earlier. In another instance, one participant of the other group described how "*it was not difficult but confusing on some occasions*". Those participants who considered the application a bit confusing and difficult did not understand the deliberate purpose of this application to exploit students' higher level skills of critical thinking in such inquiries.

Responses to question 2 were very straightforward as most of the educators (6 out of 8) considered that this tool may lead to an interesting and engaging tool for high school students. In a similar fashion, students responded positively towards question 2 as more than 70% of them were positive about their learning experiences. For instance, one of the student group participants stated that "*I really enjoyed using it. This application was pretty good and engaging, it helped you to learn about your course (science)*". The other group participants gave an interesting comment about it during the discussions as "*this type of application keeps you on focus and requires better attention but it was an interesting and enjoyable experience*".

Compared to question 2 responses, the educators responded to question 3 in a similar fashion. Educators agreed that the suggestions given in the tool were very relevant to the given topic and all these suggestions are adaptive according to the answers given to the multiple choice questions asked in the tool. One of the educators explained the relevancy of the given suggestions in such a way that "*....the given suggestions are like hints, which can guide students for generating hypotheses and their explanations*". Another one described how "*....these suggestions make them [students] think about the given topic and that is why I like the concept behind this engaging and interactive tool for learning purposes*".

On the other hand, the student groups considered the critical thinking process as a bit of a burden, but on number of occasions, participants understood it as a challenging activity to learn from. One of the students stated "*...it is a challenging task as we have to think and find out the answers ourselves instead of having straightaway answers but I really enjoyed and learnt the topic from this*". On the other hand, there were 17 students who remarked like that "*...more detail should be provided*" and "*... relevant but they (suggestions) did not explain much*". These comments showed that this application presents some challenges to the students to comprehend the given topic and make this application more engaging and interesting. However, they considered it a confusing way to generate hypotheses (see Table 3), compared to their educators. Thus, some way may be needed to convince those students about the challenging nature of this application or perhaps some further evaluation is also required to understand these challenges faced by the students.

The last two questions, as depicted in Table 1, were asked to the educators only because of their nature. With respect to how this abductive form of inquiry was a new or innovative way of learning (in response to question 4), only half of the educators supported this, while the others were confused or totally rejected this way of learning in the classroom. According to those who propose this abductive form of inquiry is an innovative way of learning, "*….this form of learning assists students to comprehend how to develop hypotheses and transform observational data into meaningful explanations*". In contrast, a few of them were against this concept because "*…this way of learning may confuse students because in a traditional way of learning, we first develop hypotheses and then collect the data afterwards but in this form of inquiry, we have to do both concurrently (collecting observational data and generating hypotheses)*".

From their replies, abductive form of inquiry not only helps to confirm both forms of data simultaneously but also guide investigators to understand both phenomena (deductive and inductive). Similarly, if an investigator starts with a hypothesis (deductive) or begins with any observation data (inductive), these are lacking in innovations because in both of inquiries, investigators tried to interpret observational data into meaningful hypotheses or vice versa [27]. These explanations may be further scrutinized with large number of experts, students or users for particular learning activity.

Responses to question 5 were promising as 75% of the educators found that mobile devices are helpful as interactive technology used in the classroom. One of the educators described how '*….mobile devices are the personal assets of our young generation and if they enjoy playing with them why can they not use them as learning tools*". However, one of the proponents of this concept elaborated that "*mobile devices could not be used in the classroom environments due to their small screens, negligence towards studies and expense (bringing their own devices)*".

Overall the responses in semi-structured group discussion were promising. Most of the educators and students both support the concept of abductive form of inquiry in the classroom after being convinced to some extent. However, there were a few suggestions given by those educators to improve 'ThinknLearn' further as "*this tool can be used as teacher-oriented along with student-centered approach*" and "*can be adaptive in such a way so that teachers can assign different inquiry learning activity in a classroom or outside if possible*". These suggestions indicate that educators were very enthusiastic to use such mobile learning tools in the classroom compared to the students. In addition, educators considered the abductive form of science inquiry an innovative and challenging activity that may enhance students' learning and inquiry skills. However, students were reluctant and confused to use this kind of approach as they believed that the hypothesis generation activity with provided suggestions may become a burden in understanding the underlying domain knowledge.

5 Conclusion and Future Work

In this paper, the comparison between educators and students' responses to the use of the mobile abductive inquiry tool 'ThinknLearn' in the classroom has been discussed. According to the data collection and analysis, this form of learning not only guides students in learning science by doing science but also provides a way that students can think in order to generate hypotheses and explanations of the given topic. The educators considered that 'ThinknLearn' has considerable mobile and software quality measures. However, while students considered this tool an interesting and engaging, some of them were not convinced with the challenge provided in terms of the hypothesis generation activity. This may need further development so that students may be comfortable with this new approach for inquiry learning activities in the classrooms.

In the future, 'ThinknLearn' can be further evaluated with other students and educators, which may help us to comprehend the use of such abductive forms of inquiry in the classroom, so that it may become an established learning practice in science education. For this purpose, we are trying to implement this tool for all science subjects at school levels. In addition, an extended version may be designed for educators so that they may assign different tasks or learning activities related to science inquiries within or outside the classroom.

References

1. Lin, M.-F., Fulford, C.P., Ho, C.P., Iyoda, R., Ackerman, L.K.: Possibilities and Challenges in Mobile Learning for K-12 teachers: A Pilot Retrospetive Survey Study. In: 7th IEEE International Conference on Wireless, Mobile and Ubiquitous Technology in Education (WMUTE 2012), pp. 132–136 (2012)
2. Scornavacca, E., Huff, S., Marshall, S.: Mobile Phones in the Classroom: If You Can't Beat Them, Join Them. Comm. of the ACM 52(4), 142–146 (2009)
3. Ogata, H., Yano, Y.: Context-aware Support for Computer-supported Ubiquitous Learning. In: IEEE International Workshop on Wireless and Mobile Technologies in Education, pp. 27–34 (2004)
4. Lai, C.-H., Yang, J.-C., Chen, F.-C., Ho, C.-W., Chant, T.-W.: Affordances of Mobile Technologies for Experiential Learning: The Interplay for Technology and Pedagogical Practices. J. of Comp. Assist. Learn., 326–337 (2007)
5. Looi, C.-K., Zhang, B., Chen, W., Seow, P., Chia, G., Norris, C., Soloway, E.: 1:1 Mobile Inquiry Learning Experience for Primary Science Students: A Study of Learning Effectiveness J. of Comp. Assist. Learn. 27(3), 269–287 (2011)
6. Vavoula, G., Sharples, M., Rudman, P., Meek, J., Lonsdale, P.: Myartspace: Design and Evaluation of Support for Learning with Multimedia Phones Between Classrooms and Museums. Comp. & Edu. 53, 286–299 (2009)
7. Grandy, R., Duschl, R.A.: Reconsidering the Character and Role of Inquiry in School Science: Analysis of a Conference. Sci. & Educ. 16, 141–166 (2007)
8. Parr, C.S., Jones, T., Songer, N.B.: Evaluation of a Handheld Data Collection Interface for Science Learning. J. of Sci. Edu. and Tech., 233–242 (2004)

9. Yarnall, L., Shechtman, N., Pennel, W.R.: Using Handheld Computers to Support Improved Classroom Assessment in Science: Result From a Field Trial. J. of Sci. Edu. and Tech. 15(2), 142–158 (2006)
10. Mulholland, P., Anastopoulou, S., Collins, T., Feisst, M., Gaved, M., Kerawalla, L., ...Wright, M.: nQuire: Technological Support for Personal Inquiry Learning. IEEE Trans. on Learn. Tech. 5(2), 157–169 (2012)
11. Oh, P.S.: Characteristics of Abductive Inquiry in Earth Science: An Undergraduate Case Study. Sci. Edu. 95(3), 409–430 (2011)
12. Ahmed, S., Parsons, D.: Evaluating 'ThinknLearn': A Mobile Science Enquiry Based Larning Application in Practice. In: 11th International Conference on Mobile and Contextual Learning (MLearn 2012), pp. 17–24 (2012)
13. Ahmed, S., Parsons, D.: Mobile Abductive Science Inquiry: Evaluating 'ThinknLearn' with High School Students. Comp. & Edu. 63, 62–72 (2013)
14. van Joolingen, W.R., Zacharia, Z.C.: Developments in Inquiry Learning. In: Balacheff, N., Ludvigsen, S., De Jong, T., Lazonder, A., Barnes, S. (eds.) Technology-Enhanced Learning Principles and Products, pp. 21–37. SpringerLink (2009)
15. Huang, Y.-M., Lin, Y.-T., Cheng, S.-C.: Effectiveness of a Mobile Plant Learning System in a Science Curriculum in Taiwanese Elementary Education. Comp. & Edu. 54, 47–58 (2010)
16. Shih, J.-L., Chu, H.-C., Hwang, G.-J., Kinshuk: An Investigation of Attitudes of Students and Teachers About Participating in a Context-aware Ubiquitous Learning Activity. British J. of Edu. Tech. 42(3), 373–394 (2011)
17. Mulder, Y.G., Lazonder, A.W., De Jong, T.: Comparing Two Types of Model Progression in an Inquiry Learning Enviornment with Modelling Facilities. Learn. and Instr. 21(5), 614–624 (2011)
18. Peker, D., Wallace, C.S.: Characterizing High School Students' Written Explanations in Biology Laboratories. Res. Res. in Sci. Edu. 41(2), 169–191 (2011)
19. Raholm, M.-B.: Abductive Reasoning and the Formation of Scientific Knowledge within Nursing Knowledge. Nursing Phil. 11(4), 260–270 (2010)
20. Wong, S.K. (B.), Nguyen, T.T., Chang, E., Jayaratna, N.: Usability Metrics for E-learning. In: Meersman, R., Tari, Z. (eds.) OTM Workshops 2003. LNCS, vol. 2889, pp. 235–252. Springer, Heidelberg (2003)
21. Sim, G., MacFarlane, S., Read, J.: All Work and No Play: Measuring Fun, Usability, and Learning in Software for Children. Comp. & Edu. 46(3), 235–248 (2006)
22. Parsons, D., Ryu, H.: A Framework for Assessing the Quality of Mobile Learning. In: 6th IEEE International Conference on Advanced Learning Technologies. IEEE Computer Society, Kerkrade (2006)
23. Lindkvist, K.: Approaches to text analysis. In: Rosengren, K.E. (ed.) Advances in Content Analysis, pp. 23–41. Sage, Beverly Hills (1981)
24. Weber, R.P.: Basic Content Analysis. Sage, Beverly Hills (1990)
25. Hsieh, H.-F., Shannon, S.E.: Three Approaches to Qualitative Content Analysis. Quali. Health Res. 15(9), 1277–1288 (2005)
26. Patton, M.Q.: Qualitative Research and Evaluation Methods. Sage, Thousand Oaks (2002)
27. Haig, B.D.: An Abductive Theory of Scientific Method. Psy. Meth. 10(4), 371–388 (2005)

Finger Velocity – A Multimodal Touch Based Tablet Application for Learning the Physics of Motion

Mattias Davidsson

CeLeKT, Department of Media Technology, Linnaeus University
351 95 Växjö, Sweden
mattias.davidsson@lnu.se
http://www.celekt.info/

Abstract. A prototype multimodal tablet application for learning the physics of motion has been developed tested and evaluated. By moving their finger across the screen the application enables the user to map its position and velocity in real-time in terms of graphs. The learning outcome of those test subjects using the application was compared to a group that did not use the application but had it shown to them, at the same time as getting an explanation of all the physics involved. There was a small but not significant difference in performance between these groups on a post-test. However, a larger (arguably significant) difference was seen between the male and female test subjects for the sub-set of questions of a more analytical nature. These were the questions targeted in this paper.

Keywords: Physics education, informal learning, mobile learning, ICT in education, embodied learning, perceptual learning.

1 Introduction

Traditionally, teaching has relied upon textual representations combined with images, and in some cases moving pictures and sound. The learner is most often a passive consumer of the material presented. The new-touch based devices such as the iPad, smartphones and android tablets makes possible a more interactive component to the learning experience. This can be seen in the ever-growing number of "apps" available for children and young learners in the apple and android marketplaces. Oftentimes these applications are still very much like the traditional learning material dressed in a digital robe, not fully taking advantage of the possibilities of the new technology. In cognitive science the research on multimodal representations of complex concepts in science stress the importance of a multitude of senses to be involved in the learning process. For example: iPad applications for learning physics are often written in a way that involves sound, text, images, and moving objects. But the interaction with the learner is still restricted to the activation of a specific movement or event. The actual movement or other physical process under study is still generated by the computer (such as

M. Kalz et al. (Eds.): mLearn 2014, CCIS 479, pp. 238–249, 2014.

the iPad) for the learner to observe and in some cases manipulate. The learner observes a moving object and might be presented with information about the motion in terms of graphs showing the speed versus time. But they do not *produce* the motion.

This work will present and evaluate a method for doing precisely this - studying the motion of your own hand in real time using a nowadays almost ubiquitous type of hardware. In [1], as well as [2], the importance of this kind of embodied multimodal learning experience is stressed. Especially when it comes to different kinds of representations of mathematical abstractions of physical processes and events. With the introduction of touch and gesture based devices the students own body can now be made a much larger part of the learning experience, adding the sense of touch and movement to the list of representations and modalities.

We will look specifically at the topic of physics education and how the new technology can be used to support the students understanding of its mathematical representations. Utilizing the possibility for a broader range of senses to be involved in the learning process would potentially increase the learning outcome. If the students could interact with, manipulate and produce the objects being studied, the idea is that they would get more emotionally involved in the learning process. If this perceptual type of learning process is turned into a game-based activity, the effect could be enlarged even further. This would also provide part of the variation in the teaching of physics that the Swedish students calls for [3], and as well as the type of variation stressed by Rose [4].

The contribution of this work is the creation and evaluation of an application specifically designed for learning the physics of motion using touch based consumer technology such as the iPad. Touching the screen, a distance-time and a velocity-time graph is produced in real time at the tip of the user's finger (see Fig. 1 and Fig. 2).

The learning experience includes embodied interactions involving several sensory modalities. The design and envisioned usage scenarios are based on theoretical argument and research as presented below.

The developed prototype application has been blind tested on year 12 students divided into two groups. In one group the students get to use the application without supervision. These students are the test group called "doers". The other group of students are shown the application and get instructions around the topics it illustrates in a more traditional, formal, classroom setting. This group acts like a reference group and is called the "watcher" group. The hypothesis being that, although the watcher group gets more formal information and instructions on the physics involved (they get to know what is "right") the doer group will perform better on a post test due to the embodied multimodal personal interaction, as was found in [5]. This research also aims to look into the proposed line of research in [5] where it was observed that many of the test persons in the doer group made gestures with their hands during the post-test. As if they were actually using the apparatus used in that study. The ability to visualize the movements in the questionnaire was proposed as a way to explain the results in the above mentioned paper.

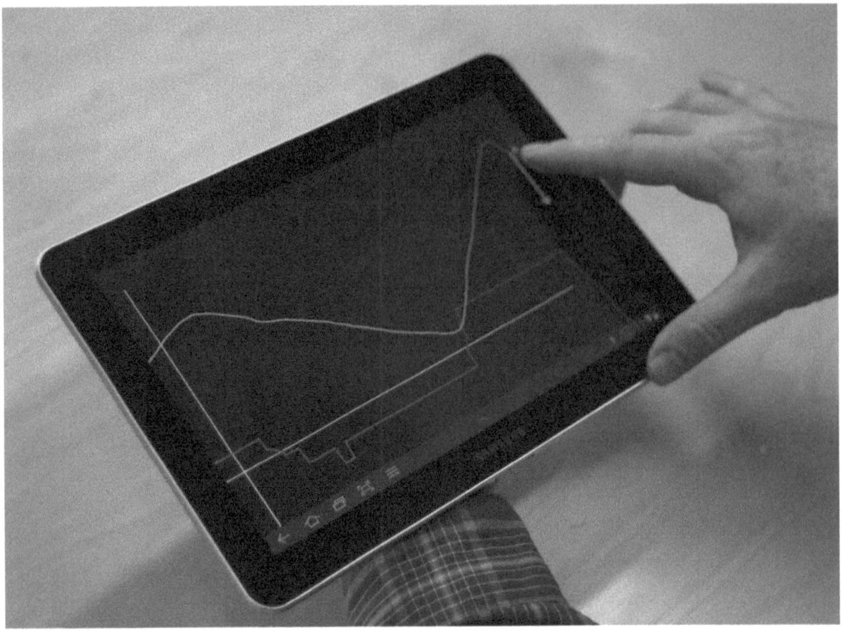

Fig. 1. An early version of the application installed on an Android tablet

2 Theoretical Background

2.1 Graphical Representations in Physics Education

The conventional illustration of motion in terms of *graphs* is distance versus time (d-t) and velocity versus time (v-t) graphs illustrating how an object's position and velocity changes with time. This study concerns learners ability to interpret and produce these kinds of illustrations, as well as finding the connections between them.

There is a multitude of sources for knowledge about learners' understanding of physics concepts connected to the physics of motion, and graphing of motion in particular. Some of these sources also contain elements of the methods and theoretical keywords for this work. In [6] it was found that the simultaneous production of graphs with the visualization of the motion did not in itself seem to have a significant effect on the students learning and understanding. It was pointed out that what was probably of importance would be the student being able to control and manipulate the motion connected to the graphical visualizations. In [7] the positive effects of the use of a motion sensor as a mean to aid students get the connection between real world motion and the abstract representations of these physical real world phenomena were studied. The method (application) tested in this work takes both these considerations into account as implemented in the application. The students do manipulate and control the motion represented by the graphical visualization. They do this by means of what could be called a motion sensor where the iPad screen senses the motion of the students finger.

2.2 Multiple and Multimodal Representations

With Multiple representations of a concept like velocity, one generally refers to the different ways of illustrating that same concept. It could be done in terms of a formula, illustrating the change in vertical speed for a free falling body. It could also be done in terms of a movie showing the actual fall, a graph showing the velocity as a function of time, or as a series of vectors of different length. Reference [8] stresses the importance of the use of multiple representations such as graphical ones together with others when it comes to aiding students in understanding complex scientific concepts. Kaput [9] argues that: the cognitive linking of representations creates a whole that is more than the sum of its parts.

Multimodal representations on the other hand generally refer to representations involving the different sensory modalities such as hearing, sight or touch, in combination with communicative modalities. Examples of communicative modalities are gestures, speech, or images. The motion of an ambulance could be communicated using moving images or a table of values of position and time as perceived through the visual modality. Adding sound would add the possibility to experience the Doppler effect as the ambulance passes. Standing at a zebra crossing, you could actually feel the ambulance passing as the pressure wave hits you, involving yet another modality in your experience of the ambulance motion.

When using the application presented here, the students involve their visual modality, in coordination with tactile/kinaesthetic modalities as they communicate and interact with the graphical representations using hand gestures. The learning experience thus involves the learners body more than just passively observing. In using his or hers sensory modalities, the learning becomes more of a learning by doing type. The intertwining of different modalities and the fact that the learner produces the object of study with their body is the key idea behind the proposed learning outcome, as drawn from the cognitive and learning sciences.

3 Related Work and Earlier Efforts

In [5] the above-mentioned importance and power of getting the students and their bodies involved in the learning process is demonstrated. They studied the effect on learning outcome by the creation and manipulation of representations involving multiple modalities and types describing the meaning of mathematical concepts of motion such as speed, distance, and time. Being able to produce and manipulate graphs showing the displacement and velocity of their hands moving as a function of time was shown to significantly increase the learning outcome. They compared pre- and post-tests with students that did not involve their own bodies in the learning process but instead watched a teacher perform the same action. The background to the study [5] was inspired by the work of Papert [10]. Papert suggested that by using their own bodies to construct symbolic representations students could be aided in their learning. A suggestion further stressed in [11]. We propose to build on and spread this kind of learning activity. This can be done in terms of touch- and gesture based applications for tablets, smartphones and digital whiteboards, in line with the conclusions in [13] and suggestions from [5]. They propose that; Providing students with personal multimodal technologies may help them to engage in learning science concepts.

The second major work that the suggested approach of this work is based upon is the study presented in [13]. The idea being that the methods presented in that paper could increase the effect presented in [5]. It would do so e.g. by enabling students to learn in a more informal setting on their own. In [13] the learners were given the task to pair graphs with other representations such as equations and written text on a computer before receiving any formal instructions. The hope was that after many trials they would see the patterns and connections between the different representations and what they actually represent. This was indeed what was observed. The students were also later observed to pick up the more formal instructions and concepts of mathematical graphical representations easier. Similarly, the suggested design in this work hypothetically would let the learners find and see patterns and connections between graphs without the need of formal instructions. Some efforts in the direction pointed out above have already been made. There are a few applications on the market, where some of the different aspects of touch capability, authentic and perceptual learning, multiple representations, ubiquity and mobility are addressed. One example is the Vernier Logger Pro desktop software and the Vernier iPad application [14]. Reference [15] has studied the use of publicly available game-like apps for iPod touch in middle school physics education. A great deal of benefit e.g. from the standpoint of the students engagement level was found.

Another study [12] further stresses the importance of engaging the learners in embodied participation. The application presented in this paper could then be part of the multitude of embodied practices sought for in [12]. However, the aspects of embodiment and informal perceptual learning have not been realized in one application. Neither does any of the above efforts include for the learners to generate and interact with the graphical representations directly using their own bodies. Hence there is no real-time multimodal feedback. Recently however, SmalLab Learning [16] has come up with a combination of Kinect type hardware and software similar to that described in this work. The effects on learning, specifically looking at the effect of embodied learning were also evaluated with overall positive results as can be seen in [17]. Most of the features and design aspects of the here presented application were derived from the research in [5] and repackaged to fit and function in touch-based hardware such as the iPad.

4 Research Questions and Hypotheses

In the work by Anastopoulou et al. [5] one clear result was shown - the students performed significantly better if they used their hands to generate graphical representations of motion in the learning process. This was a significant finding in more than one way, since the reference group got instructions and explanations on the physics involved and were guaranteed to get the right answers to the training questions, potentially giving them a head-start compared to the doer test group. Furthermore the results were solely due to the difference in score between the two groups on the more analytical questions aimed for understanding the relationships between the two different types of graphical representations. These are the questions in focus for this paper, see Hypotheses 1-3 below.

Question 1 and the corresponding Hypothesis 1 below can be seen as a effort to confirm the findings of [5]. But, it is also an extension of the same study when it comes to the age group of the test persons and the actual technology and input method used. The reference group consists of students experiencing a lecture where the same application is used and shown, and where the concepts and content are explained.

Research Question 1: Will the students using their hands in the learning process while using the application (the doers) perform significantly better than those in the control group (the watchers) on the analytical type of questions?

Hypothesis 1: The test group - the doers - will perform significantly better compared to the reference group (the watchers) on a post test on the more analytical types of questions while no such difference will be seen for the more descriptive type of questions.

Mentioned in [5] was the observation that many of the doers used their hands during the post-test, in such a manner as if they tried to recall or recreate the motion of their hand during the learning session using the application on the iPad. To test if this might be the case - that using your hands to recall/recreate the motion in a test-scenario can help you perform better - a second research question and a corresponding hypothesis is tested.

Research Question 2: Will the test persons using their hands during the post-test in such a way as to mimic the hand movement described in the questions perform better compared to the test persons that don't on the more analytical type of questions?

Hypothesis 2: Test persons moving their hands on the post-test in a similar fashion as when using the application will score significantly higher compared to those that do not on the more analytical type of questions.

In [5] there was no difference reported when it comes to the gender of the test persons. However, this is basically a default question to ask.

Research Question 3: Is there a difference in performance on the post-test results between the male and female test subjects?

Hypothesis 3: There will be no significant difference on the post-test results between male and female test subjects on the more analytical type of questions

Parallel to the above more theoretical questions, the evaluation of the developed application will also test a more practical question of pedagogical interest. Namely if it is possible for a young learner to grasp the abstract concepts of motion in a learning setting with little or no help or guidance from a teacher, only having the iPad application developed in this work to guide them. This last research question will be evaluated in a more qualitative fashion looking at how well the students managed to cope on their own during the training sessions. Will they need support to get started during the training session? Will they need help using the different features in the application? Will their results be comparable to those getting guidance from a teacher?

Research Question 4: Can this type of knowledge and understanding be obtained using an application on an iPad, without the aid of a teacher for explanations and guidance? How does the gained knowledge after this type of learning activity compare to that of a more teacher guided learning activity approach?

5 The Application – Method and Design

In [5] the method of generating and interacting with the graphs is not the same as in this study. There, the learners hand motion was measured directly using a motion sensor attached to the test persons hand. The application developed as part of this work measures the speed of the learners motion in terms of the finger moving across the screen of a tablet. The motion is then turned into a velocity- and distance versus time graph respectively (see Fig. 2).

After some early test versions implemented on an android tablet, the application used in this study was developed as a native iOS application to run on an Apple iPad. Therefore the programming language Objective-C as well as the Cocoa Touch Framework were used. All the programming was done using the IDE Xcode 4.x provided by Apple. It was build to run on devices with the major release of iOS 5. The graphical representation of the Velocity application is based on OpenGL ES.

The more technical aspects of the design and details on the functionalities are outside the scope of this paper, but some information regarding this matter can be found in [18] and in [19].

6 Test Design

The test was designed to mimic the test used [5] to as large degree as possible. However, all details could not be replicated for both technical and practical reasons. For one thing, the same hardware was not used. The quasi-experimental test was performed on 26 students of ages 17-18 studying natural sciences, divided into two sub-sets of 13 students each. The number of test subjects was increased slightly from at total of 18 in [5], and hence thought to be enough for a replication study. The pre- and post-tests discussed below were written as close to the tests used in [5] as possible. Changes were made mainly to make the language more appropriate for the age-group and hence level of education and the learning goals to be gauged in all questions remained the same.

All students first took the same pre-test. After that, one subset then got to use the application at the same time solving a set of training assignments.

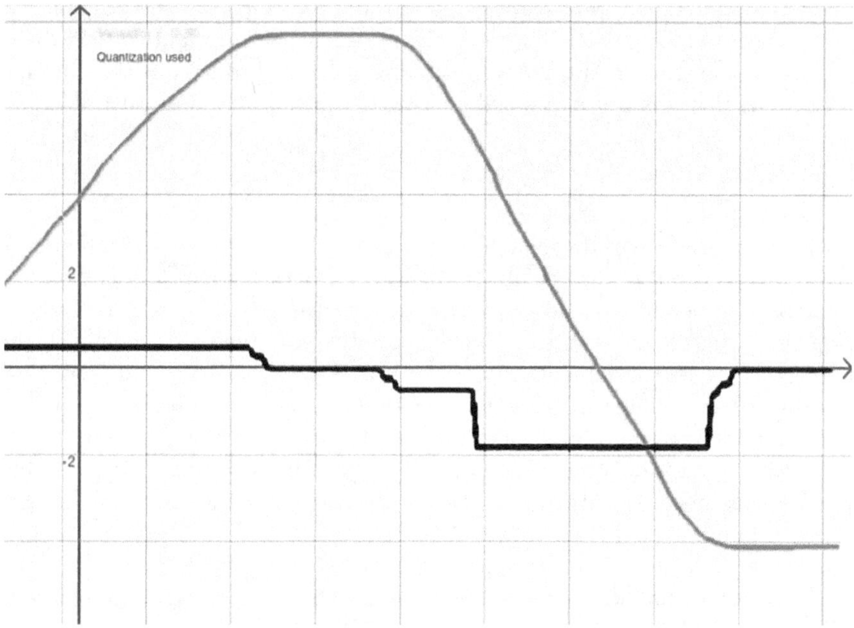

Fig. 2. A screenshot from the final version of the application used in the test. Here the two graphs; velocity, centered around the x-axis and the position graph can be seen. The x-axis shows time, and the y-axis shows the vertical position (and velocity) of the users' finger as a function of time. The feature of "quantisation"of velocity can be seen. This feature was inserted to make short time fluctuations (due to technicalities) in velocity less visible, and hence the main features of the v-t graph stand out more.

They constitute the doers group, and got no or only basic technical assistance from the test-leader. The other subset - the "watchers" - had a lesson in a more formal setting where the author used and showed the application and the students got to solve the same set of training exercises that the doers used for practice. This group also got the concepts explained to them by the teacher (author and experiment leader), and the correct answers to the training questions (see [19] for more details) given to them. In the final stage both the students in the doer and watcher group got to make one and the same post-test, gauging their knowledge and understanding (see [19] for more details). The pre-test and post-test did not contain the same questions but gauged basically the same learning goals. It should however be stated that the pre-test contained what could generally be considered easier questions (the students were expected to learn and be able to answer harder questions after the session). This had the unfortunate effect that the pre-test was probably to easy and hence potential differences in the students pre-knowledge were hard to establish.

One other flaw in the practical carrying out of the experiment was that the number of female (or male) test subjects did not end up to be the same in the doer-group compared to those in the watcher-group. This was partly due to the fact that the experiment-leader (the author) did not have the possibility to decide exactly which students were assigned to each group, and partly due to the fact that the experiment was not primarily designed to look at the difference of gender, as this was not expressed as a variable of special interest in [5].

7 Results

We will here briefly go through the results of the quasi-experiment and the analysis of the results. First of all, the statistical data for hypothesis 2 was too small (the number of test-subjects that made clear gestures with their hands during the post test were to few, and this number could not be predicted) to be able to make a clear judgment, but for sure, the hypothesis was not confirmed, nor falsified. More on hypothesis 2 can be found in [19]. The differences in pre-test scores for the different groups were all statistically insignificant. However, the pre-test was perhaps not the best method designed for gauging possible difference in pre-knowledge between the groups, since the median-values all were very close to the maximum score of the test.

Hypothesis 1 results: The difference between the doers (n=13) and the watchers (n=13) was a median score value of 16 for the doer group, and 13 for the watcher group. The maximum score on the post-test for these questions was 18.

Hypothesis 3 results: The female group (n=15) scored a median of 11 and the male group (n=11) scored a median of 16.

8 Analysis of Results

To see if the scores for the above quoted groups differed in a statistically significant manner, a Mann-Whitney U-test (MW-test) was used due to the ordinal (the distribution of scores is non gaussian and a score of 18 is *not* twice as "good" as a score of 9) rather than nominal nature of the data. The MW-test does not simply look at the difference in the median or mean values for the two groups, but instead looks at the difference in ranking. All tests are ordered (ranked) in terms of scores, and then the test calculates the chance of the ordering being due to chance instead of due to a real underlying difference in the results between the two groups. A value for the rejection of the different null-hypothesis tested, corresponding to hypotheses 1-3, was set to $p=0.05$, meaning a 5% risk of the size of the results (signal) to be due to chance. Since several hypotheses were tested, in the end a corresponding p-value will be calculated so that it will be a less than 5% chance of any of the three null-hypotheses to be rejected due to chance.

8.1 Analysis of the Results for Hypothesis 1

The post-test difference in ranking comparing the doers and the watchers gives a p-value of $p=0.14$ on a single tailed MW-test. Thus it cannot be established beyond the $p < 0.05$ limit that the difference is statistically significant.

8.2 Analysis of the Results for Hypothesis 3

The post-test difference in ranking comparing the male and female test subjects gives a p-value of $p=0.029$, on a two-tailed MW-test, and thus on its own, pointing towards there being a real significant difference in test scores between the male and female test-group.

9 Interpretation of Analysis – Reanalysis

The outset for the analysis was to test the difference between the doers and the watchers - between using the application and not using the application. However, since there was not the same number of male vs. female test subjects in the doer and watcher group, the effect from the possible difference between male and female test subjects potentially makes the results unclear. In the same way, the results that show a difference between the male and female test subjects might be due to the fact that there was a larger proportion of Doers in the male group compared to the female group. So, if the application has a positive influence on the test results this will smear the possible difference due to gender. To try and control for this, two different methods were used. In the first method, the individual scores were measured as the deviation from the mean value for that specific gender-group (male or female). Thus, the effect of using the application could be separated from that of gender. This resulted in a p-value, $p = 0.22$, once again showing no statistically significant difference between the group using the application and those that did not on the post-test. In the second method, the four (there was only four of them) male watchers were grouped together with four randomly selected female watchers. This new watcher group was then compared, using the MW-test to 4+4 randomly selected male and female doers. The process was repeated five times and an average p-value was calculated to, $p=0.27$. The two methods thus give consistent results (the second method has lower statistics and thus a slightly smaller effect is expected) that point towards gender difference contributing more to the difference in results in the original doers vs. watchers group due to the fact that the male test subjects dominated the doer-group. However, it should be stated that the results presented here does not contradict those presented in the original study by Anastopoulou et al. [5]. The actual positive effect of using the application studied here - if it exists - is most likely simply too small to be significant given the small number of test subjects.

Making a multi-hypotheses analysis of all three hypotheses (1-3) the conclusion that male test subjects perform better than female ones cannot be drawn from this work since checking several hypotheses at once increases the risk of false positives or negatives, for any-one of these. Thus an increased threshold of confidence is needed, and the result of gender difference (hypothesis 3) in this work does not pass that level.

10 Conclusions and Outlook

An application for the iPad with the aim of making it possible to learn about the physics of graphs of motion has been developed and tested. One conclusion (regarding research question four) is that it is possible for students to learn about the physics of motion without the guidance of a teacher, by instead using the tested application. At least, the students using the application does not perform worse on a post-test compared to the students that had a lecture on the topic, and was given explanations to the different concepts that the post-test aimed to gauge the students' knowledge of. On the other hand, the effect that the students using their own hands in the learning process score significantly better compared to the reference-group, as presented in [5], cannot be strengthened or confirmed.

Neither can it be claimed that the male users perform better than the female test-subjects, even though the results for that test are much closer to being significant. Further studies should be performed to settle the questions (eg. in [6] it was reported that male test subjects performed better on the type of questions asked in this paper) with more certainty. Larger statistics, an equal distribution of male vs. female test subjects in the two test-groups should be used. A "sharper" pre-test to be able to more clearly diagnose the students' pre-knowledge should be constructed to see if there is a gender difference already present at that stage.

Lastly the ideas and effects of perceptual learning as presented in [13] could be tested in a study of longer duration, where the students could use the application more extensively. The thought being that, just as learning to ride a bike - after a period of trial and error - you (your brain) simply "gets it".

Acknowledgements. Thank to Nico Reski for actually developing the application for the iOS platform. Thank to professor Marcelo Milrad for guidance and support.

References

1. Donovan, M.S., Bransford, J.D.(eds.): How Students Learn: History, Mathematics, and Science in the Classroom. The National Academies Press, Washington, D.C. (2005)
2. Linn, M.C., Eylon, B.-S.: Science learning and instruction: Taking advantage of technology to promote knowledge integration. Routledge, New York (2011)
3. Skolinspektionen. Kvalitetsgranskning, Fysik utan dragningskraft - En kvalitetsgranskning om lusten att lära fysik i grundskolan. Report. Stockholm: Skolinspektionen (Skolinspektionens rapport 2010:8) (2010)
4. Rose, C.: Accelerated Learning. Dell, New York (1985), `ftp://195.214.211.1/books/DVD-032/Rose_C._Accelerated_Learning1985en190s.pdf`
5. Anastopoulou, S., Sharples, M., Baber, C.: An evaluation of multimodal interactions with technology while learning science concepts. British Journal of Educational Technology 42(2), 266–290 (2011)

6. Beichner, R.J.: The effect of simultaneous motion presentation and graph generation in a kinematics lab. Journal of Research in Science Teaching 27(8), 803–815 (1990)
7. Lucia, M., Cicero, L.: The use of motion motion sensor can lead students to understanding the cartesian graph. In: Proceedings of the Sixth Congress of the European Society for Research in Mathematics Education, pp. 2106–2115. Service des publications, INRP (2115)
8. Ainsworth, S.: The Educational Value of Multiple-representations when Learning Complex Scientific Concepts. In: Gilbert, J., Reiner, M., Nakhleh, M. (eds.) Visualization: Theory and Practice in Science Education. Springer, Netherlands (2008)
9. Kaput, J.J.: Linking representations in the symbol systems of algebra. In: Wagner, S., Kieran, C. (eds.) Research Issues in the Learning and Teaching of Algebra, pp. 167–194. LEA, Hillsdale (1989)
10. Papert, S.: Mindstorms: Children, computers, and powerful ideas. Basic Books, New York (1980)
11. Cox, R.: Representation construction, externalised cognition and individual differences. Learning and Instruction 9, 343–363 (1999)
12. Barab, S.: Situationally embodied curriculum: Relating formalisms and contexts. Science Education 91, 750–782 (2007), doi:10.1002/sce.20217
13. Kellman, P.J.: Perceptual Learning Modules in Mathematics: Enhancing Students, Pattern Recognition, Structure Extraction, and Fluency. Topics in Cognitive Science, 1–21 (2009)
14. Vernier: Video Physics for iOS (October 19, 2011), `http://www.vernier.com/products/software/video-physics/`
15. Kelly, A.M.: Teaching Newtons laws with the iPod Touch in conceptual physics. The Physics Teacher 49(4), 202–205 (2011)
16. SmalLab Learning, SmalLab Learning website (July 15, 2012), `http://smallablearning.com/`
17. Birchfield, D., Johnson-Glenberg, M.: A Next Gen Interface for Embodied Learning: SMALLab and the Geological Layer Cake. International Journal of Gaming and Computer-Mediated Simulations (IJGCMS) 2(1), 49–58 (2010)
18. Davidsson, M.: A Mobile Application With Embodied Multimodal Interactions for Understanding Representations of Motion in Physics. In: Sánchez, I.A., Isaías, P. (eds.) Proceedings of the IADIS International Conference on Mobile Learning, pp. 263–266. IADIS Press, Berlin (2012)
19. Davidsson, M.: Creation and Evaluation of a Multimodal touch based Application for Learning the Physics of Motion (Student paper). Linnuniversitetet (2012)

Workplace Learning Using Mobile Technology: A Case Study in the Oil and Gas Industry

Mohamed Ally[1], Mohammed Samaka[2], John Impagliazzo[3], Abdulahi Mohamed[2], and Martha Robinson[1]

[1] Athabasca University
[2] Qatar University
[3] Hofstra University

Abstract. This article presents an innovative project using mobile technology to train workers in the oil and gas industry to develop their communication skills while functioning on the job. The training was delivered on mobile devices using Apps so that workers can access the training materials without having to connect to a network. Workers performance improved and workers indicated that they would like to engage in other experiences using mobile technology and that mobile learning provides flexibility for learning. Since not much research was conducted in the past on the use of mobile learning for training in organizations, more research should be conducted. Examples of research needed include how mobile sensors can be used for just in time training as workers are mobile and how to make learning pervasive using mobile technology so that workers can learn in context.

Keywords: Mobile learning, workplace learning, communication skills.

1 Introduction

Many definitions exist of mobile learning in the field. One definition of mobile learning is the use of electronic learning materials with built-in learning strategies for delivery on mobile computing devices to allow access from anywhere and at anytime [1]. Another definition of mobile learning is any sort of learning that happens when the learner is not at a fixed, predetermined location, or learning that happens when the learner takes advantage of the learning opportunities offered by mobile technologies [12]. Because of the rapid changes in technology, the definition of mobile learning is evolving. A recent definition of mobile learning that was suggested by an international standards committee is learning using information and communication technologies in mobile contexts [6]. This definition emphasize that learners are mobile and they us technologies to learn while they are mobile. This is true for the workplace where workers are mobile and they can use mobile technology for just in time learning (Figure 1).

Countries around the world are realizing the advantage of training its citizens to function productively in the 21st century so that they can remain competitive and at the same time improve the quality of life of their citizens. One example of a country

M. Kalz et al. (Eds.): mLearn 2014, CCIS 479, pp. 250–257, 2014.

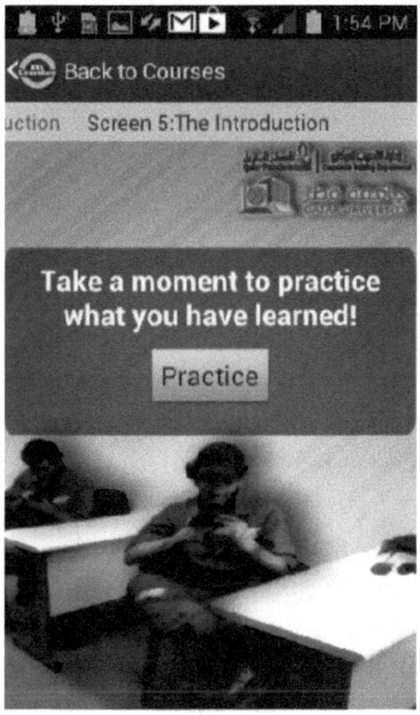

Fig. 1. Workers learning on the job

that is investing in training its citizens is Qatar. One of the goals in the Qatar National Vision 2030 is "to develop a world-class educational system that equips citizens to achieve their aspirations and to meet the needs of Qatar's society, including (1) Educational curricula and training programs responding to the current and future needs of the labor market (2) High quality educational and training opportunities appropriate to each individual's aspirations and abilities (3) Accessible educational programs for life-long learning [13]. To help achieve the above goal, Qatar is funding research projects to investigate the use of emerging training technologies to train its citizens to prepare them for the 21st century workplace. The funding is provided by the Qatar National Research Fund under the Qatar Foundation.

As the use of mobile technology increase around the world, other countries are implementing projects to take advantage of the flexibility of mobile technology to deliver education and training [9]. In some countries citizens are moving directly to mobile technology rather than using desktop and notebook computer. Citizens in some countries have multiple mobile technologies which they use for many activities. Educators and trainers need to re-think the way they develop learning materials so that citizens can use mobile technology to learn both in formal and informal situations. Use of mobile technology encourages learner-centered education so that learners are active in the learning process. Mobile learning provides flexibility in learning where learners can learn at their convenience.

This article presents a research project conducted in Qatar that developed and delivered training materials on mobile technology to train workers in the oil and gas industry to improve their communication skills in the workplace. The objectives of this research project are (1) to investigate whether the use of mobile learning improves communication skills in the workplace (2) to determine learner satisfaction with mobile learning in workplace training.

2 Benefits of Mobile Learning

With clear benefits such as affordability and portability, mobile communication technology promises to assist online learners beyond the training classroom [2]. The communication capabilities of the mobile devices allow workers to use Web 2.0 technology to form communities of practices so that they can interact with other learners in any location [5]. Also, because of the portability of the mobile devices, they can be used in different user contexts to facilitate user interaction with the environment and access to information. Leroux et al. [10] proposed a framework for the automatic detection and prediction of mobile application usage behavior patterns of the user by considering different context parameters. The detection of the usage patterns occurs automatically without the user making any explicit input. The prediction of future action is determined by the context of the user and the user past actions and behavior. As the technology become ubiquitous, there will be seamless interaction with the environment in different contexts [7].

There are many benefits of implementing mobile learning in the workplace. With the use of wireless technology, mobile devices do not have to be physically connected to networks to access information. In some cases, learners can download the learning materials as applications (Apps) to learn without having to connect to a network. Mobile devices are small enough to be portable which allow workers to take the device to any location to access information or learning materials. Because of the wireless connectivity of mobile devices, workers can interact with other workers from anywhere and at anytime to share information and expertise, complete a task, or work collaboratively on a project [3,8,14].

An important advantage of using mobile learning in the workplace is just in time training where workers can access learning materials and apply what they learn right away to promote high level learning. They can also learn in their own contexts which make the learning more meaningful. For example, if a worker wants to review the procedure to operate a piece of equipment, the worker can scan a code on the machine or take a picture of the machine that will bring up the operating procedure right away. The information could be presented in the form of text, audio, or video depending on the worker background. For workers whose English is a second language (ESL), they may want to access a video demonstration of the procedure. Another example is if a worker is out in the field and has to complete a dangerous task, the worker can access safety procedures to complete the task using the mobile technology. Learning in context encourages problem-based learning since workers experience real job problems while on the job and they can access experts for coaching while solving problems. Also, because of the communication capabilities of mobile technology, workers can interact with each other to help each other and to share information.

Because of information explosion, where it is estimated that the amount of digital information increases tenfold every five years [17], workers have to be able to access up to date training and information to keep current on the job. With mobile technology, they can access information right away regardless of time zone or location. For example, if there is a change in safety regulations to perform a task, the worker can access the updated regulations right away rather than waiting for a printed version to be delivered to the job site. Learning materials will be available in electronic repositories which allow workers to access the materials as needed. Some organizations have remote offices which make it difficult for workers to travel to obtain training. The use of mobile learning allows workers to learn at their existing locations and in context.

Research shows that individuals approach the learning process with different learning styles where there are preferences for certain learning strategies over others [11,15]. Simon [16] used the Kolb Learning Style Inventory to determine trainees learning style and investigated different training methods with the different learning styles. Simon reported that trainees whose learning style matched the training method were more successful in training outcomes. Hence, it is important to make available a variety of training strategies to match the different learning styles of workers so that they can successfully achieve the training outcomes. At the same time, organizations must cater for the new generations of workers who are mobile technology literate and adapt to technology quickly. They will demand that training be delivered on emerging technologies and that there should be flexibility in learning.

3 Methods for Research

The research for this project used mixed methods where qualitative and quantitative data were collected. The qualitative data was obtained using a questionnaire that consisted of open ended questions to determine users' satisfaction. The quantitative data was obtained using pre-test and post-test performance data. The pre-test and post-test questions were parallel questions. The questions tested the same learning outcomes. Subjects were male workers in the oil and gas industry. A total of ten subjects completed the mobile learning lessons. Fifty percent of the subjects were over 50 years of age and the other fifty percent were below 50 years. The data obtained was analyzed by a research assistant who was not involved in the data collection.

The project team consisted of a team of experts including content experts, instructional designer, and mobile technology experts. The team developed training materials on presentation skills, delivered to workers using applications (Apps) on Android phones. Before the workers started the training lessons, they completed a pre-test to determine their prior knowledge. The pre-test consisted of multiple choice, completion, and matching questions (Figure 2). To complete the training, the subjects click on the App to access the lessons. Upon completion of the training, the subjects completed a post-test that paralleled the pre-test to determine how much they learned from the mobile learning lessons. The post-test also consisted of multiple choice, completion, and matching questions. This was followed by the completion of a questionnaire to determine the satisfaction of the workers with the mobile learning experience.

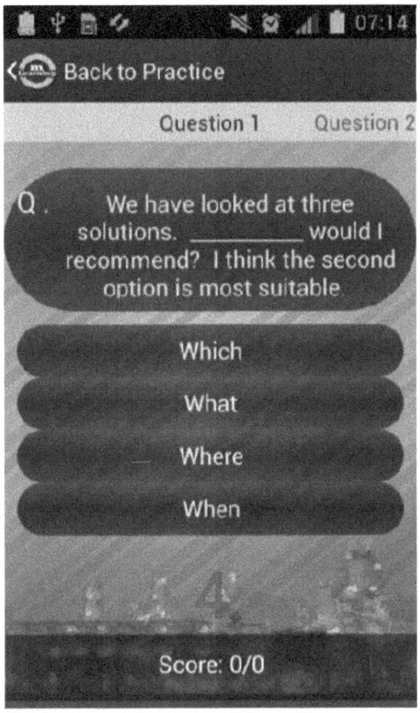

Fig. 2. Example of completion question on the post-test

4 Project Results

The average score on the pre-test was 64 percent and the average score on the post-test was 79 percent which indicated a gain of 15 percent. Ninety percent of workers improved their scores from pre-test to post-test (Figure 3). The result showed that although workers had a high prior knowledge of the lessons, they still gained from completing the mobile learning lessons. It is not surprising to see that some workers scored high on the pre-test since they may have done presentations on the job. Also, 60 percent of the workers had a university degree that may have provided the prior knowledge on presentation skills.

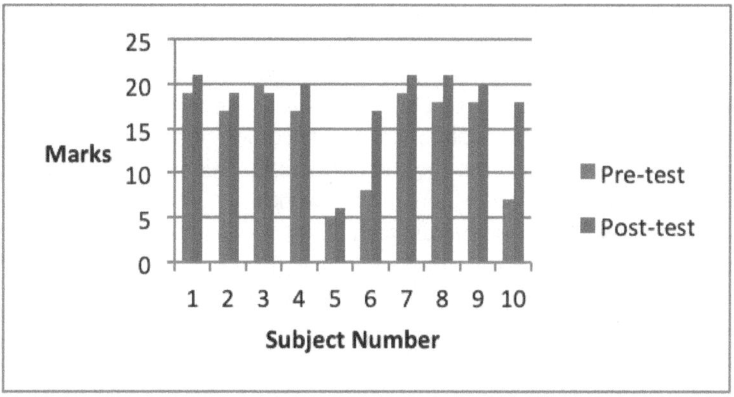

Fig. 3. Pre-test/Post-test comparison

When we asked workers about their experience using mobile phones, all indicated that they have used mobile phones before. Hence, lack of experience using mobile phones did not interfere with the workers completing the mobile learning lessons.

We asked workers why they decided to complete the presentation skills mobile learning lessons. Eight percent said that they are completing the lessons to acquire the skills for the job, ten percent said that is part of their development program, while ten percent said to acquire skills and for development. The workers were completing the presentation skills mobile lessons for their jobs, which is not surprising since communication skills is an important skill for the 21st century worker.

When asked whether mobile learning allows for anytime and anywhere learning, eighty percent of the workers strongly agree or agree, ten percent were neutral, and ten percent strongly disagree. This is the first time the workers are completing their training using mobile technology. It is encouraging to see that the majority of the workers in this study indicated that mobile learning provided flexibility in learning. When asked if they would like to take additional training with mobile technology, ninety percent strongly agreed or agreed while ten percent disagreed. Again, this is encouraging since this is the first time the workers are completing training using mobile technology.

5 Conclusion

The workers involved in this mobile learning research project were experiencing training with mobile technology for the first time. They reported that they like the flexibility that mobile learning provided learning at their own convenience. They said that mobile learning is a good method for workplace learning. The majority of research in the past was conducted in education. Since the use of mobile learning in training in organizations is in its infancy, more research should be conducted on the use of mobile learning in organization. The researchers involved in this research project will be conducing additional research to determine the effectiveness of mobile learning in the workplace.

As mobile technologies emerge and the new generations of workers enter the workforce, leading edge research should be conducted. For example, research is needed on the use of mobile sensor technology in mobile learning so that workers can learn in context and seamlessly (Jeng et al., 2010). Because workers are mobile in the workplace, it is important that mobile learning becomes pervasive where the workers can access learning materials from the environment in which they are located [18]. As the new generations of workers enter the workforce, more research should be conducted on the ways they use mobile technology for learning and how they interact while learning [4].

As information becomes digital, workers must be trained on how to access electronic information and training materials with technology. Mobile learning prepares workers for the digital world in the 21st century. There may not be a choice since all information in the future will be in digital format. Countries such as Qatar are moving in the right direction to prepare their people for the digital world.

Acknowledgment. This publication was made possible by NPRP Grant # 4-125-5-016 from the Qatar National Research Fund (a member of Qatar Foundation). The statements made herein are solely the responsibility of the authors. The authors thank Qatar University, Athabasca University and Qatar Petroleum and their associates who have provided facilities and staff to support this project. The authors also thank Martha Robinson and Abdulahi Mohamed who helped make this project possible.

References

1. Ally, M.: Using Learning Theories to Design Instruction for Mobile Learning Devices. In: Mobile Learning Anytime Everywhere, pp. 5–8. Learning and Skills Development Agency, London (2005)
2. Attewell, J.: Mobile technologies and learning: A technology update and m-learning project summary. Learning Skills and Development Agency, London (2005)
3. Brink, J.: M-Learning: The future of training technology. Training and Development, 27–29 (February 2011)
4. Chaudhuri, S., Ghosh, R.: Reverse Mentoring: A Social Exchange Tool for Keeping the Boomers Engaged and Millennials Committed. Human Resource Development Review 11(1), 55–76 (2012)
5. Clough, G.: Geolearners: Location-Based Informal Learning with Mobile and Social Technologies. IEEE Transactions on Learning Technologies 3(1) (2010)
6. ISO/IEC TS 29140-2. Information technology for learning, education and training - nomadicity and mobile technologies - part 2: Learner information model for mobile learning. International Organization for Standardization (ISO) (2011)
7. Jeng, Y.-L., Wu, T.-T., Huang, Y.-M., Tan, Q., Yang, S.J.H.: The Add-on Impact of Mobile Applications in Learning Strategies: A Review Study. Educational Technology & Society 13(3), 3–11 (2010)
8. Koole, M.: A model for framing mobile learning. In: Ally, M. (ed.) Mobile Learning: Transforming the Delivery of Education and Training, pp. 25–47. AU Press, Athabasca (2009)

9. Kukulska-Hulme, A.: Limelight on mobile learning: Integrating education and innovation. Harvard International Review, 12–16 (Spring 2013)
10. Leroux, P., Roobroeck, K., Dhoedt, B., Demeester, P., De Turck, F.: Mobile application usage prediction through context-based learning. Journal of Ambient Intelligence and Smart Environments 5, 213–235 (2013)
11. Lum, L., Bradley, P., Rasheed, N.: Accommodating Learning Styles in International Bridging Education Programs. Higher Education, Skills and Work-Based Learning 1(2), 147–168 (2011)
12. O'Malley, C., Vavoula, G., Glew, J.P., Taylor, J., Sharples, M.: Guidelines for Learning/Teaching/Tutoring in a Mobile Environment (2003), http://www.mobilearn.org/download/results/public_deliverables/MOBIlearn_D4.1_Final.pdf (accessed July 7, 2013)
13. Qatar National Vision 2030, http://www.gsdp.gov.qa/portal/page/portal/gsdp_en/qatar_national_vision/qnv_2030_document/QNV2030_English_v2.pdf (accessed July 10, 2013)
14. Samaka, M., Impagliazzo, J.: Developing a Platform for Mobile Learning Using mLearn. In: Proceedings of the International Conference on E-Technologies and Business on the Web (EBW 2013), Bangkok, Thailand, pp. 258–262 (2013)
15. Sarabdeen, J.: Learning Styles and Training Methods. Communications of the IBIMA, 1–9 (2013)
16. Simon, S.J.: The Relationship of Learning Style and Training Method to End-User Computer Satisfaction and Computer Use: A Structural Equation Model. Information Technology, Learning, and Performance Journal 18(1), 41–59 (2000)
17. The Economist: Data, data everywhere, pp.1–5 (February 25, 2010)
18. Vinu, P.V., Sherimon, P.C., Reshmy, K.: Towards pervasive mobile learning – the vision of 21st century. Procedia Social and Behavioral Sciences 15, 3067–3073 (2011)

A Vodcast Project in the Workplace

Understanding Students' Learning Processes Outside the Classroom

Laurel Evelyn Dyson

Technology and Education Design and Development Lab
University of Technology, Sydney, Australia
Laurel.E.Dyson@uts.edu.au

Abstract. A student-generated multimedia project was introduced into a first-year subject for undergraduate information technology students. Teams of students interviewed an IT professional in the workplace and video-recorded the interview. A survey of students showed statistically significant increases in students' self-reported knowledge and skills for key learning objectives. A subsequent qualitative analysis of student diaries and reflections discovered several contributing factors: the iterative nature of the activities that students undertook in order to complete the project; the multiple and evolving representations of their knowledge as they proceeded through the project; the importance of the workplace context in enhancing learning; and the affordance of the mobile devices used by the students for capturing this context on video and allowing it to be shared with other students. The research findings contribute to our understanding of how complex mobile learning projects in context-rich environments can contribute to deep learning.

Keywords: student-generated multimedia, vodcasts, video, deep learning, workplace learning, learning processes.

1 Introduction

Mobile devices – that is, portable digital devices – have been a key driver in enabling an increasingly active, learner-centred approach to education over the past decade. The affordance of these devices "as tools for complex and sustained tasks and problem solving" has been noted, as has their support for authentic learning activities, including data gathering in the field and the creation of multimedia content by students [1, p. 7]. Contextualization of learning outside the artificial environment of the classroom joins with multimedia content creation in powerful ways: "New relationships between context and production are emerging in that mobile devices not only enable the production of content but also of contexts. They position the user in new relationships with space, the physical world, and place, social space. [2, p. 23]"

This paper acknowledges this relationship between context and content through research into an assignment that required students to undertake a project in the field and create a record of their field learning in multimedia format using mobile devices.

M. Kalz et al. (Eds.): mLearn 2014, CCIS 479, pp. 258–271, 2014.

Thus context (the information technology (IT) industry workplace) and content (a vodcast, or short compressed video podcast, of an interview with an IT professional) were closely intertwined. The learners were first-year, first-semester students commencing an IT degree or IT/Business double degree. The principle learning objective was for students to acquire knowledge about the careers to which their degrees were leading them, and to enable them to make more informed choices about their majors. It had been recognized by the university that many of our students knew little about IT jobs or the IT industry, perhaps because of the extremely wide diversity of positions that IT graduates fill. The real-world context in which the learning experience was embedded was thus the world that the students would come to inhabit upon graduation.

The course in which the learning took place was a foundational communications subject that previously had focused on traditional written and oral communication skills. The introduction of the IT Careers Vodcast Project recognized that literacy practices had changed and now, in this "post-typographic world," included photography, video, sound and multimedia [3, p. 115]. Thus a secondary learning objective was the acquisition of multimedia communication skills. To make the vodcast, the students used sophisticated video and sound recording equipment provided by the university or the recording functions of their own mobile devices, such as smart phones or home digital video cameras. They worked for the most part independently although could attend an optional introductory video workshop and later in semester an editing workshop. Most chose not to participate in these but to work it out for themselves, sometimes by viewing videos they located on YouTube.

The aim of this paper is to explore the learning that occurred during this project and understand the processes that contributed to this learning. Firstly, two surveys, one before and one after the activity, revealed that significant learning took place for both key learning objectives (acquisition of IT careers knowledge and development of multimedia communication skills), based on students' perceptions. As a result of this, an exploratory qualitative study was undertaken in which student team diaries and end-of-semester reflections were examined to discover explanations for these excellent learning outcomes. This qualitative part of the study indicated several factors that contributed to learning, namely the iterative nature of the activities which students undertook to produce their vodcasts and their need to repeat many tasks more than once; the transformation of their increasing knowledge into different representational formats as they proceeded through the project; and the strongly motivating role of the workplace context and the power of video content taken on the mobile devices to capture this context and enhance learning.

The paper begins with a discussion of various studies and theoretical interpretations of the effectiveness of student-generated multimedia projects. This is followed by a description of the IT Careers Vodcast Project, and the methods used for analyzing learning and the processes which contributed to learning are described. The results are presented: firstly the statistical analysis of the student survey results and then the thematic analysis of selected student diaries, supported by selected comments from student reflections.

The paper contributes to our understanding of student-generated multimedia learning. In particular, it goes beyond existing research of the observed learning processes of small, in-class projects and provides insights into how teams of students learn when studying independently of their teacher, outside the classroom. It further demonstrates the potential of this learning approach even in very large courses. It is hoped that this example of a successful, sustainable project may inspire other academics to adopt this pedagogical approach.

2 Deep Learning in Student-Generated Multimedia Projects

Many educational benefits have been identified with projects in which students create multimedia content. One study of student-generated video demonstrated that students were more engaged, displayed greater autonomy in their learning, adopted a strong sense of self-esteem and ownership, and developed both oral and IT literacy skills [4]. Another study showed that the authentic voice of the students emerged as they expressed their creativity, sense of humour and individuality [5].

Student-generated multimedia projects represent a form of "learning by doing" [6]. Students take part in knowledge discovery as they plan and produce meaningful multimedia artefacts for sharing with their peers [7]. In order to do this successfully, they must have a deep understanding of the topic: to explain a concept one must know it [5], [8, 9]. Some authors go so far as to suggest that it is the very awareness of their peer audience that spurs students on to produce their best work in order to maintain the respect of their fellow students [10].

However, obviously the multimedia itself and the practice of producing it are highly motivating, too, particularly for students living in a world where the convergence of recording functions in mobile devices, combined with user-generated content and file-sharing platforms to host it, are the norm [11]. Multimedia projects recognize that students who have been exposed to technology for most of their lives require new pedagogical methods to engage them [12]. Users are now "actively engaged in shaping their own forms of individualised generation of contexts for learning" [2, p. 23]. Because these tasks are inherently motivating, students work hard and even poorer students often do well [13]. Highly motivated learners, such as those engaged in multimedia projects, typically adopt a deep approach to learning. Deep learners "understand ideas and seek meanings … have an intrinsic interest in the task and an expectation of enjoyment in carrying it out [14].

In order to understand more precisely why student-generated multimedia projects result in deep learning, various researchers have been investigating the learning processes in which students engage as they undertake these projects. Some researchers have focused on social knowledge-building: since these tasks are typically complex, they lend themselves to team work and team-based idea generation, negotiation of meaning, perspective taking and group problem solving [15]. Metacognitive processes are involved, that is, "strategies the learner uses to achieve specific learning goals, like planning and organising, allocating attention to relevant and irrelevant factors, looking for relationships and patterns, monitoring comprehension, identifying and testing procedures, evaluating outcomes, and reflecting on learning" [16, p. 35].

Other research focuses on the evolving representational forms in which students' knowledge is expressed at each stage of the learning process. Studying trainee teachers in a classroom making "slowmations" (a type of simple, stop-go animation), researchers observed students moving through a predetermined sequence of five stages or activities which followed a digital storytelling procedure: 1) Researching the topic; 2) Planning, including creating a storyboard; 3) Constructing models; 4) Manipulating and photographing the models; and 5) Editing the photographs and adding text and voice to create the slowmation [9], [17]. Produced at each stage was a particular representation of the concept being studied, respectively: 1) Research notes; 2) Storyboard; 3) Models of the concept; 4) Photographs of models; and 5) Narrated animation (slowmation). In "creating a representation, students make meaning as they are thinking about the relationship between what they are making (the 'representation') and the concept or object they are trying to represent" [9]. Each representation has an affordance, or role, that makes students consider the topic in a particular way, depending on its mode of representation – whether notes, diagrams, photographs or the finished multimedia product. Meaning is "multiplied" as it is transferred from one representational mode to another, resulting in deep learning. The progression is not strictly linear as students often return to previous representations to check their knowledge before moving forward again to the next stage.

The research approach adopted in the current study aimed at discovering what learning processes students were adopting in this self-directed project, rather than imposing a predetermined learning strategy as [9], [17] had done. Because the learning process was largely determined by the IT student teams, the artefacts or representations of knowledge at each stage of learning were also open to discovery. A further point of difference from both [16] and [17] was the field-based aspect of the IT Careers Vodcast Project, which meant that learning processes were highly contextualized, and thus understanding the role of context in the learning process, and the role of mobile technology in facilitating learning in the field, was an additional important line of inquiry. Since students were mobile and could not be observed directly, a different approach to capturing data needed to be adopted.

3 Design and Implementation of the IT Careers Vodcast Project

Prior to the launch of the IT Careers Vodcast Project an assignment had been trialled whereby students researched IT careers in teams of nine students each. One student from each team went to the workplace to interview an IT professional and reported back their findings to their team, who prepared a written report for their tutor. The teams then gave a presentation to all the remaining students enrolled in the subject, some 300. The assignment was seen as valuable but there were several issues: only one in nine students enjoyed the benefit of meeting someone working in IT and visiting their workplace; the teams were too large and "loafing" occurred, with most work being done by the student who had conducted the interview; and organizing the mass presentation for sharing the results with 300 students was logistically difficult.

Moreover, there was no explicit linking of the careers being researched to the majors and study paths that students should follow in order to prepare for a particular career in IT.

A redesign of the assignment focused on addressing these issues (Figure 1). Firstly it was decided that all students should have the valuable experience of meeting the professional and seeing what the IT workplace was like. Secondly, teams were reduced to occasionally three but mostly four students each. Thirdly, to facilitate sharing of the interview results, mobile technology was introduced: students were given the brief to video-record their interviews, show an edited recording to their tutorial class of 30 students as part of a formal presentation towards the end of semester, and then further edit the video to create a 5-8 minute vodcast for upload to the learning management system (LMS) where all students enrolled in the subject could access it. The assignment shifted from a traditional fieldwork activity, in which oral and written communication skills were developed, to mobile-supported fieldwork, in which students built their oral, written *and* multimedia communication skills to create a sharable and engaging multimedia product. The vodcasts could then form part of a growing repository of IT career resources while at the same time, as externalized representations of student understanding [18], they could promote learning conversations amongst the students as well as between teacher and students during the tutorial presentations.

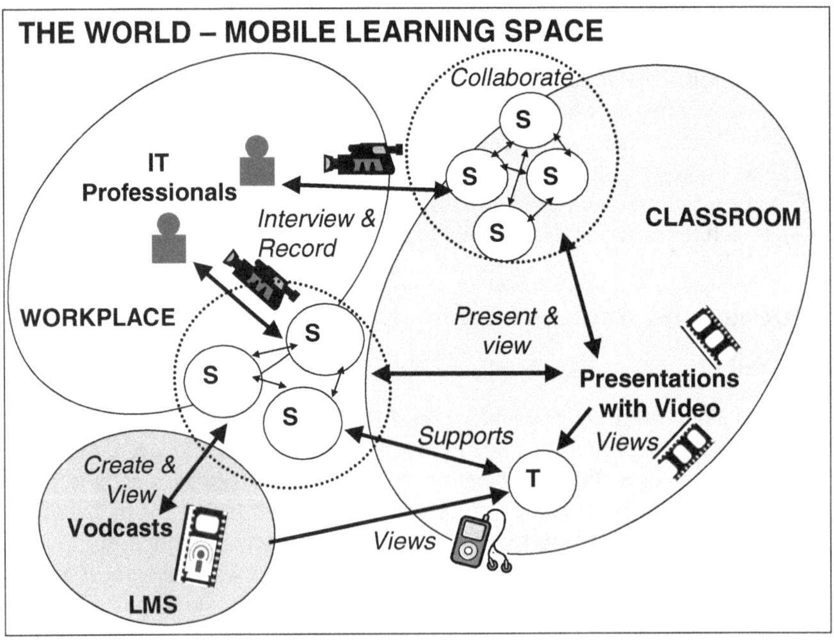

Fig. 1. Dynamics of the IT Careers Vodcast Project (T – teacher; S – student; LMS – Learning Management System)

Finally, in order for students to understand how the careers they had investigated related to the different IT majors offered by the university, the tutors gave students a short presentation about the content of the majors in the last week of semester and students then mapped all the careers they and their peers had investigated to the majors. That concluded the IT Careers Vodcast Project for the students. Students undertook the assignment over a 10 week period, beginning in the fourth week of semester and finishing in the final teaching week.

The redesigned assignment was first implemented in the second semester of the Australian academic year when enrolments following the mid-year intake are typically smaller for this subject (96 students in this case), thus making the project more manageable in its first iteration. When it was found to be successful, some small changes were made before offering it on a continuing basis in the course. To gain some idea of the scale, enrolments typically range from 320-430 in the larger semester to 95-130 in the smaller semester.

4 Research Methods

4.1 Quantitative Analysis of Students' Learning

To evaluate students' learning in the vodcast assignment the second semester it was offered, students were surveyed anonymously about their knowledge of IT careers, the major which they intended to follow to achieve their chosen career, and their multimedia skills. The survey consisted of three statements about careers and two questions about multimedia skills which they rated using 5-point Likert scales: 1) I know what career I would like to follow at the end of this degree; 2) I am well informed about different IT career options; 3) I know what area of study or IT major I want to undertake in this degree; 4) How do you rate your knowledge and skill of making video recordings; 5) How do you rate your knowledge and skill of video and/or audio editing?

The same survey was presented at the beginning of semester ($n = 316$; response rate = 92%) as at the end of semester ($n = 275$; response rate = 80% for all items except that one student failed to answer Q 3, thus $n = 274$ for that item). This provided a measure of changes in their self-perceptions over the course. As no other teaching about IT careers had been provided, the evaluation is a valid measure of their perceived changes in knowledge, i.e., learning, while undertaking the various activities associated with the project (making a vodcast, listening to their peers present their careers findings, watching other vodcasts, and mapping careers to majors). The means for each question were compared (end versus beginning of semester) using a 1-tailed independent samples t test and SPSS software. Since the samples are large, the assumptions of the t test regarding continuity and normality of distribution hold [19].

4.2 Qualitative Analysis of Students' Learning Processes

Following the findings from the student surveys, research was conducted the subsequent year to discover why the IT Careers Vodcast Project had been so effective in teaching the students about careers. As the students completed their work by themselves and away from the classroom (unlike [4], [7], [9]), their learning processes could not be directly observed and another method had to be adopted. The new cohort of students was asked to report on their learning by means of a weekly team Diary and end-of-semester team Reflection. The hardcopy Diary and Reflection was submitted by each team of students at the end of semester and a random selection was then entered into a spreadsheet, totalling 45 diaries in all (almost half the diaries submitted for the subject that semester).

In the Diary, students recorded their activities during the project, how they organized their work and how they overcame any obstacles. A further random selection of 16 diaries was made from the initial 45 and a thematic analysis was conducted of students' entries over the 10 weeks of the project: the information was sorted into themes based on student activities, with the resulting artefacts or representations also recorded or inferred, if necessary, from the stated activities. The thematic analysis was conducted by a research assistant and the author, with cross-checking to minimize subjectivity and improve reliability of themes identified. This analysis revealed the key activities undertaken by students, the artefacts they produced, and the way in which their knowledge was represented and transformed as the project progresses.

Following this analysis, the 45 Reflections were examined for insight into how mobile learning contributed to students' learning, specifically the impact of the workplace context and the contribution of mobile technology in capturing students' knowledge of the careers studied and sharing this knowledge with other students. Information on these aspects of the project was gathered from students' comments in response to the Reflection questions, "What were your key learnings about IT Careers? How did the Careers Project contribute to your learning about IT careers?" and "What were your most interesting experiences during the Careers project?"

5 Achievement of Learning Outcomes

The results of the *t* test comparing students' perception of their knowledge before and after undertaking the IT Careers Vodcast Project are presented in Table 1. It can be seen that, for all items rated by students, the mean rating at the end of the semester is significantly higher than at the beginning of semester: significance is at least at the 0.005 level. For items 2-4, the significance is in fact higher, at the 0.001 level.

Thus, according to students' self-evaluation of their knowledge about IT careers and their multimedia recording and editing knowledge or skills, the project was extremely successful. The two main learning objectives of the project were realized, based on students' self-reporting.

Table 1. Students' Self-Reported Learning

Statement/Question	Suvey	Mean Score (out of 5)	Analysis using 1-tailed *t* test		
			t (*df*)	p	*Significance Level*
1. IT career to be followed	Post	3.65	2.865 (589)	0.002	$p < 0.005$
	Pre	3.43			
2. IT career options	Post	3.95	10.007 (589)	0.000	$p < 0.001$
	Pre	3.29			
3. IT major	Post	3.83	3.995 (588)	0.000	$p < 0.001$
	Pre	3.49			
4. Video recording skills	Post	3.67	3.419 (589)	0.0005	$p < 0.001$
	Pre	3.40			
5. Video and/or audio editing skills	Post	3.58	4.600 (589)	0.000	$p < 0.001$
	Pre	3.19			

6 Students' Learning Processes

6.1 Iteration of Learning Activities

The results of the thematic analysis of students' team diaries are given in Table 2. The main activities appear in the order in which most teams recorded their commencement of the associated tasks (for example, all but one team stated that they undertook team organization activities right from the beginning). However, main activities often overlap, with teams inevitably commencing preparation for their video before the interview, and the filming of video footage taking place at the interview. Not all students recorded all activities or tasks listed: for example, only 8 out of 16 teams explicitly noted that they had performed background research; while 4 teams had set up a Facebook page and 1 team used Google Docs to manage team discussions and contributions; and only 4 teams stated that they had used brainstorming as a creative thinking technique. A minority of teams chose to edit their video straight into the vodcast format and show this at their class presentation instead of a less finished video.

Some of the students' activities are of a pragmatic nature, particularly Team Organization and Project Planning. Though developing skills in these areas was not the main aim of the assignment, they are highly valuable and, indeed, necessary skills for any IT graduate to possess. Thus the opportunity for students to practice and learn these skills, often by trial and error, was a valuable part of the project. Students' comments in their diaries sometimes show their awareness of this, as was the case with one team: "[Week 4] People don't get along well. - Still figure out how to make it better. … [Week 5] Try to strengthen the relationship (first time, so need more time to work in a team). - Go out for drinks. [Week 7] Conflict of time between members - negotiate between members - Compare timetable then finally have a good time."

Table 2. Key Project Activities and Outputs

Main Activities	Tasks	Artefacts or Outputs
Team Organization	Form team Exchange contact details Email or message team members Assign roles and allocate work Set up Facebook page, Google Docs, etc. Build relationships Establish rules for team operation	Contact list of members Emails, messages Agreed roles Online collaborative space Team rules
Project Planning	Brainstorming Plan project Discuss/choose career/IT professional Contact/email IT professional Make appointment	Ideas for proceeding with project Project plan Choice of career/ IT professional Contact details Appointment
Background Research	Research an IT career Research the IT professional, and his/her organization Incorporate into interview questions, video or presentation	Research notes or downloads
Interview	Plan interview; interview techniques; roles of team Draft and refine interview questions Practise interview Conduct interview Inspect workplace	Interview plan Interview questions (draft, final) Interview answers Observations of workplace
Video Production	Book/choose video equipment Learn how to make a video, test equipment or practise View past examples of vodcasts Design video (length, look, etc.) Record interview Develop script Locate editing software Edit video Edit audio track Incorporate video into presentation	Equipment booking Video design Rough interview footage Script Video (draft, final)
Presentation	Plan presentation Create presentation slides Script presentation Assign roles Rehearse presentation Present slides and video or vodcast to class Watch other groups' present and comment	Presentation outline/plan Presentation slides (draft, final) Rehearsal Oral and multimedia presentation
Vodcast Production	Locate editing or file conversion software Compress/shorten video to make vodcast Upload vodcast to LMS View other teams' vodcasts	Vodcast

One of the most obvious features of the project was the large number of tasks that student teams needed to undertake in order to complete the project successfully. In addition, many of the diaries showed how teams had to revisit tasks more than once, with students creating drafts before the final artefact was arrived at. This was very noticeable with the development of interview questions: for example, one team had a delay in being allocated their interviewee: "[Week 6] Each team member individually wrote a few questions for generic IT professional as we wait for [the university] to assign the group a professional. … [Week 7] assigned an IT professional to our group, we then wrote further questions pertaining to his specific field, being Software Development and 3D Effects. [Week 8] The group meet on campus to choose a final list of questions from a bank of questions we individually formulated." This reiterative nature of some of the tasks can reasonably be interpreted as contributing to the outstanding learning outcomes from the project.

Most teams encountered technical difficulties, often due to poor planning, for example not testing or practising with the video equipment before arriving in the IT professional's workplace for the interview, or not participating in the video workshop that was offered to all students on a voluntary basis (only one team records that they sent a team member to the workshop, who then reported back to his team on what he had learnt). This meant that a lot more work had to be done during the editing phase. Few students seemed to have had video experience but were prepared to work it out, for example: "[Week 8] The microphone given wasn't working so we had to use the default camera mic. - We learned that we should have checked the equipment before using it. [Break Week] Edited a bit of the vdieo interview, but not yet finished. The raw format from the camcorder wasn't compatible with the video editing software. - Found a program and learned how to convert it to a recognisable format e.g. avi. [Week 9] Finished off the editing of the video. [Week 11] Rewatched video to see if there are anymore improvements to be made. - Checked to see if there is anything wrong. [Week 12] Tried to fix the audio of the video. Audio was difficult to fix". However, these students found the video-making the most engaging aspect of the assignment. Some teams turned their technical problems to advantage, as the team who had to go back to the workplace to conduct and record the interview a second time: "With the second interview, we used multiple angles and it can be considered new skills have been learnt." This team actually went to the IT professional's workplace three times in total as, the week after the initial interview, they went to take "photos, and additional videos for the vodcast and meet other staffs." This supports the view that repeated attempts and more time on the tasks probably contributed to deeper learning. Certainly, changes in students' perception of their knowledge and skills of video recording and editing over the course of the project were highly significant (Table 1).

6.2 Evolving Representations of Student Knowledge

In addition to the practice that students received through their oft repeated attempts at tasks – the successive drafting of their interview questions and presentation slides, and their planning and design of various aspects of the project – there is also evidence of an evolution in the way that their knowledge about the IT career was presented. Extracting some of the main representations from the multitude of artefacts shown in Table 2, and concentrating on the five main activities that contributed directly to their learning about IT careers and development of multimedia skills, one can see a variety of representational formats, or modes, and an increasingly sophisticated expression of the students' knowledge as they progress from their choice of career and interviewee, to the interview (oral and audio) and workplace visit (visual), to the video production (video), class presentation (oral, multimedia), and finally to the production of the vodcast (Table 3). It was not obvious from the diaries, but was very clear from viewing the students' vodcasts, that these were highly crafted representations, with "intros," "outros," titles, captions, background music, humour, and special effects. This diversity of representations and the multiplication of meaning contained within this evolving representation of careers knowledge may well contribute to the students' achievement of deep learning, as put forward by [17].

Table 3. Representations of Careers and Multimedia Knowledge

Main Activities	Major Representations
Project Planning	Choice of career/ IT professional
Background Research	Research notes or downloads
Interview	Interview questions and answers Workplace observations
Video Production	Video footage
Presentation	Presentation slides Oral and multimedia presentation
Vodcast Production	Vodcast

6.3 Context and Content in Mobile Learning

Students' group reflections on "What were your key learnings about IT Careers? How did the Careers Project contribute to your learning about IT careers?" often showed that they appreciated the context of the learning activity. Students reported that the project allowed them to learn aspects of a career that would probably not be part of any textbook or website job description, such as dress codes, behaviour in the workplace, workplace culture, the work environment, and how job structures varied between small and large firms. As one team wrote, "Learnt about daily life – couldn't be googled! – More direct way of learning – great experience."

All but 2 of the 45 reflections examined stated that it was a highly motivating way of learning when responding to the question, "What were your most interesting experiences during the Careers project?" A major factor was the workplace learning context: for example, the "most interesting experience during this project was to go to an IT professionals workplace and gain a first hand experience of their work environment and see what they do on a day-to-day basis"; and "The interview was the highlight of this whole experience. To see someone talk so enthusiastically about IT was very uplifting and made me happy to be doing IT."

One question that must be asked is whether students learnt from watching the videos shown in the class presentations and the vodcasts viewable from the LMS, or was the learning chiefly related to their own experience in the field? In other words, did the affordance of the mobile devices (video cameras or mobile phones) for capturing and sharing information about the career contribute to the learning outcomes? Though for most students the visit to the IT professional's workplace was a compelling experience, a number of students' reflections noted that they also learnt from their peers and this allowed them to learn about more than just the one career they had investigated themselves. For example, one team wrote that, "We were unaware of many of the IT jobs that had been studied. It always looked like a great workplace environment and almost all of the jobs had opportunity for advancement"; while another team stated that their most interesting experience in the project was "Watching other people's vodcasts and learning from them." Thus, the significant increases in students' learning of careers and career options, as reported earlier, did not just derive from the single encounter with one IT professional. The video content captured the context of the workplace-based interview and conveyed this to other students.

7 Conclusion

The IT Careers Vodcast Project has been extremely successful in giving IT students the opportunity to become better informed about careers in their field of future employment and in developing the multimedia communication skills they need increasingly today. In fact, students' self-reported learning on these two key learning objectives has been shown to be statistically significant. The project has moreover been an excellent vehicle for students to develop soft skills, such as teamwork and project management.

Despite the difficulty of assessing mobile learning which is undertaken by students in the field, independently of their teacher, a method was adopted using students' diaries and reflections to understand the learning processes and why such good learning outcomes had been achieved. This revealed three main contributing factors. Firstly, a practice and "time spent" effect, through reiteration of tasks and the large amount of effort and time dedicated to achieving the desired goals. This was particularly noticeable in students' development of multimedia skills. Secondly, there was a progression in the way students' knowledge of the career was represented, and these representations showed an increasing sophistication in terms of the multimedia format [17]. Thirdly, the context of learning was extremely motivating to most students, and this was enhanced by the ability of the video taken by the mobile devices to communicate the workplace context in a much more powerful way than is usually possible with more traditional media such as text.

There may well be other features of the assignment which contributed to learning but for which there is insufficient space to elaborate. For example, there is an ample body of literature detailing the power of collaborative learning dating back at least to the 1980s [20], and teamwork was certainly a prominent feature of the assignment, and one which many students enjoyed. In addition, the efficacy of project-based learning has been well documented [21] and is an example of that "learning by doing" referred to earlier, which leads to knowledge discovery and deep learning. More research is planned in the coming semester with postgraduate students undertaking a similar style of project on another topic. This should clarify further how students learn in these student-generated, multimedia, mobile learning projects.

Acknowledgments. Thanks are due to Marijke Wright of the University Careers Service for being the first to suggest that our information technology students would benefit from a careers project in which students met IT professionals in the workplace, and then for her continued and dedicated support for its implementation. The expertise of Andrew Litchfield, who provided multimedia advice, and Sophie Abels and Barbara Courtille, who assisted with data processing and analysis, is also gratefully acknowledged. Finally, the reviewing of a draft of the paper by colleagues Jessica Frawley and Babak Abedin and their helpful suggestions for improvement are acknowledged.

References

1. Herrington, A., Herrington, J.: Authentic Mobile Learning in Higher Education. In: Proc. Australian Association for Research in Education (AARE) 2007 Conference, Fremantle, pp. 1–9 (2007)
2. Pachler, N., Bachmair, B., Cook, J.: Mobile Learning: Structures, Agency, Practices. Springer, New York (2010)
3. Davies, J.P.: DOA: Educaiton in the Electronic Culture. Scarecrow Press, Lanham (2003)
4. Schuck, S., Kearney, M.: Capturing Learning through Student-Generated Digital Video. Australian Educational Computing 21(1), 15–20 (2006)
5. Frydenberg, M.: Principles and Pedagogy: The Two P's of Podcasting in the Information Technology Classroom. In: Proc. ISECON, Champagne, vol. 23, pp. 1–10 (2006)
6. Dewey, J.: Experience and Education. Touchstone Books, New York (1938)
7. Rocha, A.M.M., Coutinho, C.P.: Web 2.0 Tools in High Schools in Portugal: Creating Screencasts and Vodcasts for Learning. US-China Education Review A 1, 54–62 (2011)
8. McGarr, O.: A Review of Podcasting in Higher Education: Its Influence on the Traditional Lecture. Australasian Journal of Educational Technology 25(3), 309–321 (2009)
9. Hoban, G., Nielsen, W.: The 5Rs: A New Teaching Approach to Encourge Slowmations of Science Concepts. Teaching Science (September 2010)
10. Wheeler, S., Yeomans, P., Wheeler, D.: The Good, the Bad and the Wiki: Evaluating Student-Generated Content for Collaborative Learning. British Journal of Educational Technology 39(6), 987–995 (2008)
11. Dyson, L.E.: Student-Generated Mobile Learning: A Shift in the Educational Paradigm for the 21st Century. anzMLearn Transactions on Mobile Learning 1, 15–19 (2012)
12. Tapscott, D.: Growing up Digital: The Rise of the Net Generation. McGraw-Hill, New York (1998)
13. Wikan, G., Mølster, T., Faugli, B., Hope, R.: Digital Multimodal Texts and Their Role in Project Work: Opportunities and Dilemmas. Technology, Pedagogy and Education 19(2), 225–235 (2010)
14. Prosser, M., Trigwell, K.: Understanding Learning and Teaching: The Experience in Higher Education. Open University Press, Buckingham (1999)
15. Lee, M.J., McLoughlin, C., Chan, A.: Talk the Talk: Learner-Generated Podcasts as Catalysts for Knowledge Creation. British Journal of Educational Technology 39(3), 501–521 (2008)
16. McLoughlin, C., Lee, M.J.W., Chan, A.: Using Student Generated Podcasts to Foster Reflection and Metacognition. Australian Educational Computing 21(2), 34–40 (2006)

17. Hoban, G., Loughran, J., Nielsen, W.: Slowmation: Preservice Elementary Teachers Representing Science Knowledge through Creating Multimodal Digital Animations. Journal of Research in Science Teaching, 1–25 (2011)
18. Sharples, M.: Disruptive Devices: Mobile Technology for Conversational Learning. International Journal of Continuing Engineering Education and Lifelong Learning 12(5/6), 504–520 (2003)
19. Hinton, P.R.: Statistics Explained, 2nd edn. Routledge, London (2004)
20. Webb, N.M.: Information Processing Approaches to Collaborative Learning. In: Hmelo-Silver, C.E., Chinn, C.A., Chan, C.K.K., O'Donnell, A.M. (eds.) The International Handbook of Collaborative Learning, pp. 19–40 (2013)
21. Thomas, J.W.: A Review of Research on Project-Based Learning. Autodesk Foundation, San Rafael (2000)

M-Workplace Learning @ ITC-ILO

Design Thinking for Mobile Learning Interventions at Work

Alessia Messuti[1,2], Tom Wambeke[1], Marco Kalz[3], and Barbara Bruschi[2]

[1] International Training Centre of the International Labour Organization, Turin, Italy
[2] Università degli Studi di Torino, Dipartimento di Filosofia e Scienze dell'Educazione, Turin, Italy
[3] Open University of the Netherlands, Welten Institute, Heerlen, The Netherlands
{a.messuti,t.wambeke}@itcilo.org, marco.kalz@ou.nl, barbara.bruschi@unito.it

Abstract. The paper describes an applied research project intended to facilitate the learning at/for work of learning and development professionals through mobile learning. In particular, the audience was narrowed to professionals of the International Training Centre of the International Labour Organization, a UN agency based in Turin, Italy. The methodology adopted moved beyond the traditional instructional design towards design thinking, which takes into account five phases (discovery, interpretation, ideation, experimentation and evolution) that help identifying a concrete challenge until building a practical solution. The results covered in this paper represent the first three phases of the design thinking process: discovery, interpretation and ideation. Building on these findings, three workplace learning challenges were identified to be solved through mobile learning: improving knowledge sharing and effective communication at work, facilitating appropriation of learning experiences at work, increasing motivation towards learning at work through game concepts.

Keywords: adult learning, mobile learning, workplace learning, continuous learning, informal learning, mobile methods, mobile tools, design thinking.

1 Introduction

Workplace learning has become increasingly important in the development of skills in a time of increasing competition as it ensures that workers' skills are constantly renewed and adapted. However, in such context it is difficult to find suitable time for attending formal training sessions and workplace-learning initiatives are often geared towards providing quick access to content-driven instruction. As a result, professional staff are relying more and more on informal initiatives where learning happens at the moment of need.

This paper describes an applied research project intended to facilitate the learning at/for work of learning and development professionals through mobile learning strategies. In particular, the project focus is to seek whether the use of mobile methods and tools can support social and informal learning at the workplace. The audience of

M. Kalz et al. (Eds.): mLearn 2014, CCIS 479, pp. 272–286, 2014.

the research project was narrowed to learning and development professionals working for the training center of the International Labour Organization (ITC-ILO), but we expect that results of this project can inform other organizations.

The International Labour Organization (ILO) is the oldest UN agency, working towards the goal of decent work for all. The unique tripartite structure of the ILO gives an equal voice to workers, employers and governments to ensure that the views of the social partners are closely reflected in labour standards and in shaping policies and programs. In 1964 the ILO identified the need to train people on how to put its values into practice in a real world context, and so it created the International Training Centre (ITC-ILO) in Turin, Italy as its vocational training institute, the only one of its kind in the world. The centre's mission is to be the leading global provider of learning and training for the world of work.

The centre's learning, knowledge sharing and capacity-building activities and programs for governments, workers and employers' organizations are focused in the fields of: rights at work, enterprise, employment and skills development, social protection, social dialogue, labour law and labour administration, workers' and employers' organizations, gender and non-discrimination, sustainable development and governance. With 14.000 participants each year from 192 countries, around half of the centre's work takes place at Turin campus in Italy, while the rest of its training activities take place on the field in developing regions [1]. As a result, ITC-ILO staff is usually traveling to deliver training across the globe, making it more and more difficult for them to attend formal training sessions at the workplace.

Therefore the macro objective of the "M-Workplace Learning @ ITC-ILO" research is to identify effective training modalities that suit the needs of a specific group of workers. The overall project was informed by previous experiences and projects conducted since 2011 by the ITC-ILO which pointed out the potential of mobile methodologies as a means to facilitate learning and knowledge sharing with ITC-ILO beneficiaries, as well as to allow the extended participation of staff in professional, work-based communities [2-3].

2 Methodology

The methodological approach adopted for this research progressively moved beyond the traditional instructional design towards design thinking methodologies. Design thinking is a structured approach to generate ideas and provide innovative solutions to practical needs [4]. It is a human-centred, collaborative and experimental approach, which begins by understanding the needs and motivation of people, translating them into frameworks and opportunities, solutions and prototypes. However design thinking was originally meant for big corporations, it is possible to use this approach to address any challenge, especially those that focus on the design and development of learning experiences. The approach takes into account five phases that help identifying a concrete challenge until finding and building a practical solution:

- Discovery: framing a practical challenge and defining the audience;
- Interpretation: reviewing the history of the challenge, find common themes and translate them into actionable statements;

- Ideation: generating as many ideas as possible to serve the identified needs
- Experimentation: creating working prototypes
- Evolution: gathering feedback, determine if the solution met the goals, discuss what could be improved, measure success and document the process [4]

The content of this paper includes results from the first three phases of the design thinking process: discovery, interpretation and ideation. In the next sections we discuss the approach taken into detail.

2.1 Discovery

As previously mentioned, the macro objective of the research project is to identify effective learning and training modalities that suit the needs of ITC-ILO learning and development professionals. In order to define the context of the research, an extensive research was conducted between November 2013 and February 2014, including a study of a comprehensive learning needs analysis of staff, a desk review and a series of interviews with learning and development professionals working for different departments. A comprehensive learning needs analysis was conducted at ITC-ILO in 2013 consisting out of three different phases: an online survey to assess individual learning needs, several team meetings to assess team learning needs and focus groups to gather important elements for the organizational learning strategy. The analysis overall objective was to build a learning culture at ITC-ILO workplace settings.

After having analysed data and comments, the first noteworthy fact that emerged from the survey was that 35% of staff did not engage in any form of training in 2012; and almost 50% of staff did not attend any language training even if the staff generally perceives language competencies as essential to the work at the ITC-ILO. Generally speaking, staff indicated that more investment in alternative learning modalities other than formal training would have been welcome such as: refresher courses, more career development opportunities, more e-Learning opportunities, more informal lunchtime sessions, more staff exchanges among sectors.

A qualitative literature review was done to gain an in-depth-understanding of the "mobile learning" and "workplace learning" concepts, and to identify good practices of using mobile technology for continuous learning at work. The qualitative review was done by using different channels such as academic social networks (Mendeley and Academia.edu), content curation tools (Scoop.it), books and manuals.

As a result, mobile workplace learning has been framed according to the definition by Pimmer & Pachler: "the processes of coming to know, and of being able to operate successfully in, and across, new and ever changing contexts, including learning for, at and through work, by means of mobile devices" [5]. The definition highlights the importance of mobile learning for knowledge application rather than for knowledge acquisition by stressing the practical aspect of being able to operate successfully across contexts. In addition to this, a review of the five moments of learning need introduced by Gottfredson and Mosher [6] reveals that only three moments are most appropriate for mobile learning: 1) when trying to remember, 2) when things change, 3) when something goes wrong. This implies that mobile learning is most suitable for performance support rather than being aligned with more traditional modes of learning.

A qualitative field study was conducted to gain an in-depth understanding of (potential) mobile learning practices at ITC-ILO. For this purpose, 26 staff members (42% men and 58% women), including senior managers, activity managers, programme assistants, and interns from different departments were interviewed. Visual facilitation techniques were used to facilitate the discussion around the topics of workplace learning, motivation towards training at work, mobile behaviours and interest in connecting mobile technology with learning experiences.

A visual template was used to engage staff during the interviews by writing or drawing. Interviewees were firstly asked to write down their exact job title and then, to reflect on how they could describe it in other words (e.g. what does "activity manager" means to you?). This question was useful to frame the roles, activities, motivations and frustrations of people towards workplace settings. Later on, interviewees were asked to think if attending training was also part of their work life and encouraged to translate their work-related tasks in images by using selected Power Point ClipArts. The result was a collage where the size of images was proportional to the relevance of the tasks performed on a regular basis at the workplace. Staff development training was often present in small percentages (see fig. 1). These collages allowed the visualization of interviewees' work life and made them reflect on the reasons why training is not so relevant, or on the fact that despite their job title, a big portion of their work life is dedicated to time-consuming activities such as email management.

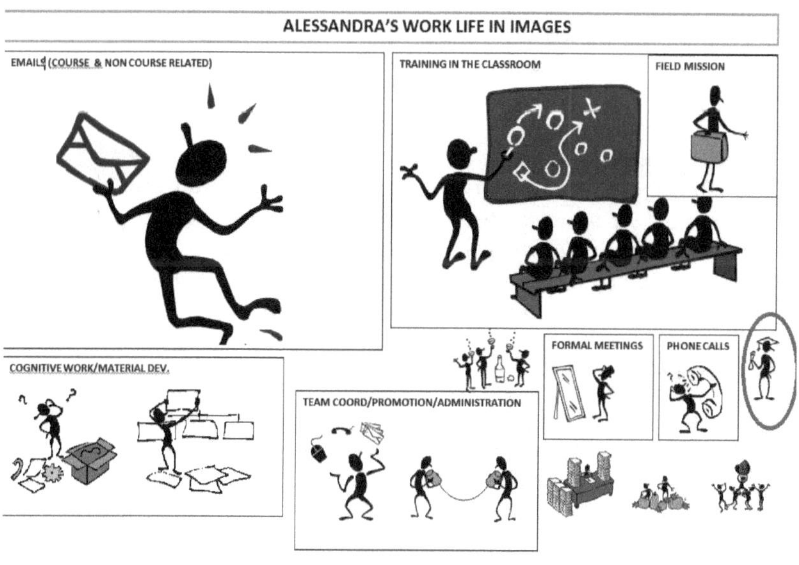

Fig. 1. Visual representation of work-related tasks where the percentage of staff development activities is rather low

Generally, interviewed staff stressed the fact that learning at work mostly happens through sharing with others, through targeted interaction (individual meetings) and systematic follow-up actions rather than only relying on formal learning interventions. Limited attendance to staff development activities was often connected to time constraints and busy schedules. Despite different learning preferences, they overall agreed on the fact that work-related training should be short in duration but regular, inspirational and collaborative.

Finally, interviewees were asked to give a definition for mobile. Generally speaking, the concept of mobile was associated with a device (mobile phones, tablets, laptops, cameras etc.) and it was difficult to perceive it as a standing alone concept. Although when pushed to think beyond the tool, mobile was overall associated to easy life and support, accessibility, flexibility, new generation, future but also chaos. Interviewees were also encouraged to reflect on mobile technology in terms of applications e.g. social media. The most popular ones were Facebook, Twitter, Linkedin, Google Drive and Google Calendar, Dropbox, YouTube, Skype. In these terms, mobile meant better communication and organization, accessible information, and entertainment but also time consuming aspects and privacy concerns.

"*I am concerned about privacy: which kind of documents may I put on the cloud?*"

"*I already use my mobile to listen to podcasts of Spanish news. Mobile learning could be useful to me to improve my social media skills*"

"*I am not a very technological person. Technology is time consuming and I prefer speaking rather than writing*"

In order to measure mobile behaviours at the workplace, interviewees were asked to match icons of devices to work-related tasks visualized on the collage (see fig. 2). This reflection helped understand how mobile is already embedded in their working life and encouraged the reflection on how can it be connected to staff development activities as well.

"*In the field I have a lot of dead moments in between assignments. If I am for example in a boring context, I could have easily used my time for training.*"

"*I would have done the security in the field training more eagerly at home on my Ipad instead of staying in the office until late*"

"*I would have appreciated a short assignment after some of the F2F trainings I took in the past, e.g. an sms interaction to gather feedback*"

2.2 Interpretation

Building on the discovery findings, the interpretation phase consisted in finding main themes and translating them into actionable statements. Mobile learning has to be "on the spot" meaning that it has to be framed in a specific context, a concept not acknowledged by "anywhere, anytime". According to Parsons [8], it has to be taken into account that individuals are not always able to learn and that they rarely learn while physically moving. But it is true that they tend to take their learning tools with them to the most appropriate places. Therefore, it is important to understand how it will be possible to strategically use already available learning tools and integrate them into workplace learning initiatives.

Fig. 2. To measure mobile behaviours, interviewees were asked to match icons of devices to work-related tasks and encouraged to reflect on how mobile can be connected to staff development activities

In order to understand to what extent mobile methods and tools can be integrated into ITC-ILO staff development activities, the adapted version of Engerstrom's framework [9] for analysing mobile learning was taken into account as it describes the dialectical relationship between technology and learning by including the following elements:

- the learners
- the learning goal
- the learning tools
- the learning context
- the communication among learners
- the learners' attitude/behaviour (motivation)

These six elements constitute a mobile learning ecosystem where each component interplays with the others, and cannot be considered in isolation.The match between the above mentioned framework components with the discovery phase findings resulted in the following three questions:

1. How will it be possible to create relevant mobile learning spots that augment informal learning and knowledge sharing (learning goal) by means of effective and

improved interaction (communication) within and outside the organization (learning context)?
2. How can personal mobile devices (learning tools) contribute to the personalization and appropriation of learning experiences, where the "m" that usually stands for "mobile" can easily be perceived as "me"?
3. How will it be possible to increase motivation (learners' attitude) towards learning at work through game concepts?

2.3 Ideation

Building on the interpretation phase, main themes or actionable statements were identified. These led to considering three workplace-learning challenges that could be solved through mobile learning.

1. Improve knowledge sharing (furthermore "KS") and effective communication at the workplace;
2. Facilitate appropriation of learning experiences at the workplace;
3. Increase motivation towards learning at the workplace through game concepts.

The ideation phase consisted in designing practical solutions that took into account ITC-ILO staff learning needs and that were applicable within ITC-ILO staff development initiatives.

Table 1. Challenges vs. Ideas for future development

Challenges/Ideas	**Idea 1: The Community of Practice for the glocal workplace**	**Idea 2: The CompassScape**	**Idea 3: Greening the Campus workshop**	**Idea 4: The e-Campus Game-board**
KS & effective communication	X	X		
Appropriation of learning	X	X	X	X
Motivation towards learning				X

Table 1 provides an overview of challenges and future development and implementation projects that can impact these challenges. We will discuss these approaches in the next section.

3 Challenges and Ideas for Future Development

3.1 Community of Practice for the 'Glocal' Workplace

In 2012 ITC-ILO started exploring the use of Yammer to improve cross-departmental communication but has not been rated as a successful tool for interaction and knowledge sharing at work. Interviewed staff stated that they did not see a clear strategy for using social networks within the organisation as ITC-ILO is perceived as a rather small organization. They additionally stated that they are already managing popular social media e.g. Facebook, Linkedin, Twitter, therefore they are resistant to use "yet another tool". Security concerns related to the online sharing of institutional documents on web 2.0 were highlighted. On the other hand, in 2013 ITC-ILO launched the institutional e-campus (http://ecampus.itcilo.org/), a Moodle-based virtual learning system increasing participants' access to learning opportunities beyond the classroom. The e-campus includes six learning modalities: tutored courses, self-learning, webinars, blended courses, academies and communities of practice (furthermore CoP).

Although, the e-campus has been conceived as a platform to better reach and train participants all over the world, it is also a relevant tool for internal learning and knowledge exchange. In particular, the CoPs present the potential of facilitating KS and interaction at work as they are built inside an institutional platform, and focus on work-related topics. Until now the e-campus CoPs have not been explored yet. They can be recognised as an institutional channel to connect with colleagues and external professionals interested to the same domain. If CoP on ITC-ILO e-campus constitute an open network, they can also be used as a marketing tool to attract potential participants on ITC-ILO e-Learning catalogue.

After having conducted interviews on mobile behaviours and mobile learning strategies, it was noted that some staff is already familiar with lots of mobile devices for personal reasons (organization, entertainment, communication, and information) and look forward to make a link between technology and learning activities, others are considered rather resistant to technology even though they perceive its social importance. Mobile learning therefore generates both interest towards effective use of technology at work and frustration in terms of time consuming aspects related to new technology usage. In particular, there is a growing interest towards the effective use of such devices for training delivery inside and outside the classroom settings.

By focusing on the ITC-ILO e-campus CoPs rather than on an external web 2.0 tool, the focus of KS at work is shifted on a strategic institutional platform that is more and more becoming of central importance for ITC-ILO in terms of making learning opportunities available anytime and anywhere. A CoP on "Learning, Technology and Knowledge Management" was already created in 2013 but it was put to no use so far. The idea is to use it to better explore how mobile learning methodologies and technologies may be integrated into ITC-ILO learning modalities.

One of the challenging questions in the CoP is: "How do you make learning accessible anytime and anywhere?"

The Learning, Technology and Knowledge Management community will constitute an open network, accessible both from ITC-ILO staff and an external audience. This will facilitate the connection with the external world.

The primary target of the CoP is constituted by the 26 staff members who participated to the research project interviews. The secondary target is constituted by any other ITC-ILO staff interested in covering the topic. The tertiary target is constituted by external audience e.g. in March 2014, the University of Turin (Italy) organised a rounded table around the topic of mobile and education. The event gathered about twenty learning and development local professionals and researchers, including two ITC-ILO staff interested in exploring mobile learning potentials and ways to collaborate together. These twenty people will be invited to participate in the CoP.

The CoP will be launched through sharing the results from the interviews conducted for this research project which has the purpose to frame mobile learning concepts and behaviours. The results will be shared in a visual format. The objective of the CoP is to discuss and progressively learn how to integrate mobile technology into training delivery and validate mobile learning methods. In the last two years, the ITC-ILO Distance Education Learning Technology Applications (DELTA) has already produced substantial guidance on mobile methodologies and tools that facilitate learning beyond the classroom, before, during or after a face-to-face event. This experience will be used to make sure that ITC-ILO colleagues have access to such information in a contextualised environment.

In particular, this material will be located in the CoP "Toolkit box". In addition to this, the CoP will have a News forum, a main discussion forum and a database listing all the mobile methods. The main discussion forum will be used to encourage discussion on how can mobile learning be integrated before, during and after face-to-face by providing practical tips and sharing best practices e.g. how can we inspire learners before a face-to-face (furthermore "F2F") session by means of mobile devices? How can we engage interaction during a F2F session by means of mobile devices? How can we gather feedback during a F2F session by means of mobile devices?

ITC-ILO staff will also be encouraged to rate the mobile methods listed in the database and share why they would or would not use such methods in the delivery of training activities. This task will be helpful for evaluating the relevance and efficacy of the developed mobile methods, increasing the overall quality of the mobile learning toolkit and encouraging ITC-ILO trainers to test mobile learning strategies.

Participation to the CoP initiative will be recognised through a visual and social tool, namely Open Badges by Mozilla Foundation [10]. Open badges are a digital representation of interests and learned skills, and can be displayed online e.g. the e-campus personal profile or social media, and offline as they are images.

Badges are considered mobile tools: "Get recognition for learning that happens anywhere. Then share it on the places that matter". Participants to the CoP will be entitled a "m-Learning" badge, facilitating the visualization of whom within the organisation is already familiar with such methodologies.

The evaluation of the CoP initiative will take place after three months from the launch and will take into account the qualitative and quantitative elements.

3.2 CompassScape

The CompassScape derives from the integration of two terms: the Compass and the Learnscape. The Compass is a project of the ITC-ILO centre's DELTA unit. DELTA is made up of a team of specialists who combine expertise in learning and knowledge sharing methodologies with professional backgrounds in international development.

The Compass uses the metaphor of a navigational instrument to guide trainers through a repository of participatory learning, training and knowledge sharing methods. It orients the design of learning and training activities and provides inspiration in the search for useful, effective and relevant learning and training methodologies. It contains sixty participatory knowledge sharing methods and technologies to be used in any workshops or meeting.

These methods were developed and validated in workshops all over the world, conducted by the ITC-ILO and are currently available online (http://compass.itcilo.org/). The website is regularly updated with information on new methods or examples and best practices to better describe them.

Building on the information gathered by the DELTA unit on the use of Compass, there is a general resistance to use new methods and ITC-ILO staff keeps using always the same ones e.g. the World-café or Fishbowl, rather than learning and exploring the use of new learning strategies. On the other hand, observation in the field led to a particular reflection on the use of methods within ITC-ILO workplace settings. In February 2014, the Information Communication Technology Services (ICTS) requested to be trained by DELTA on the Compass methods and their potential use for internal decision-making procedures. Eight ICTS members attended an interactive session facilitated by DELTA where they were involved in identifying suitable Compass methods that would match typical ICTS scenarios. As a result, Compass methods are perceived to be relevant not only for trainers but also for other ITC-ILO staff involved in problem-solving and internal decision-making.

Finally the learnscape concept focuses on how does the architecture influence learning and training activities and how can physical spaces be optimized to facilitate the appropriation of learning experiences. The DELTA unit is currently conducting research on this matter by trying to transform the traditional campus training settings in an environment conducive to learning The CompassScape initiative was born by matching workplace learning issues and influence of physical space on learning.The micro-objective of the CompassScape initiative was framed in two questions: 1) how will it be possible to raise awareness towards the Compass methods? 2) how will it be possible to encourage the exploration of more methods?

The CompassScape was conceived as a specific itinerary that identifies areas inside the ITC-ILO campus where specific methods can be visualized and implemented. In particular, the objective is to identify under-used spaces such as the park or open spaces that until now serve as decoration rather than having a strong connotation for learning activities. In other words the CompassScape is the result of the integration between optimization of physical spaces and effective mobile learning strategies, by allowing learners to learn in a specific context in an informal way, through recognition of visual patterns and facilitating the appropriation of learning experiences through mobile devices. In this context, mobile learning is intended in its broad sense as learning beyond the classroom, extending the learning experience across informal spaces, and encouraging "out-of-the-room" reflection.

The CompassScape addresses primarily ITC-ILO staff involved in training activities (trainers and programme assistants involved with learning spaces organization); secondly ITC-ILO staff who is not regularly involved in training activities e.g. ICTS, and thirdly training participants involved in training of trainers activities who can get inspired by methods and apply them in their own context. The methodological approach adopted for the CompassScape is Harold Jarche's framework for professional development: seek-sense-share [10]. This framework consists of three steps: seeking important information and getting up-to-date, making sense and personalize the use of information, and finally sharing ideas and experiences with the network.

According to this framework, a time-framed initiative will be organized around the following phases:

- Exploring the ITC-ILO environment and identifying methods through visual patterns;
- Making sense of information through elaboration of visual data, use of mobile devices to access rich-media information e.g. videos through augmented reality.
- Sharing feedback through mobile devices by social media or software for engaging with a dispersed audience e.g. Polleverywhere

The implementation of the CompassScape initiative will consist in the identification of a set of methods from the Compass database. These methods will be visualized through patterns and associated to specific places within the campus. The idea behind is to encourage matching the implementation of certain methods in areas other than classroom settings or inspire staff to use new methods through visualization, reflection and personalization of information.

A map will be created where to locate a set of labels displaying visual methods. These labels will be enriched by information accessible through mobile devices e.g. a QR code directing to a short video description of the method. In addition to this, there will be instructions on how to share feedback about the methods e.g. sending SMS, use of microblogging and social media.

The CompassScape initiative will be tested through a time-framed initiative of 2 weeks, followed by an evaluation. The evaluation will take into account the following elements:

- Quantitative data e.g. number of staff registered to participate into the initiative, number of staff actively participating into the initiative by sharing feedback;
- Qualitative data e.g. analysis of comments and feedback, and formulation of suggestions of improvements.

3.3 Mobile Learning for Improving Environmental Behaviours

The idea is to concretely use mobile learning methods that were developed in the past 2 years by DELTA unit for selected projects and contextualise them into workplace learning initiatives. Moreover, this initiative links back with the CoP to be created in the framework of this research and intended to facilitate recognition and validation of mobile learning methods.

The ITC-ILO has been working intensively to improve its green soul in the recent years. For this reason a voluntary-based committee (the GCAG) was created in 2011.

Its objective is to undertake an environmental review with external auditors, produce, monitor and evaluate an action plan, review the curriculum work to include environmental-related courses and activities, inform and involving staff, participants and service providers, and adopt an environmental code of conduct. Among the various initiatives, in February 2014 the centre was awarded the International Eco-Schools Award Certificate and Green Flag for its achievements in improving the campus environmental performance and in promoting a sustainable approach to learning.

The GCAG realized that, despite all these efforts, the ITC-ILO staffs and the services know little about the green improvements and that greener personal behaviours would make the difference. On this basis, the GCAG designed a short workshop addressing staff and services.

The shape of the workshop will enable participants to explore existing information and report back on the several green initiatives undertaken on campus. Once participants have shared the current green picture of the ITC-ILO, they will be exposed to a wide range of informative and inspiring resources that will constitute the basis for a constructive discussion on how everybody can contribute to a greener campus. By adding a mobile component to one of the workshop sessions, it will be measured how learning can be extended beyond the classroom settings and amplify learning objectives.

The Greening the Campus Workshop was already designed and resulted in one day training, by including the following components: green IT strategy, green facilities, energy management, waste management, green procurement, green monitoring system and green learning. Mobile learning methodologies will be selected to suit the pre-learning, learning and follow-up learning phases. Those methods were already categorized by DELTA according to benefits, learning need moments and category (deliver content, assign tasks, gather feedback, provide support). The evaluation of this initiative will be made through the comparison with sessions implemented without using mobile learning methods in order to measure whether mobile can really be an added value and have an impact on overall learning objectives. We will also build on related projects and experiences at other locations for this topic [12].

3.4 The e-Campus Game-Board: Applying Game Concepts to e-Learning Solutions

The ITC-ILO e-campus project starts from a forward looking vision that will help the centre to position itself better in the e-Learning landscape.The e-campus supports different type of activities and not exclusively e-learning, namely self-guided e-learning, tutor-based e-learning, on-line communities of practices, short promotional open educational resources and relevant ILO knowledge products. To convince colleagues moving to a centralized e-Learning platform, a series of incentives were presented such as decreasing investment in staff time when managing activities in the central e-campus; the benefit of using actual features that are currently not available in individual department platforms (webinars, mobile learning,); higher visibility in communication and promotion in contrast with decentralized learning products.

In order to accompany the transition phase to a centralized platform, a series of training activities were planned targeting managerial and administrative staff to get familiar with the e-campus features. Within this initiative, the concept of mobile learning will constitute an additional learning component supporting staff training activities during the transition phase. In the discovery phase, it was noted that mobile learning is not only associated with mere technology but also with learning across contexts, flexible and informal learning modalities, and finally tools that enable learning at the moment of need enabling knowledge application, reflection and collaboration.

The e-Campus game-board initiative will focus on addressing workplace learning needs (building internal e-Learning capacity) through making available to ITC-ILO staff a tool that facilitates decisions about e-Learning processes. Building e-Learning courses can be time-consuming and costly, therefore the e-Campus game-board will not constitute a technical training on how to use features on the online platform or getting to better know the online environment. It will rather be focusing on selecting the right e-Learning modality and its suitable components. For example, focusing on sharing information or changing performance, keeping it simple or building a complex structure etc. The primary target of this initiative will be trainers belonging to different department; secondly staff involved with practical use of the online platform, dealing with content update, participants' registration or technical issues.

As part of the e-campus project, an appropriate learning management system was selected through a benchmarking exercise, which opted for an open source platform, namely Moodle. The e-Campus game-board initiative will be informed by already existing literature on how to integrate the Bloom's Taxonomy into Moodle learning management system. In particular, inspiration came from the visual Moodle teachers' guide created by Joyce Seitzinger [11], which allows them to compare the functionalities and pedagogical advantages of some standard Moodle tools with the different levels of Bloom Taxonomy, and finally leading to a comprehensive decision about e-Learning structure.

In order to create a customised e-Campus version, the following steps will be taken into account:

- Analyse the e-campus delivery modalities
- Create scenarios for each e-Learning modality and build a questions' guide
- Link Bloom's Taxonomy levels with e-Learning scenarios
- Create a list of e-campus functionalities that serve the different e-Learning scenarios

Bloom Taxonomy levels and e-Learning delivery modalities will be identified through different coloured cards, containing questions, visuals or description. Through the e-Campus game-board, staff will be encouraged to reflect on how to reach learning needs and build suitable e-Learning. The e-Campus game-board is meant to work out of formal training sessions, during departmental decision-making processes related to e-Learning delivery. Moreover, it can be used as a tool to describe how the e-Campus works to external ITC-ILO partners that want to jointly deliver e-Learning.

4 Discussion and Conclusion

In this paper we attempted to show ITC-ILO efforts in creating a learning culture through a human-centred and personalised approach which tackles concrete workplace learning needs.

By integrating the results from the staff learning needs analysis, targeted interviews and observation on the field, it was possible to collect information around the ways staff engages with workplace training activities and inspiration about their use of mobile technology.

Understanding and interpreting their needs was a crucial step that helped rethink workplace learning delivery, shifting from traditional towards informal opportunities. Additionally, the paper outlines potential connections of mobile learning strategies with staff development learning activities, where the intention is to make strategic use of available mobile technology to augment continuous learning at work. In order to improve knowledge sharing, facilitate appropriation of learning experiences and increase motivation towards learning at the workplace, four ideas were identified: 1) the community of practice for the glocal workplace, 2) the CompassScape, 3) Greening the Campus workshop, 4) the e-Campus game-board. As part of the design thinking methodological approach used within this research, ideation will be followed by experimentation of working prototypes. The implementation of the above mentioned activities will take place approximately between June and November 2014. The outputs that will result from the implementation will be the task of a future paper.

Acknowledgement. The "M-Workplace learning @ ITC-ILO" research project is co-funded by the Italian Fondazione Goria and the International Training Centre of the International Labour Organisation (ITC-ILO) through the "Master dei Talenti per la Società Civile" fellowship programme. The total duration is of eighteen months starting in November 2013 and to be implemented as follows: twelve months at the Distance Education Learning and Technology Applications (DELTA) unit of the ITC-ILO in Turin, and the final six months at the Welten Institute of the Open University of the Netherlands.

References

1. http://www.itcilo.org/en/the-centre
2. http://mobile.itcilo.org
3. Parker, J.: Mobile Learning for Africa. Politecnico di Milano, Milan (2011)
4. IDEO, Riverdale Country School: Design Thinking for Educators Toolkit, pp. 11–14 (2013)
5. Pimmer, C., Pachler, N.: Mobile Learning in the Workplace: Unlocking the Value of Mobile Technology for Work-Based Education. In: Ally, M., Tsinakos, A. (eds.) Increasing Access to Mobile Learning, ch. 14, pp. 193–203. Commonwealth of Learning and Athabasca University, Vancouver (2014)
6. Gottfredson, C., Mosher, B.: Innovative Performance Support: Strategies and Practices for Learning in the Workflow. McGraw-Hill, New York (2011)

7. Sharples, M., Taylor, J., Vavoula, G.: Towards a Theory of Mobile Learning, Centre for Educational Technology and Distance Learning, University of Birmingham (2005)
8. Parsons, D.: The Future of Mobile Learning and Implications for Education and Training. In: Ally, M., Tsinakos, A. (eds.) Increasing Access to Mobile Learning, ch. 16, pp. 217–229. Commonwealth of Learning and Athabasca University, Vancouver (2014)
9. Sharples, M., Taylor, J., Vavoula, G.: A Theory of Learning for the Mobile Age. In: Andrews, R., Haythornthwaite, C. (eds.) The Handbook of E-Learning Research, London, pp. 221–247 (2007)
10. http://www.openbadges.org
11. Jarche, A.: Seek Sense Share, Creative Commons (2012), http://www.jarche.com
12. Kalz, M., Börner, D., Ternier, S., Specht, M.: Mindergie: A pervasive learning game for pro-environmental behaviour at the workplace. In: Wong, L.-H., Milrad, M., Specht, M. (eds.) Seamless Learning in the Age of Mobile Connectivity. Springer, Heidelberg (2014)
13. Seitzinger, J.: Moodle Tool Guide for Teachers, Creative Commons (2010), http://www.cats-pyjamas.net/

Context-Aware Mobile Professional Learning in PRiME

Christoph Greven, Mohamed Amine Chatti, Hendrik Thüs, and Ulrik Schroeder

Learning Technologies Research Group
RWTH Aachen University
Ahornstraße 55
52074 Aachen, Germany
{greven,chatti,thues,schroeder}@cs.rwth-aachen.de

Abstract. Technology Enhanced Learning (TEL) in professional and organizational settings is increasingly gaining importance. The high availability of mobile end devices and their ability to support learning across contexts open up new perspectives for effective professional learning and knowledge management. The BMBF project Professional Reflective Mobile Personal Learning Environments (PRiME) addresses the challenge of mobile learning in context and realizes a seamless learning framework which connects learning and work processes. PRiME enables the mobile professional learner to harness implicit knowledge and supports continuous knowledge creation and reflection at three different layers: the personal learning environment (PLE), the personal knowledge network (PKN), and the network of practice (NoP).

Keywords: mobile learning, professional learning, knowledge management, context, reflection.

1 Introduction

The widespread use of mobile technologies has led to an increasing interest in mobile learning. Mobile learning can be seen as a subset of technology-enhanced learning (TEL) that includes the usage of mobile devices to enable learning anywhere and at any time. Mobile learning is thus an interaction or activity of an individual which uses a mobile device, capable of having a reliable connection to communicate with a mobile learning platform, with the main goal to consume or create knowledge. Context is a central topic of research in that area. In fact, a major benefit of mobile devices is that they enable learning across contexts. Context-aware mobile learning applications leverage the context information of the learner to provide personalized and motivating learning experiences. Context-aware learning is increasingly important at the workplace, where learning is normally done on-the-fly; this means it is very fast and mostly integrated in the work process.

Several researchers have proposed theoretical work around mobile learning and proposed models on how to leverage emerging mobile technologies in teaching and learning (Koole, 2009; Park, 2011; Sharples, Taylor and Vavoula, 2010). However, comparatively little research has been done to focus on the context attribute in mobile

M. Kalz et al. (Eds.): mLearn 2014, CCIS 479, pp. 287–299, 2014.

learning at the workplace and only few research on context-aware mobile professional learning exist. In this paper, we present the conceptual and first implementation details of the Professional Reflective Mobile Personal Learning Environments (PRiME) project. PRiME provides a new vision of learning at the workplace defined by the seamless integration of mobile learning and knowledge management concepts into one solution toward a new model of mobile professional learning in context. In PRiME, learning is no longer regarded as an external training activity separate from the work flow, but rather as a learner-controlled evolving activity embedded directly into work processes.

The remainder of this paper is structured as follows. Section 2 gives an introduction to the term mobile learning in context. In Section 3, we address the relationship between professional learning and knowledge management. Section 4 presents the conceptual and implementation details of PRiME. Finally, section 6 gives a summary of the main results of the paper and outlines perspectives for future work.

2 Mobile Learning in Context

It has been widely recognized that context is one of the most distinctive features in mobile learning (Specht, 2009; Wang, 2004). The term context-aware was brought and defined by Schilit et al. (1994) to describe the location, identities of nearby people, objects and changes to objects. A very popular definition is given by Dey (2001, p. 5) who defines context as "... any information that can be used to characterize the situation of an entity. An entity is a person, place, or object that is considered relevant to the interaction between the user and the application, including the user and the applications themselves". According to Dey at al. (2001), there are three types of information that need to be gathered to describe any specific context - places, people and things. Specht et al. (2011) view context as any kind of information to determine, specify, or clarify the meaning of an event.

Context can be categorized into two types: extrinsic and intrinsic context. The extrinsic context deals with the user's current state of the environment. It may be the user's current position, the time of the day or the interval in which the user learns, or the object the user currently deals with and how this information can support his learning. It may also contain the relation to learners which was described by Costa et al. (2006). Intrinsic context information deals with the inside of a user. For Costa et al., "intrinsic context information inheres in a single entity" (2006, p. 3), an entity can either be a device or a person. This context information contains for example the user's level of attention or the level of motivation.

Harnessing context in a mobile learning experience has a wide range of benefits including personalization, adaptation, intelligent feedback, and recommendation. Adding context information to a set of applications may lead to presenting more personalized data to the learner and to more personalized applications. Context may contain information about the current situation, in which the learner is in, his or her pace, the pre-knowledge, interest, peers, motivation, etc. Context-aware mobile learning applications leverage this context information of the learner to provide personalized and motivating learning experiences.

As mobile devices are becoming ubiquitous and the workplace is increasingly mobile, there is an interest in the educational applications of mobile technologies in a professional learning context. Mobile professional learning refers to the use of mobile devices at the workplace for the purpose of learning while on the move.

3 Professional Learning and Knowledge Management

In an organizational environment, professional learning and knowledge management have attracted attention over the past years and are meanwhile important tasks to increase competitive advantages of an organization. In practice, however, the two fields have evolved down separate paths. While knowledge management concentrates on knowledge collection and distribution, professional learning focuses on formal learning and training of the employees. This tightened perspective can still be read from today's companies' organizations. Knowledge management and professional learning are commonly related to two different departments, namely IT and human resources.

Over the past few years, companies and researchers are starting to recognize relationships and intersections between the two fields and to explore the potential and benefits of their integration (Dunn and Iliff, 2005; Grace and Butler, 2005; Hall, 2001) Chatti et al. (2012) go a step further and point out that professional learning and knowledge management can be viewed as two sides of the same coin and stress the need for the seamless integration of the two concepts into one solution for the purpose of increasing individual and organizational performance. The authors introduce the Learning as a Network (LaaN) theory as a bridge between professional learning and knowledge management. LaaN starts from the learner and views learning as the continuous creation of a Personal Knowledge Network (PKN). In LaaN, a knowledge worker is a lifelong learner who is continuously creating and optimizing her PKN. LaaN enables the seamless integration of learning and work. The view of learning as the continuous creation of a PKN makes learning and work so intertwined that learning becomes work and work becomes learning. Professional learning in LaaN is no longer regarded as an external online training activity separate from the work flow, but rather as a learner-controlled evolving activity embedded directly into work processes.

4 PRiME

The joint research project Professional Reflective Mobile Personal Learning Environments (PRiME) is conducted by the Learning Technologies Research Group of the RWTH Aachen University and DB Training, Learning & Consulting of the Deutsche Bahn AG. It is funded by the German German Federal Ministry of Education and Research (Bundesministerium für Bildung und Forschung - BMBF) with a runtime of three years, finishing in June 2016.

PRiME draws together the concepts behind the LaaN theory and context-aware mobile learning. It provides a framework for the integration of professional learning and knowledge management within a mobile learning and working environment. The learning process in PRiME is a spiral and cyclic conversion of individual and organizational learning at three different levels of knowledge creation and reflection.

PRiME addresses the following objectives:

- Provide an innovative professional learning approach, where informal and network learning converge around a self-directed learning environment. This approach is grounded in the Learning as a Network (LaaN) theory.
- Design a work-integrated framework that links mobile job activities and self-directed learning in context.
- Develop and evaluate mobile learning applications to support mobile learning in context.
- Support continuous knowledge creation and reflection at three levels: (a) the personal learning environment (PLE) level where professional learners can annotate learning materials on their mobile tablet devices; (b) these materials can be shared, commented, and rated by peers at the personal knowledge network (PKN) level; (c) the new generated learning materials can then be shared and used within the company at the network of practice (NoP) level.
- Develop and evaluate learning analytics tools and methods (e.g. dashboards, recommendation, intelligent feedback, context-based search) to support reflective learning at the workplace.

4.1 Scenarios

Mobile professional learners represent the primary target group of PRiME. As a proof of concept, we addressed service technicians at Deutsche Bahn as a possible target group. These include car inspectors and the corresponding training developers, trainers, and specialist authors working in the field of car inspection service of the long-distance passenger transport DB Fernverkehr AG. In the following sections, the different occupational images and areas of activities as well as possible scenarios are explained in some detail.

Car Inspector

The car inspector is a mechanic that performs rail-worthiness checks on trains. He repairs small-scale damages on trains and decides about trains' dispositions on extensive problems. Either he works in a marshaling yard where trains are parked on holding tracks until their next usage or he attends a train on duty to ensure a smooth production flow. A car inspector is normally on his own after he got some work instructions when beginning his working shift. The field of work covers various different technical systems and machines and hence a lot of expertise and knowledge is required. Being an all-rounder, he has to master different tasks, such as handling newest and very old models, common and extraordinary procedures, and domestic and foreign machinery. In order to take accurate decisions in specific work contexts, either he has the required knowledge or he has to refer to related materials. Due to the complexity of knowledge that has to be acquired, remembering all needed details is impossible. On the other hand, getting access to auxiliary material is not really realizable due to the non-manageable physical amount of material, normally available as hardcopies. The following use case describes a possible flow of workflow of a car inspector with PRiME support.

Carl, being part of the car inspection service team, is using the PRiME system. He has access to the learning materials (called bundles in PRiME) from a workshop he attended recently. The bundles are covering knowledge assets regarding his field of application that has been created by a training developer. Furthermore, Carl searches the system for some bundles that might be useful for his next working day when he will need to deal with a specific machine. He finds several matching bundles that he subscribes to in order to receive changes and news related to these bundles. Carl faces a problem while he is on the field to repair a machine. He checks the related bundles as they are also available offline on his tablet. He finds a small learning unit (called snippet in PRiME) that might assist him with his current problem and follows the instructions provided in the snippet. While repairing the machine, Carl comes up with a new solution. He uses the video camera of his tablet to record the process. He then extends the related snippet with the captured video as annotation and saves it for future use. Whenever he is using the snippet again he gets his annotation associated with it. Back in his office, Carl enhances his annotation with some more information explaining the steps of his new solution in more details. He then shares his annotation with his colleagues who also subscribed to the same bundle. They are informed about the new annotation via a personalized aggregation interface (called *Newsstream* in PRiME). Carl's contribution further initiates a discussion and his colleagues start to post comments and other annotations related to the same snippet. Some colleagues consider Carl's annotation to be very helpful and give it a very good rating. Carl himself is also informed about those activities in his *Newsstream* and he is glad that he could help his colleagues. Some weeks later the specialist author who published the snippet became aware of Carl's highly rated solution and asks him for his permission to implement it in a new version of the snippet. All the bundles which contain the enhanced snippet will be updated. The updated bundles are then communicated to all subscribers through their *Newsstreams*. Carl's contribution is now part of the organizational-wide learning materials.

Specialist Author

The specialist author is responsible for the creation of new learning resources. Doris is a specialist author. She creates guidelines, instruction rules related to car inspections. She uses the PRiME *Snippet Creator* as an authoring tool to easily create multimedia (text, image, audio, or video) learning units. These snippets then be stored in the PRiME system and can be used by training developers as backbones for trainings and workshops (refer to the training developer scenario). Moreover, Doris uses the PRiME *Snippeter* to automatically convert an existing Word document containing working instructions to a set of snippets.

Training Developer

As compared to the car inspectors and the specialist authors, training developers are not technical experts and belong to another department at Deutsche Bahn AG. The training developers are responsible for the selection, aggregation, and creation of trainings from existing learning materials (bundles) created by specialist authors. Corinna is a training developer. She uses the PRiME system to create bundles that may be used in workshops or trainings. She got the task to prepare new bundles to be used by a trainer in a workshop with car inspectors. Corinna uses a PRiME authoring tool (*Bundler*) to search, filter, and aggregate existing bundles and snippets to create a new bundle. She also reuses parts of bundles she created in a previous work. The PRiME *Bundler* enables Corinna to easily structure and arrange content in her new bundle with simple drag and drop actions. When she finishes, the new bundle is published and can be used by the trainer in the workshop and subscribed to by the car inspectors who are interested in it. Corinna is informed as soon as someone rates or comments on her bundle. She also receives feedback from the trainer who used her bundle in his workshop. Based on the feedback, she enhances her bundle with further content, rearranges some parts, and deletes content which is not required. Changes of her bundles are again communicated to all subscribers.

Trainer

The trainers are responsible for the organization and execution of the professional technical trainings and workshops with the technicians. They use the learning materials (bundles) prepared by the training developers. These trainings are commonly done in traditional face to face classroom settings, often lasting for several days. During that time, participants interrupt their normal work to attend the training. Besides knowledge about the topic, the trainer should have the didactical skills required for an effective training. George is a trainer and is in charge of practical trainings and workshops with car inspectors, covering facilities and techniques of car inspection services. He is asked to arrange a workshop soon addressing the topic maintenance of passenger trains. Therefore, he makes use of the PRiME system to select relevant bundles required for his workshop. The bundles contain different snippets as textual instructions, diagrams, and demonstration videos. With the help of the PRiME export function, George is able to export the bundles as a simple presentation slides. In the workshop, he collects feedback to the bundles and forwards them to the training developer. After the workshop, the car inspectors get access to the bundles and can subscribe to them for continuous learning, as presented in the car inspector scenario.

4.2 Implementation

As presented in the scenarios above, learning and knowledge creation in PRiME is a continuous process involving professional learners, their networks, and the organization. As depicted in Figure 1 PRiME divides the learning and working process into three layers, namely the Personal Learning Environment (PLE), the Personal

Knowledge Network (PKN), and the Network of Practice (NoP). In the following sections, we discuss in detail the work and learning activities in relation to each layer and how these activities are supported by the PRiME tool set.

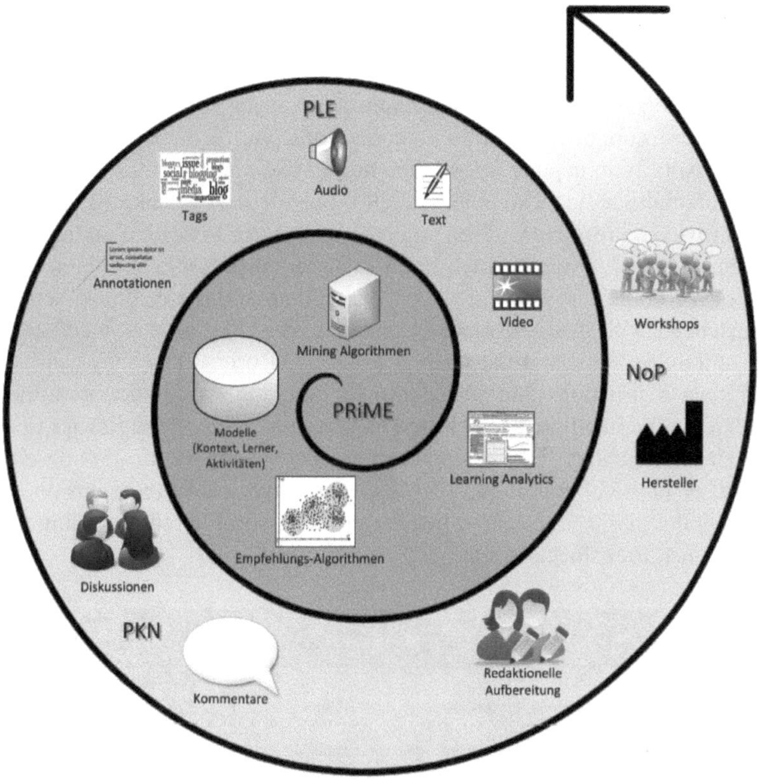

Fig. 1. Continuous creation of knowledge in PRiME considering the three reflection layers: Personal Learning Environment (PLE), Personal Knowledge Network (PKN), and Network of Practice (NoP)

Personal Learning Environment (PLE)

The Personal Learning Environment (PLE) represents the knowledge home of the professional learners (in our scenarios car inspectors but also specialist authors, training developers, trainers) enabling them to create their individual learning environments by assembling the knowledge assets which are relevant for their working context. In PRiME, these include bundles, snippets, and private annotations. A snippet is an atomic learning unit that can take the form of a text, image, audio, or video. A bundle is a set of snippets representing a learning material. An annotation is a multimedia enrichment of a snippet or a bundle.

The mobile PLE can be accessed at work at any time to solve current tasks. Car inspectors can use their mobile PLE to search and subscribe to relevant bundles, visualize the content of snippets and bundles, take notes, and make annotation. These activities are supported by a set of native mobile applications. The Android-based application *Reader* enables car inspectors to search for, subscribe to, and annotate bundles. As shown in Figure 2, the left side of the *Reader* visualizes a bundle and its associated snippets. Swiping from the right border fades in a list of annotations available for the selected content on the left. Personal annotations can be added via the buttons at the top of the annotation column and appear at the beginning of the list. Annotations are comments, questions, corrections, etc. to a bundle or snippet that offer the possibility to capture knowledge during the work process. They cover various types of multimedia (text, image, audio, or video) that make the creation very easy and provide a great expressiveness at the same time. Instead of taking a note on a loose sheet of paper which normally gets lost, car inspectors can take a photo of a machine or record a short video of a procedure. Annotations are context-sensitive and can be extended with meta information, such as recording time or location. At first, they are strictly personal and not visible to any other user. Thus, they can hold some personal work instructions that are helpful for future tasks. Furthermore, the car inspector can use the intelligent search functionality provided by the *Reader* to discover context-relevant bundles. For example, some knowledge is location-based due to machinery, or physical conditions such as noise might result in exclusion of media types containing audio. The search result can further be filtered according to author, topic, keywords, time, location, etc.

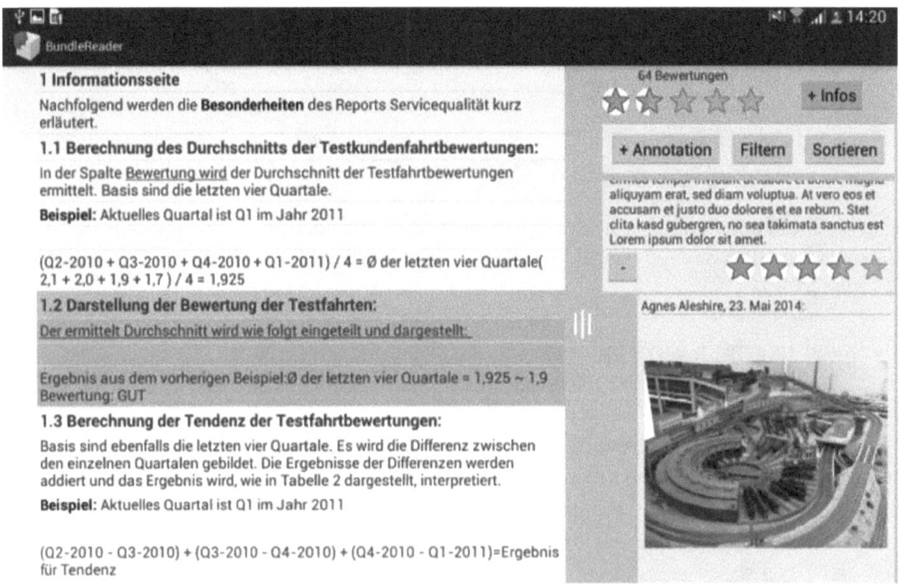

Fig. 2. Reader application for knowledge access and personal annotations

Car inspectors at work often do not have the time to write extensive annotations and link them to a specific snippet or bundle. Instead they can also use the PRiME system to simply take a photo, record a video, write a short text, or create a chart and store it for future use. These notes can then be used as a basis for annotations. A kind of media gallery collects all of the created notes and synchronizes them with a server-sided personal repository. Figure 3 shows the *Notepad* application that is realizing this functionality. The list holds all notes captured by a car inspector with optionally a short comment and keywords. By simply pressing a button in the top right corner new notes can be added. Selection from storage like SD card is possible as well as using the device-internal tools to record multimedia like the camera application.

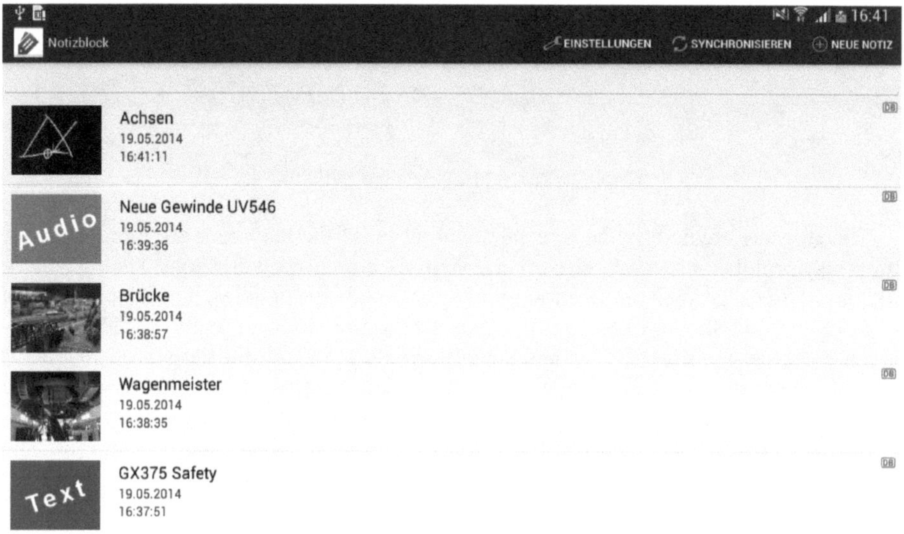

Fig. 3. Notepad for fast and easy collection of multimedia

As presented in the scenarios, specialist authors are responsible for creation and maintenance of guidelines and instruction rules. Today, most of such documents are already available in digital formats, e.g. as pdf, word, PowerPoint. The PRiME *Snippeter* Web application offers means to import such documents and automatically convert them to snippets. Figure 4. shows the import of a docx file. After the document has been uploaded, the *Snippeter* splits up the file into snippets and rebuilds its hierarchical logical structure into tree-structured bundles. The resulting bundles are then presented to the specialist author and can be modified. This includes merge and split of snippets. Whenever it is also possible to extract metadata from the source file, these are attached to the generated snippets. If the results are still not satisfactory, the specialist author may add metadata manually via the *Snippeter* interface. When the import is done, an initial version of the document is available in PRiME as a bundle. In addition to converting an existing document to snippets, the specialist author may use the *Snippet Creator* Web application to create snippets from scratch.

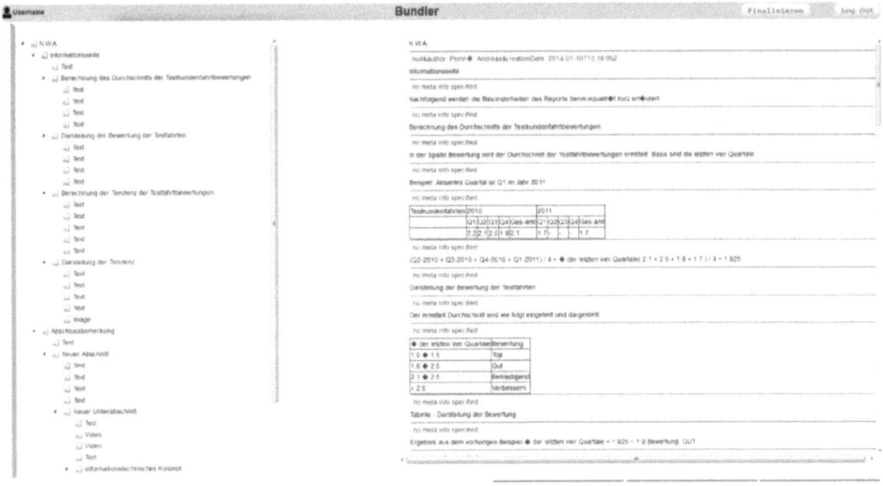

Fig. 4. Snippeter after importing a docx file into PRiME

The snippets created by the Specialist author build the backbone for bundles that a training developer can compose for a specific training or workshop. Their primary tool is the *Bundler* Web application to aggregate and edit bundles. The functionalities of a *Bundler* are shown in Figure 5. The left column is used to search for already existing snippets. Different filters and search criteria help to limit the search results to only show context-relevant snippets. In the middle column, a tree-like view helps to structure and arrange snippets at various bundle levels. Simple drag and drop actions can be used to include search results from the left column into the bundle tree. The right column shows a document-like view of the aggregated bundle.

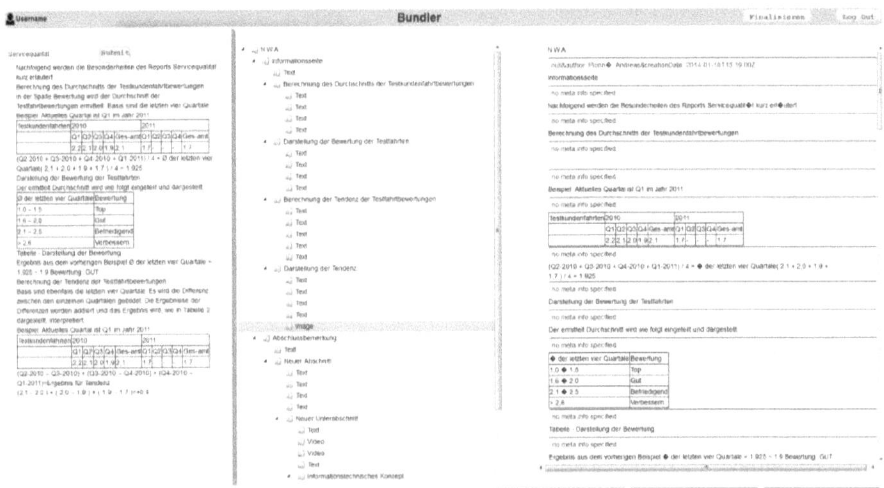

Fig. 5. Bundler to create and arrange new, reusable knowledge bundles

The *Bundler* can also be used by the trainer to search for relevant pre-assembled bundles for his training or workshop. The *Bundler* further offers different export modules that allow the trainer to convert bundles to traditional formats, such as pdf, word, PowerPoint that can be used as handouts in the workshop.

Personal Knowledge Network (PKN)

The Personal Knowledge Network (PKN) represents the networking layer in PRiME. This layer supports sharing and interaction with selected peers, thus enabling collaborative knowledge creation. In the case of car inspectors, previously taken personal annotations in the *Reader* can be shared with selected peers or groups that can be personally defined. Annotations can then be seen by all members of the publishing group who can give ratings and might reply to these annotations with their own ones. This way, expert discussions can emerge resulting in collaborative creation and maturing of knowledge. Car inspectors who apply the knowledge in their daily tasks have the possibility to give valuable feedback to aid the specialist authors in improving the produced snippets. Because over time the available knowledge is rapidly growing, recognizing changes in the system becomes harder. In PRiME, car inspectors, specialist authors, training developers, and trainers who subscribed to a specific bundle continuously receive notifications on the annotations, ratings, and changes made to the bundle. Car inspectors can follow the discussion and discover quality snippets/bundles. Specialist authors, training developers, and trainers can get continuous feedback that can be used in the enhancement of their snippets or bundles. This is achieved through the native Android-based *Newsstream* application which provides an aggregated view of recent activities, as shown in Figure 6. Furthermore, car inspectors use the *Newsstream* to receive recommendations according to their preference and activities in the system.

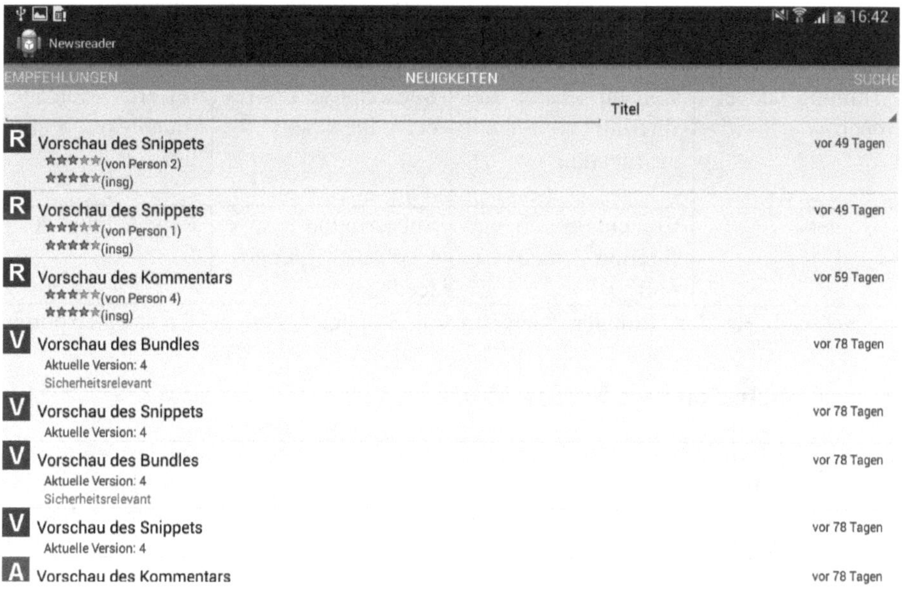

Fig. 6. Newsstream informing about activities related to favored knowledge

Network of Practice (NoP)
The Network of Practice (NoP) represents the organization layer in PRiME. This layer supports the propagation of the knowledge created at the personal network level to the entire organization. Quality knowledge that emerges as a result of the continuous interaction between all PRiME users at the PKN level builds the cornerstone for the enhancement of the organization-wide guidelines and instructions rules. The specialist author can use the highly rated annotations as a base for new snippets or versions of the same. For this he uses the *Snippet Creator*. The training developer and the trainer can use the *Bundler* for the creation of new bundles to be used in the next trainings and workshops. This continuous knowledge creation and reflection process ensures that the organization knowledge is accurate and up to date.

The following table summarizes the mapping of the different roles and most important functionalities of the PRiME toolkit to the three reflection layers, namely PLE, PKN, and NoP.

Table 1. Mapping roles, functionalities, and reflection layers in PRiME

	PLE	PKN	NoP
Car inspector	Multimedia data collection[1], knowledge access[2], offline availability[2], personal annotations[2]	Subscription[4], sharing of annotations[2], receiving annotation [2], rating[2], news aggregation[4], receive recommendations[4]	
Specialist author	Data import[5], content processing[3], content creation[6]	Subscription[4], discovery of need for changes[4], receive feedback[2]	rework content (snippets)[6], Publication[6]
Training developer	Content search and filtering[3], content aggregation[3]	Subscription[4], receive feedback[2]	rework content (bundles)[3], Publication[3]
Trainer	Content search and filtering[3], data export[7]	Subscription[4], give feedback[2], receive feedback[2]	Rework training [3]

[1]Notepad, [2]Reader, [3]Bundler, [4]Newsstream, [5]Snippeter, [6]Snippet Creator, [7]Exporter

5 Conclusion and Future Work

In this paper, we addressed the challenges of mobile learning in context as well as the convergence of professional learning and knowledge management. We presented the theoretical, conceptual, and implementational details of the Professional Reflective Mobile Personal Learning Environments (PRiME). The main goal of PRiME is to offer seamless learning across times, locations, and social contexts combining the work and learning processes into one. Learning in PRiME is the result of a continuous knowledge creation and reflection at three layers, namely personal learning environment (PLE), personal knowledge network (PKN), and network of practice (NoP). Different mobile applications have been introduced to support the various activities related to each of these layers. The project is still in the first phase of development. Hence, comprehensive evaluations are not yet possible. So far, the applications cover basic functionalities but will be continuously extended and incorporate context such as social connections as friends or collegues, learning time or behavior, physical environment, etc. Thus, future work will include the implementation of personal dashboards to support self-reflection and awareness, as well as different learning analytics methods that leverages the context information to provide effective recommendation and intelligent feedback to the PRiME users.

References

1. Chatti, M.A., Schroeder, U., Jarke, M.: LaaN: Convergence of Knowledge Man-agement and Technology Enhanced Learning. IEEE Transactions on Learning Technologies 5(2), 177–189 (2012)
2. Costa, P.D., Almeida, J.P.A., Pires, L.F., Guizzardi, G., van Sinderen, M.: Towards Conceptual foundations for context-aware applications. In: Proceedings of the Third International Workshop on Modeling and Retrieval of Context (MRC 2006)
3. Dey, A.K.: Understanding and using context. Personal and Ubiquitous Computing 5(1), 4–7 (2001)
4. Dunn, P., Iliff, M.: At Cross Purposes—Why E-Learning and Knowledge Management Dont Get Along. Learning Light. E-Learning and Knowledge Management (2005), `http://www.learninglight.eu/Register1/Learning%20Light%20E-learning%20and%20Knowledge%20Management.pdf`
5. Grace, A., Butler, T.: Learning Management Systems: A New Beginning in the Management of Learning and Knowledge. International Journal Knowledge and Learning 1(1/2), 12–24 (2005)
6. Hall, B.: Learning Management and Knowledge Management: Is the Holy Grail of Integration Close at Hand? white paper (2001), `http://www.jacqueslecavalier.com/linked/lkmwp-080301.pdf`
7. Koole, M.L.: A model for framing mobile learning. Mobile Learning: Transforming the Delivery of Education and Training, 25–47 (2009)
8. Park, Y.: A pedagogical framework for mobile learning: Categorizing educational applications of mobile technologies into four types. The International Review of Research in Open and Distance Learning 12(2), 78–102 (2011)
9. Schilit, B., Adams, N., Want, R.: Context-aware computing applications. In: First Workshop on Mobile Computing Systems and Applications, WMCSA 1994, pp. 85–90 (1994)
10. Sharples, M., Taylor, J., Vavoula, G.: A theory of learning for the mobile age. In: Medienbildung in neuen Kulturräumen, pp. 87–99 (2010)
11. Specht, M.: Learning in a Technology Enhanced World, pp. 33–34. OCÉ, Maastricht (2009)
12. Specht, M., Ternier, S., Greller, W.: Mobile augmented reality for learning: A case study. Journal of the Research Center for Educational Technology 7(1), 117–127 (2011)
13. Wang, Y.-K.: Context awareness and adaptation in mobile learning. In: Proceedings of the 2nd IEEE International Workshop on Wireless and Mobile Technologies in Education, pp. 154–158 (2004)

Nurses' Work-Based Mobile Learning

Dorothy (Willy) Fahlman

Athabasca University, Canada
willyf@athabascau.ca

Abstract. Informal learning is the most prevalent mode of learning in workplaces. Additionally, mobile devices are increasing used as learning tools for informal learning. However, in the healthcare workplace, the use of mobiles for work-based informal learning of nurses is still relatively new. This paper reports on a qualitative study that explored Canadian registered nurses' mobile use for individual and collaborative informal work-based learning. Results suggest that the participants primarily used individual informal work-based mobile learning. This learning was influenced by the context of the healthcare workplace.

Keywords: Work-based mobile learning, informal learning, nursing, RN, individual, collaboration, constructivism.

1 Introduction

In the 21st century knowledge-based economy, there is global competition for advanced knowledge and skills development. This leads to an ever-increasing demand for work-based education and training that engages learners and supports workforce development. The workplace is a rich setting for both formal and informal learning; however, most learning is informal in nature within all spheres of work [1]. Furthermore, mobile device use has become ubiquitous including penetrating the workplace organizations. Consequently, work-based mobile learning (WBML) is a rapidly emerging field combining the approaches of work-based learning and mobile learning [2]. While mobile devices are becoming more commonplace in the healthcare workplace, their use for work-based professional learning in nursing is still in its infancy [3]. This presentation explores registered nurses' (RNs) individual and collaborative informal WBML, including factors that promote or inhibit collaborative learning within a community of professional practice.

2 Background

In Canada, there are over 270,000 RNs [4]. As members of a self-regulated profession, RNs are required to maintain competency and continued professional development using self-directed learning activities and reflective practice [5]. Unfortunately, accessibility and availability issues with formal work-based education are creating challenges for RNs to engage in learning. Subsequently, RNs are seeking more autonomous and

M. Kalz et al. (Eds.): mLearn 2014, CCIS 479, pp. 300–303, 2014.

flexible means for meeting their professional learning needs [6]. Informal learning in the workplace provides an appealing alternative for seamless learning. To this end, mobiles are being used as learning tools by RNs in their workplaces [7].Their communication/multimedia capabilities including audio, text, images, and video open the possibilities to informal WBML and collaboration within a community of learners.

3 Theoretical Framework

The study's theoretical framework draws on the theories of informal learning in the workplace in addition to aligning with the constructivist lens that includes individual and social WBML. Based on the model of informal and incidental learning put forth by Marsick, Watkins, and Lovin [8], informal learning occurs outside of the formal, institutional sponsored classroom-based activities and is built on experience that is linked to meaningful job activities. It often happens in non-routine circumstances, focuses on action and problem-solving using reflection, and is influenced by the workplace context. Informal learning can be self-directed and intentional, incidental or unplanned learning that becomes conscious after an experience, and tacit learning that is neither intentional nor conscious [9]. Although informal learning may occur individually or collaboratively, this model values learning through communities of practice grounded on social interaction and social construction of knowledge. Likewise, WBML incorporates the perspectives of cognitive and socio-cultural constructivism; whereby, learners can construct new knowledge individually or through collaboration and social discourse that builds mutual understanding [2].

4 Methodology

Qualitative descriptive inquiry explored data obtained from semi-structured interviews. Ten Canadian regulated RNs who self-reported frequent mobile use for individual and collaborative informal WBML were purposively selected:

- Gender: female – 8, male – 2;
- Birthdate: born 1981-2000 - 3; born 1965-1980 - 4; born 1946-1964 – 3;
- Years employed as an RN: 3-20
- Work-related setting: hospital - 6, long term care - 1; community nursing - 1; private clinic – 1; home care – 1;
- Location: rural – 3; population centre – 7;
- Position: staff nurse – 6; management – 2; educator -1; community nurse -1;
- Mobile device used: smartphones – 8; tablets - 2; iPod – 1; PDA – 1;
- Length of mobile device usage varied from under 3 months to 5 years.

The interview questions were piloted. The interviews were conducted by phone and recorded. Transcripts were inductively analyzed, condensing data into codes then categories, and finally themes for further interpretation. Trustworthiness was enhanced using member checking, peer debriefing, and data source triangulation.

5 Results and Discussion

The study RNs reported primarily participating individually in informal WBML. They used their mobiles for "just-in-time" access to evidence-based resources for self-directed learning when challenged with new situations, procedures, and/or treatments. They searched online databases and the Web, read online resources, and/or viewed online videos, webcasts, or podcasts. They stated mobiles provided flexibility in time and place of learning, convenience, and learner-centred control resulting in perceptions of increased self-confidence and self-efficacy for professional practice.

Although the study RNs were motivated to engage in collaborative informal WBML, there were perceived workplace constraints influencing their participation. Some RNs expressed employer's restrictions and/or bans on mobiles due to the potential blurring of boundaries for personal and work use. Additionally, there were connectivity and bandwidth issues in remote/rural areas. Also, concerns were articulated about needing to "bring your own device" (BYOD) to work and incurring personal data plan costs, and privacy/confidentiality issues related to security of health information. These factors limited peer-to-peer informal WBML.

Irrespective of the constraints, the study RNs did provide a few instances of social-cultural informal WBML including group texting for problem-solving, sharing resources via email, and accessing listservs to access information, but refrained from actively interacting. Also, several participants shared information on their mobiles in-person with their on-site colleagues for further discussion and reflection.

These findings are consistent with Wihak and Hall [10] who posit that self-directed learners prefer individuals modes of informal learning. However, the pervasive nature of the workplace influenced their perceptions and conceivably their preferences. As Marsick, Watkins, and Lovin [8] argue, the organization culture influences work-based informal learning "permeating every phase of the learning process -- from how the learner will understand the situation, to what is learned, what solutions are available, and how the existing resources will be used" (p.67). Likewise, the context shapes the social-cultural processes for meaning-making and the potential for the practitioner's transition from periphery participation to full participation within a community of professional practice [11].

6 Conclusion and Recommendations

The study RNs primarily engaged in self-directed "just-in-time" individual informal WBML. They proactively used their mobiles to acquire knowledge and develop skills as a means for improving professional nursing practice regardless of the workplace constraints. However, a major study limitation must be noted: all participants were Canadian diploma-prepared practicing RNs enrolled in an online Bachelor of Nursing program at a unimodal distance education institution. As such, they may have been more engaged with technology than other RNs.

Mobiles can be used to pedagogically support and enhance informal WBML experiences. Furthermore, mobile use should not be restricted in the healthcare workplace as it can be argued that these constraints are neither practical nor feasible [12]. Moreover, the importance of promoting the intertwining process of work and informal learning using mobiles within a community of nursing practice must not be ignored. Therefore, the following recommendations should be considered:

- Visionary leadership with a clear strategy including stakeholders' input to generate mutual trust and respect with informal WBML.
- Analyze BYOD vs. employer-supplied mobiles' cost/benefits for the best fit for the workplace and RNs.
- Incorporate ongoing managerial and technical support for enhancing collaborative informal WBML and co-learning within a community of practice.
- Implement security measures such as Mobile Device Management (MDM) controlling all mobile activities or Mobile Application Management (MAM) governing only specific apps, and also designate secure websites for social informal WBML.
- Develop policies, procedures, and education related to privacy and confidentiality to ensure appropriate and professional use of mobiles.

References

1. Livingstone, D.W., Mirchandani, K., Sawchuk, P.H.: Introduction: Critical perspectives on learning and work in turbulent times. In: Livingstone, D.W., Mirchandani, K., Sawchuk, P.H. (eds.) The Future of Lifelong Learning and Work, pp. 1–10. Sense Publishers, Rotterdam (2008)
2. Pimmer, C., Pachler, N.: Mobile Learning in the Workplace: Unlocking the value of mobile technology for work-based education. In: Ally, M., Tsinakos, A. (eds.) Perspectives on Open and Distance Learning: Increasing Access through Mobile Learning, pp. 193–204. Athabasca University Press (2013)
3. Burden, K., Schuck, S., Aubusson, P.: Ethical professional mobile learning for teaching and nursing workplaces. In: Pachler, N., Pimmer, C., Siepold, J. (eds.) Work-based Mobile Learning: Concepts and Cases, pp. 277–304. Peter Lang, Oxford (2011)
4. Canadian Institute of Health Information: Regulated Nurses: Canadian Trends, 2007 to 2011 (2013), `https://secure.cihi.ca/estore/productFamily.htm?locale=en&pf=PFC2016&lang=en` (2013)
5. Canadian Nurses Association: Competencies (2014), `http://cna-aiic.ca/en/becoming-an-rn/rn-exam/competencies`
6. Kenny, R.F., Park, C., Van Neste-Kenny, J., Burton, P.A., Meiers, J.: Using mobile learning to enhance the quality of nursing practice education. In: Ally, M. (ed.) Mobile Learning: Transforming the Delivery of Education and Training, pp. 51–75. AU Press, Edmonton (2009)
7. Fahlman, D.: Examining informal learning using mobile devices in the healthcare workplace. Can. J. Learn. Technol. 39 (2013)
8. Marsick, V.J., Watkins, K.E., Lovin, B.: Revisiting informal and incidental learning as a vehicle for professional learning and development. Elabor. Prof., 59–76 (2011)
9. Schugurensky, D.: The forms of informal learning: Towards a conceptualization of the field. University of Toronto, Toronto (2000)
10. Wihak, C., Hall, G.: Work-related informal learning. Centre for Workplace Skills (2011)
11. Lave, J., Wenger, E.: Situated learning: Legitimate peripheral participation. Cambridge University Press (1991)
12. Pimmer, C., Brysiewicz, P., Linxen, S., Walters, F., Chipps, J., Gröhbiel, U.: Informal mobile learning in nurse education and practice in remote areas—A case study from rural South Africa. Nurse Educ. Today (2014)

Review of Trends in Mobile Learning Studies in Mathematics: A Meta-Analysis

Helen Crompton[1] and Diane Burke[2]

[1] Teaching & Learning, Old Dominion University, 145 Education Bldg. Norfolk, VA 23529
[2] Education Department, Keuka College, 141 Central Ave, Keuka Park, NY 14478

Abstract. The field of mobile learning is growing an exponential rate. To best understand this field, it is crucial to gain a collective understanding of the research that has taken place. There are a number of recent studies which looked at collective trends in mobile learning across subjects. However, these results do not provide insight into specific subject areas. This meta-analysis reveals the trends in mobile learning in mathematics with a comprehensive analysis and synthesis of 48 studies from the year 2000 onward. Major findings include that most of the studies focus on effectiveness, followed by learning design. Mobile phones are currently the most widely used device used. Also, the use of mobile devices for mathematics learning is most common in elementary (5-11 years old) school settings.

Keywords: mathematics, literature, mobile learning, mlearning, meta-analysis, research.

1 Introduction

Mobile learning is now receiving increasing attention as a way to transform and rejuvenate educational practices [1, 2, 3]. To better understand the field of mobile learning, the research needs to be compiled to gain a complete picture of the field. This review of the literature will provide information on the trends and patterns of the development and the usage patterns of mobile technologies and learning. This will provide a springboard for future studies.

There are those who have completed a recent mobile learning meta-analysis with a learner focus [e.g.4, 5] and those who have focused on the device and software [e.g. 6]. Although these meta-analyses provide a rich source of information on mobile learning, they do not provide a wealth of data on a particular subject. Therefore, the meta-analysis can only be used as generalizable information. The purpose of this study is to conduct a meta-analysis on mobile learning as it relates to mathematics. The researchers will parse out pertinent information to provide a comprehensive examination of this topic. Other recent meta-analysis [e.g. 4, 6, 5] have used only peer reviewed articles. This is problematic as journals on average take eight to nine months to be published.

M. Kalz et al. (Eds.): mLearn 2014, CCIS 479, pp. 304–314, 2014.

With the rapid progression in mobile technologies a study of only peer-reviewed journal articles will provide dated information from the prior year. To ameliorate this issue and provide the most up-to-date information, the researchers of this study include conference proceedings and also dissertations as well as peer reviewed publications. This study is beneficial to researchers, policy makers, educational leaders, and other stakeholders as the use of mobile learning in mathematics is elucidated. Three questions are used to drive this meta-analysis on mobile learning in mathematics: 1) In studies involving mobile learning and mathematics, what were the major research purposes, methodologies, and outcomes addressed in the studies? 2) In studies involving mobile learning and mathematics, what were the mathematical concepts, educational levels, and educational context of the studies? 3) In studies involving mobile learning and mathematics, what were the mobile devices used and the geographical distribution of the studies?

2 Literature Review

2.1 Definition of Mobile Learning

As the field of mobile learning has developed and devices have advanced, there have been a number of ephemeral definitions of mobile learning. Earlier definitions have often been technocentric or have named a particular device which quickly becomes dated [7]. Nonetheless, trends have emerged from these definitions that highlight the four central constructs of mobile learning as pedagogy, technological devices, context, and social interactions [7]. Using these constructs, Crompton defined mobile learning as "learning across multiple contexts, through social and content interactions, using personal electronic devices" [3 p. 4]. This is the definition selected for this article as it best encompasses the meaning of the term.

Determining which devices are included in m-learning has also been a topic of debate among scholars [8, 9]. For this study, Crompton's [1] criteria have been used (see p. 48) to define what qualifies as a mobile device. She proffers that the device must be portable and have a prompt on off button. The latter is extremely problematic with traditional laptops as they take a while to start and they are typically not left on stand-by mode to use quickly. For this reason, laptops are not included as mobile devices in this study.

2.2 Trends in Mobile Learning Research

As mobile learning is a relatively new field of study, there is a paucity of studies that collectively analyze mobile learning research. In 2009 Frohberg, Goth, and Schwabe [10] conducted a critical analysis of mobile learning projects. The researchers focused on six criteria: context, control, tools, communication, subject, and objective. Using a framework to systematically analyze and position mobile learning projects, Frohberg et al. [10] reported screening 1469 publications to finally analyze 102 publications.

Using, the Mobile Learning Analysis Framework [11, 12], Frohberg et al. [10] found that although mobile devices were primarily for communication, they found few connections to the research regarding communication or collaboration. The researchers also found that the majority of the studies supported novice learners.

Hung and Zhang [13] conducted a study of mobile learning research trends from 2003 to 2008. Text mining techniques were used to provide basic bibliometric statistics, trends in frequency of topics, predominance of topics by country, and preference for each topic by journal. To summarize the findings, the researchers found that: 1) mobile learning articles increased from eight in 2003 to 36 in 2008; 2) effectiveness, evaluation, and personalized systems were the most popular area of study; 3) Taiwan conducted the most mobile learning studies; and 4) mobile learning is utilized by *early adopters* [14].

A meta-analysis has been selected for this study to answer the three research questions. This was also the methodology chosen by two other recent studies. Wu et al. [3] conducted one of the most recent meta-analysis of mobile learning. This study analyses 164 studies from across content areas. The researchers' foci were on the research purposes, methodologies, outcomes, devices, learners, and disciplines. From this study, Wu et al. [3] were able to present new findings from the research on each of the topics they studied; for example, mobile phones and PDAs were the most commonly used devices in mobile learning research.

Despite the wealth of information the Wu et al [3] study provides, very little on individual subjects and topic areas can be parsed out of that information. Therefore, the meta-analysis in this study will focus on the research on mobile learning in mathematics to provide a detailed review of the trends in this field.

3 Method

A systematic review and analysis was conducted from a data pool consisting of computerized bibliographic databases (e.g. Ebscohost, Proquest, and ERIC). The methodology was based on the rigorous protocol developed by scholars [15, & 16]. In addition, this protocol was used by Wu et al. [3] in a meta-analysis on general mobile learning studies. For this study, the inclusion/exclusion criteria, data sources and search strategies, and data coding and analysis can be found below.

3.1 Inclusion/Exclusion Criteria

The inclusion and exclusion criteria can be found in Table 1. To be included in this meta-analysis, each study had to meet all the inclusion criteria.

Table 1. Inclusion Exclusion Criteria

Inclusion Criteria	Exclusion Criteria
Must include mobile learning as a primary condition	The mobile device must not be laptops
Must have mathematics as focus of learning	The mobile device must not be Netbooks
Must include identifiable learner level	The mobile device must not be calculators (that just calculate or graph)
Must include mobile devices while learning	The mobile device must not be audience response systems (e.g. clickers)
Must include educational activities while implementing mobile learning	
Must include primary research (not a collective review of past studies)	
Must be publicly available	
Must be peer reviewed if the article is published in a journal	

3.2 Data Sources and Search Strategies

This meta-analysis included studies located through a comprehensive search of publicly available literature, primarily through manual electronic searches of the following databases: ERIC, EBSCOHOST, ProQuest and Dissertation Abstracts, Wiley International Science, Elsevier Direct, JSTOR, and Sage Journal On-line . Manual searches were conducted in the International Journal of Emerging Technologies in Learning, ELEED, Journal of Mobile Teaching, IEEE, and the International Journal of Mobile and Blended Learning. Search terms included the following keywords: "Mathematics," "Math," and "Maths," with "mobile learning," "m-learning," "hand-held," "tablets," "I-pads," Ubiquitous learning," "wireless learning," "location-aware," context-aware," and "situated learning."

The initial search resulted in 19,267 articles. This large number uncovered that the search terms were being interpreted broadly by the search engines. A review of the articles revealed that the changing meaning of terms over time resulted in inaccurate findings. Therefore this study went from 2000 and onwards. The search from the year 2000 gained a total of 754 articles. After removing duplicates and those that did not meet our initial criteria of including mobile learning and mathematics, 58 studies remained. Two researchers then independently confirmed the inclusion/ exclusion criteria for each study. The inter-rater agreement for coding was 94.8%.

Disagreements between the two coders were resolved through discussion and further review of the disputed studies. In total, 48 studies met all the inclusion criteria listed in Table 1 and were used in the analyses.

3.3 Data Coding and Analysis

The research questions of this study framed the eight features that were coded for analysis: 1) research purpose, 2) research method (e.g. observations, case study), 3) learning outcomes (i.e. positive, negative, neither, and other), 4) mathematical concepts based on the National Council of Mathematics (NCTM) Standards, 5) educational levels (e.g. elementary, middle, high school, higher education, and special education), 6) educational context (i.e. formal, informal, and non-formal), 7) mobile devices (e.g. mobile phones, tablets), and 8) countries of study.

The research purposes were classified as one of two types: 1) application-dominant with device-minor, 2) device-dominant with application-minor. Type one focuses on student learning and then the device (e.g. the researcher wants to focus on the student and see how the program influences learning. Type two focuses on the device before student learning (e.g. the development of a mobile application). The educational levels were Pre-Kindergarten (2-4 years old), elementary (5-11 years old), middle school (1-14 years old), high school (14-18 years old), and higher education (18 years and older). The educational contexts of the studies were coded following Crompton's [7] categories of educational context, formal, non-formal, and informal. Formal is intended learning in a typical educational setting (e.g. a taught class in a school), non-formal is non-intended learning (e.g. determining sale percentages in a shop), and informal learning, which is intended learning in an atypical setting (e.g. a lesson taking place in a playground).

4 Results

4.1 Distribution of Research Purposes

Each of the studies was classified into one of three categories according to the research purpose: (1) evaluating the effects of mobile learning, (2) designing a mobile system for learning or (3) investigating the affective domain during mobile learning. As seen in Figure 1, evaluating the effects of mobile learning was the most common research purpose (69%), followed by designing a mobile system for learning (25%) and investigating the affective domain during mobile learning (6%).

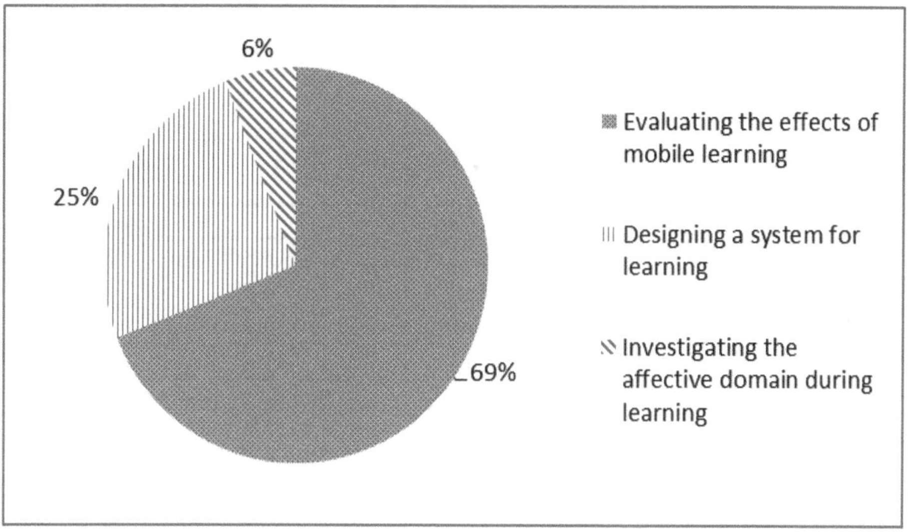

Fig. 1. Distribution of Mobile Learning Studies by Research Purposes

4.2 Distribution of Research Methods

The research was classified into two types: 1) application-dominant with device-minor. 2) device-dominant with application-minor, Type one focuses on student learning and then the device (e.g. the researcher wants to focus on the student and see how the program influences learning. Type two focuses on the device before student learning (e.g. the development of a mobile application). Type one studies focusing on the student learning first comprised 77% of the studies. Type two studies focusing on the device before the student comprised 23% of the studies.

Table 2 provides a summary of the classification of mobile learning studies by methodology. The number in brackets is the number of studies that specified using that methodology (or type of data collection). A single study may have specified a number of different methodologies which were listed once for each type specified. Table 2 indicates that for purpose 1, (evaluating the effects of mobile learning) researchers primarily relied on surveys followed by case studies with four or less on the other methodologies. For purpose two (designing a mobile system for learning), there was no dominant methodology. For research purpose three (investigating the affective domain during learning), only three methodologies were used in three studies. The information provided in the articles was varied from researchers who were specific in their methodology e.g. quantitative t-test, to those who defined their approach in broader terms e.g. quantitative.

Table 2. Classification of Mobile Learning Studies by Methodology

Application-Dominant with Device-Minor		Device-Dominant with Application-Minor
Purpose One	**Purpose Three**	**Purpose Two**
Survey (8)	Constant comparison (1)	Observation (1)
Grounded theory (1)	Grounded theory (1)	Interviews (1)
Quasi-experimental (2)	Survey (1)	Questionnaire (1)
Qualitative coding (1)		Design-based research (1)
Case study (7)		Test scores (1)
Test scores (4)		Case study (1)
Design-based research (3)		Quasi –Experimental (2)
Observations/Field Notes (4)		Action Research (1)
Experimental (4)		Focus groups (1)
Alternative treatments design (1)		Usability Analysis (1)
Quantitative (2)		Experimental control group (2)
Quantitative t-test (1)		
Mixed methods (4)		
Comparison linear modeling (1)		
Interviews (3)		

4.3 Distribution of Research Outcomes

From the analysis 75% of the studies reported positive outcomes. Fifteen percent reported neutral outcomes, no studies reported a negative outcome and 12% reported outcomes that were not related to effects on student learning.

4.4 Distribution of Mathematical Concepts

The majority of the studies (72%) did not indicate the specific NCTM standard being taught. Algebra and Numbers and Operations were the standards most often stated (9%) followed by Geometry (7%). Problem-solving (4%) and Reasoning and Proof (2%) were the remaining standards being studied. Communications, Connections,

Representation and Data Analysis and Probability were not represented in any of the studies. Figure 3 displays the NCTM mathematical concepts with the academic level of the students.

4.5 Distribution of Educational Levels

Elementary schools were most often the setting of the research studies (41%), followed by middle schools (24%). High schools (20%), higher education (11%) and special education (2%) were the remaining settings. No studies reported Pre-K settings.

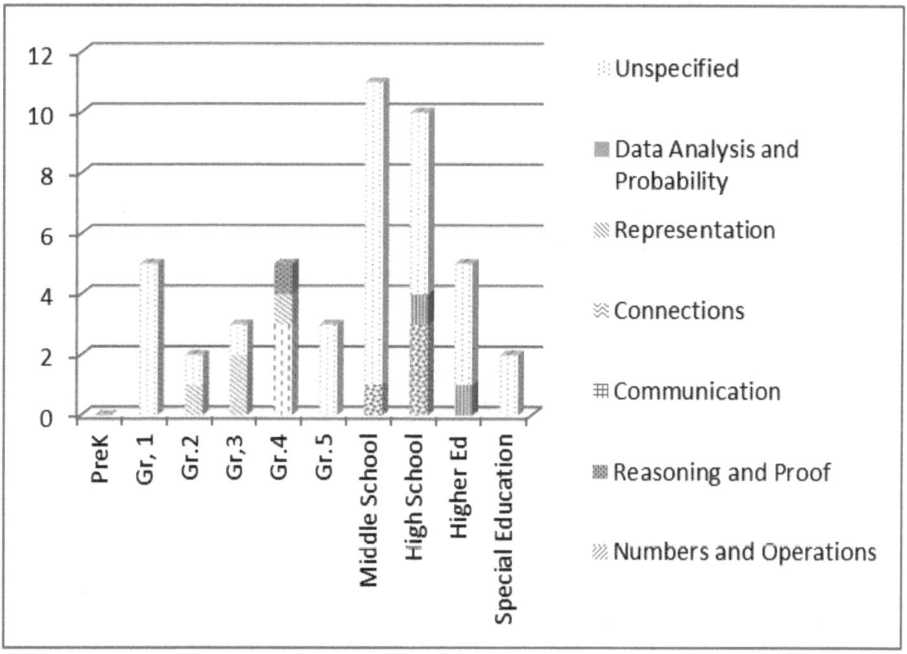

Fig. 2. Distribution of Educational Concepts and Levels

4.6 Distribution of Types of Educational Contexts

Based on research [7], three different categories of educational context were identified: formal education, non-formal education and informal education. The majority of the studies took place in formal educational context (85%), while the remainder (15%) took place in informal contexts. No studies were conducted in non-formal contexts.

4.7 Distribution of Types of Mobile Devices

Mobile phones were reported as the most frequently used mobile device (33%). Tablets (18%) were next in frequency followed by iPads (14%) and iPods (14%). Gameboys (2%) were last. Not all studies specified a type of mobile device (18%) and just used the term mobile device.

4.8 Distribution of Countries of Study

The United States was the country with the highest number of studies (42%) followed by Israel (15%) and Taiwan (10%) and the Caribbean (6). The United Kingdom, Chile, So Africa and Sweden each comprised 4% and Spain, The United Arab Emirates, Malaysia, Nigeria and India each comprised 2%.

5 Discussion

This meta-analysis provides a valuable synthesis for studies on mobile learning in mathematics. Each finding is delineated below.

Most of the studies on mobile learning in mathematics focus on effectiveness, followed by mobile learning design for the primary research purpose. As seen in Figure 1, 69% of the 48 studies stated that evaluating the effectiveness of mobile learning in mathematics was the primary research purpose. This finding is similar to the result found by Wu, et al [3], who both studied mobile learning in general. As mobile learning has become more common in educational settings, there is an increasing interest in the effectiveness of mobile learning in relation to student outcomes. A growing body of research regarding the effectiveness of mobile learning will help in answering this question.

Most studies of mobile learning in mathematics adopted surveys as primary research methods. As seen in Table 2, surveys were the primary research method (9) followed by case studies (8) regardless of the research purpose. The findings reveal a wide variance across the research methodologies with no one dominant methodology. In an analysis of the various type of studies, qualitative methodologies are more predominant than quantitative. This is an interesting difference from the findings of Wu et al. [3]. In their meta-analysis on mobile learning across all subjects they found quantitative methods to be dominant.

Most studies of mobile learning in mathematics report positive outcomes. Figure 2 shows that 36 of the 48 mobile learning in mathematics studies present positive outcomes. This finding correlates with the Wu et al. [3] findings about mobile learning in general. A word of caution needs to be stated here as it can be argued that less research is published with a negative outcome than those that are positive.

The majority of the studies did not indicate the specific mathematical concept being taught. Although the mathematic concepts were not always identified, typically a grade level was presented. This could mean that the authors of the research manuscripts thought that the reader understood what is expected of students at that grade level.

It could also lead to the inference that the mathematics was not the focus of the study but the technologies used. Finally, it could be that the researchers were developing mathematical skills and abilities across the subject and not confined to a particular strand in mathematics.

The use of mobile devices for mathematics learning is most common in elementary school settings. Mobile learning is most frequently used in elementary mathematics settings (41%) followed by middle school (24%). This is an interesting difference from the findings of Wu et al. [3]. In their meta-analysis on mobile learning across all subjects they found higher education to be the most frequent context for mobile learning studies. However, when the focus is on mathematics the research takes place primarily in an elementary setting.

Formal educational contexts were the most often places in which mobile learning and mathematics took place. The majority of the studies took place in formal educational context (85%), while the remainder (15%) took place in informal contexts. No studies were conducted in non-formal contexts.

Mobile phones were the most widely used devices for mobile learning in mathematics. Mobile phones were the device cited most often as the mobile learning tool used in the mathematics studies. They accounted for 33% of the devices identified. However, the iPad (14%) and the iPod Touch (14%) combined almost met the usage rate of mobile phones. The generic term, tablet, appears in 18% of the studies and 18% of the studies do not identify a specific device. As new technologies emerge, in the future this may be split into sub categories of mobile learning, such as wearable technologies.

Research on mobile learning in mathematics is geographically diverse; however, the majority of the studies have been conducted in the United States. An analysis of the 48 research studies included in this meta-analysis indicated that 42% of the studies were conducted in the United States. However, studies included in this analysis represented 12 different countries found on five different continents—North America, South America, Europe, Asia and Africa. This is similar to the findings of Hung and Zhang [13], who studied mobile learning across the subjects. They found Taiwan as the most common country followed by the USA. The growing global access to mobile technologies has made the use of mobile devices in educational settings more prevalent around the world and it is appropriate that the research would be world-wide.

6 Conclusion

Scholars conducting recent meta-analysis of the field of mobile learning [e.g. 10, 13, 3] provided valuable information in helping the research community in better understanding mobile learning trends. However, specific information about a particular subject area cannot be extracted from that information. This study provides a unique meta-analysis on mobile learning within the field of mathematics.

This unique study provides new information to the academic field of mobile learning in mathematics. In summary, this meta-analysis of mobile learning and mathematics provides researchers and practitioners with a better understanding of how mobile learning is being used in the study of mathematics. The reported results can assist researchers in building upon previous research and adding to gaps in the research. Practitioners can benefit from reviewing what is happening in the classroom and discovering ways to use mobile learning with their own students.

References

1. Crompton, H.: Mobile learning: New approach, new theory. In: Berge, Z.L., Muilenburg, L.Y. (eds.) Handbook of Mobile Learning, pp. 47–57. Routledge, Florence (2013a)
2. Hwang, G.J., Tsai, C.C., Yang, S.J.H.: Criteria, strategies and research issues of context-aware ubiquitous learning. Educational Technology & Society 11(2), 81–91 (2008)
3. Wu, P.H., Hwang, G.J., Su, L.H., Huang, Y.M.: A context-aware mobile learning system for supporting cognitive apprenticeships in nursing skills training. Educational Technology & Society 15(1), 223–236 (2012)
4. Capretz, L.F., Alrasheedi, M.: A meta-analysis of critical success factors affecting mobile learning. In: IEEE International Conference on Teaching, Assessment, and Learning for Engineering, TALE (2013)
5. Wu, W.H., Wu, Y.C., Chen, C.Y., Kao, H.Y., Lin, C.H., Huang, S.H.: Review of trends from mobile learning studies: A meta-analysis. Computers & Education 59, 817–827 (2012)
6. Pereira, O.R.E., Rodrigues, J.J.P.C.: Survey and analysis of current mobile learning application and technologies. ACM Computing Surveys 46(2), Article 27 (2013), doi: `http://dx.doi.org/10.1145/2543581.2543594`
7. Crompton, H.: A historical overview of mobile learning: Toward learner-centered education. In: Berge, Z.L., Muilenburg, L.Y. (eds.) Handbook of Mobile Learning, pp. 3–14. Routledge, Florence (2013b)
8. Caudill, J.G.: The growth of m-learning and the growth of mobile computing: Parallel developments. International Review of Research in Open and Distance Learning 8(2), 1–13 (2007)
9. Traxler, J.: Learning in a mobile age. International Journal of Mobile and Blended Learning 1(1), 1–12 (2009)
10. Frohberg, D., Goth, C., Schwabe, G.: Mobile learning projects – a critical analysis of the state of the art. Journal of Computer Assisted Learning 25, 307–331 (2009)
11. Taylor, J., Sharples, M., O'Malley, C., Valvoula, G., Waycott, J.: Towards a task model for mobile elearning: A dialectical approach. International Journal of Learning Technology 2, 138–158 (2006)
12. Sharples, M., Taylor, J., Vavoula, G.: A theory of learning for the mobile age. In: Andrews, R., Haythornwaite, C. (eds.) The Sage Handbook of Elearning Research, pp. 221–247. Sage, London (2007)
13. Hung, J.L., Zhang, K.: Examining mobile learning trends 2003-2008: A categorical meta-trend analysis using text mining techniques. Journal of Computers in Higher Education 24, 1–17 (2012), doi:10.1007/s12528-011-9044-9
14. Rogers, E.M.: Diffusion of innovations, 5th edn. Free Press, New York (2003)
15. Hossler, D., Scalese-Love, P.: Grounded meta-analysis: A guide for research syntheses. Review of Higher Education 13, 1–28 (1989)
16. Ke, F.: A qualitative meta-analysis of computer games as learning tools. In: Ferdig, R.E. (ed.) Handbook of Research on Effective Electronic Gaming in Education, pp. 1–32. Information Science Reference, Hershey (2009)

Towards an Indigenous Model for Effective Mobile Learning

Travis Timoko

Te Wānanga o Aotearoa
Te Awamutu, New Zealand
Travis.Timoko@twoa.ac.nz

Abstract. This research will leverage off the popularity of emerging technologies with young people to identify (mobile) learning practices that produce positive experiences for students. The research will take place and be guided by a learning environment that adheres to tikanga Māori practices. The expectations and mobile learning experiences of tutors and students will be obtained; this data will then be analysed to construct a mobile learning model that adheres to the needs of students within tikanga Māori learning environments. Adaption of a tikanga based teaching and learning philosophy for the development of formal mobile learning is being used as a basis for professional development and learning design. It is anticipated that a mobile learning framework that can produce positive teaching and learning experiences will contribute to improving educational outcomes for Māori and other students.

Keywords: Indigenous, mobile learning, model.

1 Introduction

Within the New Zealand Aotearoa educational context several areas are converging that lend impetus to the development of a mobile learning model that aligns to the indigenous culture (Māori), these include: proliferation of mobile technologies [1], uptake of mobile devices by youth and Government focus on better education outcomes for target populations [2], in particular Māori and Pacifica students between the ages of 16 and 17 [3].

It is from the potential of integrating these broad (and loosely defined areas) that this research looks to explore further. More precisely, it seeks to find an effective way to use mobile technology to engage young people in formal learning to produce positive learning experiences in accordance with tikanga Māori (Māori traditions and customs). Positive experiences could include something as simple as having fun or learning something new from the internet, through to something more complex like the construction of a video presentation as a piece of assessment towards specified learning criteria.

As a relatively new and continuously evolving approach, mobile learning presents exciting opportunities for teaching and learning, with many practitioners exploring and defining this area. The following Figure (Fig. 1.) presents several key elements

M. Kalz et al. (Eds.): mLearn 2014, CCIS 479, pp. 315–320, 2014.

for the identification of a tikanga based mobile learning model. The effectiveness of mobile learning is continually undergoing review. One such approach is the M3 Evaluation Framework, established by Vavoula & Sharples [4] to address challenges they had drawn from the literature. The M3 evaluation framework provides an approach to undertaking the proposed research, as well as contributes to the generation of data at various junctures and levels and will therefore play a central evaluation role within this research.

Kaupapa Māori is not a new term with its origin reaching back thousands of years [5], however modern definitions and applications are continuously evolving. Smith [6] highlights three factors that gave voice to Māori resistance to the prevailing Western approaches to research of the late 1960's as; the establishment of the Waitangi Tribunal, the revitalisation of language through Te Kohanga Reo movement and the spaces created in social science from critical and reflexive approaches to research. Indicative of the resistance was distrust, borne out of Māori experience of Western approaches to research that (mis)represented Māori knowledge. Redressing these concerns, Māori researchers sought to validate and legitimise Māori knowledge in the conduct of social science research. The critique of the prevailing paradigm created the conditions by which culturally sensitive approaches to research were developed enabling the emergence of a more sympathetic Māori approach [6]. These new ways of thinking and the frameworks offered up for applying the thinking emerged under the rubric of Kaupapa Māori research [6]. The kaupapa Māori approach will guide all aspects of this research.

Tikanga Māori, as briefly stated previously, relates to the traditional customs and practices of the indigenous people (the Māori) of Aotearoa New Zealand. These customs and traditions are practiced formally and informally, often whenever and wherever Māori come together.

Te Wānanga o Aotearoa (TWoA) is the largest Māori tertiary education provider in Aotearoa New Zealand. It was established as a wānanga pursuant to section 162 of the Education Act 1989. A wānanga is characterised "by teaching and research that maintains, advances, and disseminates knowledge and develops intellectual independence, and assists the application of knowledge regarding ahuatanga Maori (Maori tradition) according to tikanga Maori (Maori custom)" [7].

Accordingly, TWoA has established its own approaches to teaching and learning, and recently this has been established through ako wānanga a "philosophy, theory and applied educational practice of teaching and learning" [8]. Stemming from the ako wānanga philosophy is the TAWA framework, an acronym that captures the key elements of ako wānanga: namely Te Hiringa (spirituality, passion and motivation), Ako (living, learning and teaching), Whanaungatanga (respectful relationships and connections), Aro (reflective practice). This research will use the TAWA framework to design, develop and deploy formal mobile learning experiences for students.

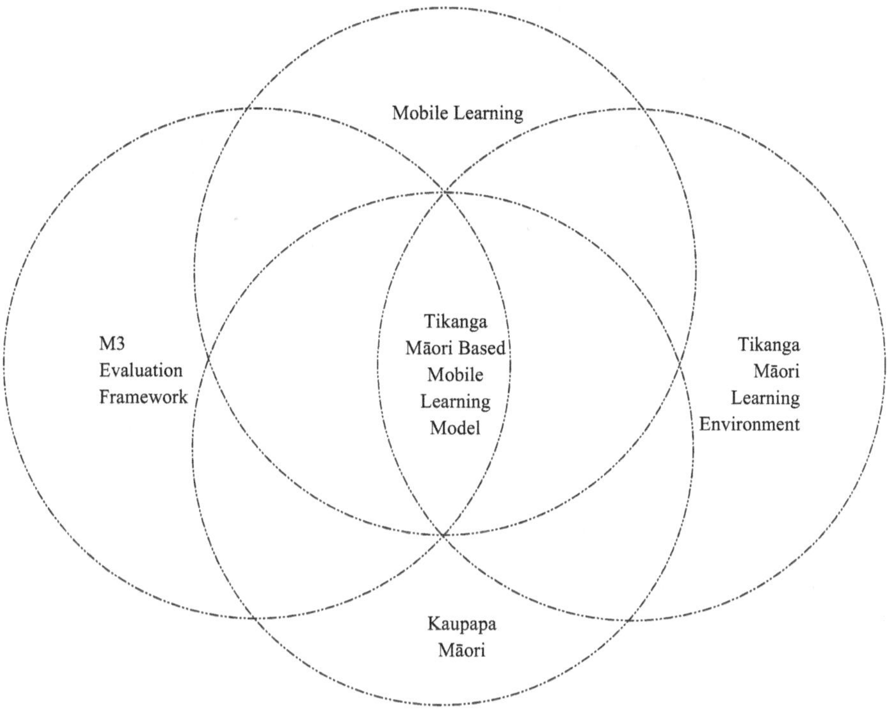

Fig. 1. Conceptual framework for the identification of a tikanga Māori based mobile learning model.

1.1 Research Questions

The research questions to be explored are:

- Question one: How do teachers/tutors and students currently apply mobile learning/emerging technology for teaching and/or learning purposes?
- Question two: What are key stakeholders (teachers, students, principal/s and parents) expectations and requirements of a mobile learning experience?
- Question three: How does tikanga Māori contribute to mobile learning models, approaches or pedagogies?
- Question four: What contribution does a tikanga Māori based mobile learning activity make to student experiences?

1.2 Proposed Methodology

This research will follow a four phase approach towards the development of an indigenous model for mobile learning. The first phase focuses on a review of the literature of the above mentioned elements to provide an understanding of the concepts and how they might be able to come together for the purpose of effective teaching and learning.

The second phase would focus on working with stakeholders to design, develop, deploy and implement a mobile learning experience that adheres to tikanga Māori practices (via the TAWA framework). Phase three would employ an iterative process of deploying the mobile learning experience, observations (but also including interviews, focus group interviews/discussions, video observations and video diaries), improvements and then redeployment. The purpose of the iterative process assists to saturate themes that emerge from the mobile learning experiences that produce positive learning outcomes for students. The fourth phase would be the documentation of the effective practices that would contribute to an indigenous model of mobile learning.

2 Initial Explorations

An initial mobile learning experience has been developed and deployed using the TAWA framework, although thorough analysis of data has not yet taken place. However, several key themes have begun to emerge from initial observations, including staff willingness to participate in mobile learning, professional development needs in relation to learning design for formal mobile learning experiences, access to emerging/mobile technology and unexpected levels of engagement by students with the content and activities.

2.1 Some Early Observations

Although all stakeholders expressed a willingness to be involved in the development and delivery of a mobile learning experience to produce formal learning outcomes, converting the enthusiasm into practice did require a shift in approach to get the first cohort underway. The willingness to be involved stemmed from familiarity and the obvious potential inherent within mobile devices/smartphones, with each of the participating stakeholders owning at least one device and using the mobile devices personally for formal or informal learning. However, the leap from traditional classroom practices to the use of mobile devices for teaching and learning proved to be beyond what was comfortable, so a compromise was made and a stakeholder was identified (former tutor of information technology) to co-design and guest facilitate the mobile learning experience.

This presented the first two observations in relation to willingness to participate but some discomfort in developing and delivering a mobile learning experience. As a result of this observation the TAWA based training package was extended from covering several key concepts via email communications and several meetings during available time, to a full day session that builds familiarity through covering formal learning, the TAWA Framework and key mobile learning features and practices into the training.

Leading up to the deployment of the first cohort funding was sort to obtain a class set of mobile devices, however funding was unable to be secured. As such a third element/observation of BOYD emerged, as a matter of necessity and practicality.

Students and stakeholders brought a number of various mobile devices to capture artifacts from their learning experiences.

The mobile learning experience was developed using the TAWA Framework in relation to a unit standard to demonstrate and apply knowledge of a personal computer system. TAWA was used to ensure the learning avoided traditional abstracted and decontextualized approaches highlighted by Herrington, Reeves & Oliver [9], while taking full advantage of mobile devices and learning [10], [11], [12], [13], [14], [15], [16]. This mobile learning experience materialized, in part, as the disassembling of several broken personal computers by students to identify components of a personal computer system, the identification and understanding of each of the components and the capturing of each component through video recording or photos from the mobile devices.

During these latter activities higher levels of engagement than were initially anticipated emerged as displayed with students completely absorbed with disassembling the broken computers, capturing the evidence of each of the components and the high levels of energy with continual questioning and inquiry to the facilitator by the students in relation to the activities.

Moving forward, further iterations of mobile learning experiences will direct focus on two of the preliminary themes to emerge, being; professional development of tutor/s in the use of mobile devices for teaching and learning within the TAWA Framework and the levels of engagement observed. It is anticipated that improved professional development for tutors will enable them to deliver the mobile learning experience. With a focus on the levels of engagement, it will be interesting to observe whether similar levels emerge with different curriculum content and potentially different contexts as well.

3 Conclusions

This research has set out to leverage off the popularity of emerging technology for students within a tikanga based learning environment. A conceptual model has been constructed to convey the key elements guiding the research. A tikanga based teaching and learning philosophy has been adapted to design a mobile learning experience that can contribute to formal learning outcomes for students. From the first cohort several themes relating to willingness to participate by stakeholders, professional development relating to mobile learning and higher than expected levels of engagement have emerged which will form the basis of future iterations of the study. It is anticipated that a mobile learning model based on tikanga Māori practices that produces positive learning experiences will contribute to improving educational outcomes for Māori and potentially other students.

References

1. Herrington, J.: New technologies, new pedagogies mobile learning in higher education. Papers/Faculty of Education, University of Wollongong 91 (p. 1 online resource (138 p.)) (2009)

2. National Government, Youth services vital to reducing welfare dependence (2012), http://www.national.org.nz/Article.aspx?ArticleID=38353 (retrieved May 7, 2012)
3. Ministry of Education. Briefing to the Incoming Minister Education Minister (2011), http://www.minedu.govt.nz/theMinistry/PolicyAndStrategy/BriefingToIncomingMinister.aspx (retrieved)
4. Vavoula, G., Sharples, M.: Meeting the Challenges in Evaluating Mobile Learning: A 3-Level Evaluation framework. International Jounral of Mobile and Blended Learning 1(2), 54–75 (2009)
5. Smith, L.T., Reid, P.: Maori Research Development: Kaupapa Maori Principles and Practices A literature Review. International Research Institute for Maori and Indigenous Education (2000)
6. Smith, L.T.: Decolonizing methodologies: Research and indigenous peoples. Zed Books, University of Otago Press, Dunedin, N.Z. (1999)
7. Education Act (1989), http://www.legislation.govt.nz/act/public/1989/0080/latest/DLM183668.html (retrieved)
8. Te Wānanga o Aotearoa. Ako Wānanga: Philosophy, theory and practices (Supporting whānau transformation through education in practice and reality). Te Wānanga o Aotearoa, New Zealand (2012)
9. Herrington, J., Reeves, T.C., Oliver, R.: A guide to authentic E-learning. Routledge, New York (2010)
10. Buffington, M.L.: Podcasting Possibilities for Art Education. Art Education 63, 6 (2010)
11. El-Hussein, M.O.M., Cronje, J.C.: Defining Mobile Learning in the Higher Education Landscape. Education Technology and Society 13(3), 12–21 (2010)
12. Johnson, L., Smith, R., Willis, H., Levine, A., Haywood, K.: The 2011 Horizon Report, Austin, Texas (2011)
13. Lee, M., Chan, A.: Pervasive, lifestyle-integrated mobile learning for distance learners: An analysis and unexpected results from a podcasting study. Open Learning 22(3), 201–218 (2007)
14. McGarr, O.: A review of podcasting in higher education: Its influence on the traditional lecture. Australasian Journal of Technology 25(3), 309–321 (2009)
15. Shumack, K.A., Gilcrest, M.: Easy Podcasting for the Busy Teacher. Journal of Technology Integration in the Classroom 1 (2009)
16. Traxler, J.: Defining, Discussing, and Evaluating Mobile Learning: The moving finger writes and having writ... .International Review of Research in Open and Distance Learning 8(2) (2007)

Pedagogical Affordances of Smart Mobile Devices Integrated with Web 2.0 Tools to Enhance English Language Teaching and Learning

Nurhasmiza Sazalli

University of Exeter
United Kingdom
nbas201@exeter.ac.uk

Abstract. I report on the provisional findings of an ongoing research project investigating the pedagogical affordances of mobile learning in combination with Web 2.0 tools for the learning of English for English as Second Language (ESL) learners. Using Design Based Research (DBR) as an approach to conduct this study, this paper presents the finding from an iteration that has completed so far. The initial design framework for the study was developed from the literature. It was tested and developed through a series of iterations and the impact of each iteration was evaluated using interviews and qualitative data analysis. One of the most important finding reported in this paper is the impact of a sense of social obligation whereby participants felt under pressure from their peers to post and to participate. This social obligation effect can have both positive and negative consequences for learning. Future research will focus on exploring ways in which pedagogical designs for m-learning with social networking can take this social obligation effect into account in order to avoid its negative consequences and make best use of its positive consequences.

Keywords: mobile learning (m-learning), social network, Web 2.0 tools, Design Based Research, Smart mobile devices.

1 Introduction

The use of Smart mobile technology allows all the tools of Web 2.0 to be accessed anytime and anywhere. As mobile technology can be used to be integrated in teaching and learning process, educators need to understand how they can be effectively used to support various kinds of learning (Kukulska-Hulme, A and Shield, 2008) and develop effective methods and materials for learning [3]. The teachers in the study by (Purcell & Buchanan, 2013:p50) also believed that new technologies should be incorporated into classrooms and schools, as long as they enhance the lesson plan and encourage learning [4]. In this paper, I report on a study using DBR to investigate how best to integrate the use of Smart mobile technology (Smart phones and tablets) with web-based social networking (Facebook) in the teaching and learning of English as a second language with adult students. Participants of the iteration was a group on

M. Kalz et al. (Eds.): mLearn 2014, CCIS 479, pp. 321–327, 2014.

Malaysian undergraduates studying at University of Exeter, taking various course such as Business Studies, Engineering, Physics, and Law. These participants who range between 21 to 28 years old, volunteered to take part in this research.

2 Research Aim

Reporting work in progress, this research aimed to find the affordances of integrating the tools of Web 2.0 and Smart mobile devices for the teaching and learning of English for English as a Second Language (ESL) students and under what circumstances do the affordances work best. As the study investigated the use of the combination, it hoped to investigate to what extent learning through Smart mobile devices and Web 2.0 tools support collaborative learning.

3 Method

Adopting a DBR methodology, this study involved designing, developing and evaluating a number of educational interventions for students studying English language via Smart mobile devices and Web 2.0 tools. As defined by Wang & Hannafin (2005: p6), DBR is "a systematic but flexible methodology that aims to improve educational practices through iterative analysis, design, development, and implementation, based on collaboration among researchers and practitioners in real-world settings, and leading to contextually-sensitive design principles and theories" [5]. It involves designing interventions that are tested, evaluated, refined and adjusted (Cobb, Confrey, DiSessa, Lehrer, & Schauble, 2003) [2]. These practices reflect DBR's continuous cyclical and iterative characters, which aim to produce design principles, learning theories, interventions of curricular products, instructional tools, and or practical solutions which can be continuously refined and improved.

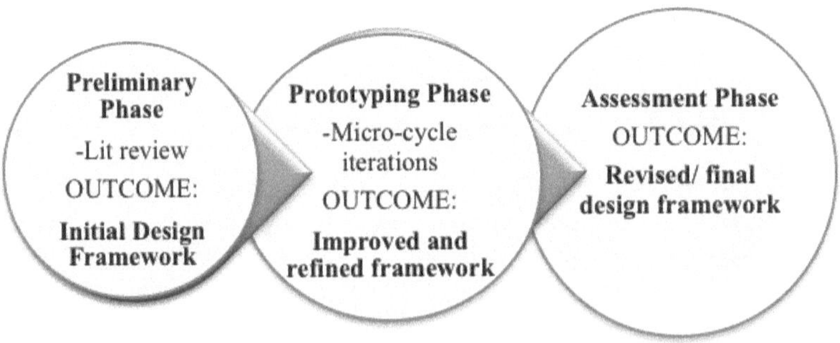

Fig. 1. Diagram 1. Three main stages in Design Based Research

This study was motivated by few conjectures. The first conjecture was language learning experience could be enhanced with the use of Smart mobile devices and Web 2.0 tools, and their integration into a tertiary education curriculum for ESL learners could promote collaborative learning among students. The second conjecture was learners' uses of the technologies were shaped by the learning activities that they were engaged in and teachers' roles were important to facilitate students' understanding. The third conjecture referred to the pedagogical affordances of mobile learning and social networking that could enhance learning. Possible affordances of Smart handheld devices in combination with Web 2.0 tools were individuality, portability, variety use of multimedia, social interactivity, connectivity, instant information finding and sharing, and anytime and anywhere access. As shown in Diagram 1, this study has three main phases. Preliminary Phase was completed; the study is now in the Prototyping Phase.

The Preliminary Phase acted as a theoretical and empirical foundation of the whole study. In this stage, comprehensive review of literature was conducted to clarify the key research terms used in the research, finding the affordances of Web 2.0 and Smart mobile devices for language learning from the literature, and understanding the theoretical principles that underpinned most mobile learning projects and related them to the needs of English as Second Language (ESL) learners. The outcome for this stage was a development of an initial design framework (Design Framework 1) as shown in Diagram 2 below. Then, I tested and developed the framework through a series of two iterations. Each iteration, being a micro-cycle of research with formative evaluation as the most important research activity, aimed at improving and refining the interventions to produce Design Framework 2, and Design Framework 3. To test the affordances of Smart mobile devices and Web 2.0 tools for learning in every iteration, all the tasks designed prompted the participants to use the technologies and were done in groups because this study hoped to find how they learn collaboratively. In Iteration 1, the respondents were given tasks that required them to use their Smart mobile devices like Smartphones or tablets and the social networking site, Facebook to search for information, contact their group members to discuss and plan their work and share information for their presentation on their Facebook Group. There was no time restriction of when the students should do the task with their group mates so long that they could complete them before the day they were assessed. They were free to use other tools of Web 2.0 that they found useful to complete their task. The outcomes of their group work were group presentations. Participants' perceptions of the impact of each intervention were evaluated using semi-structured qualitative interviews (Bryman, 2012) [1].

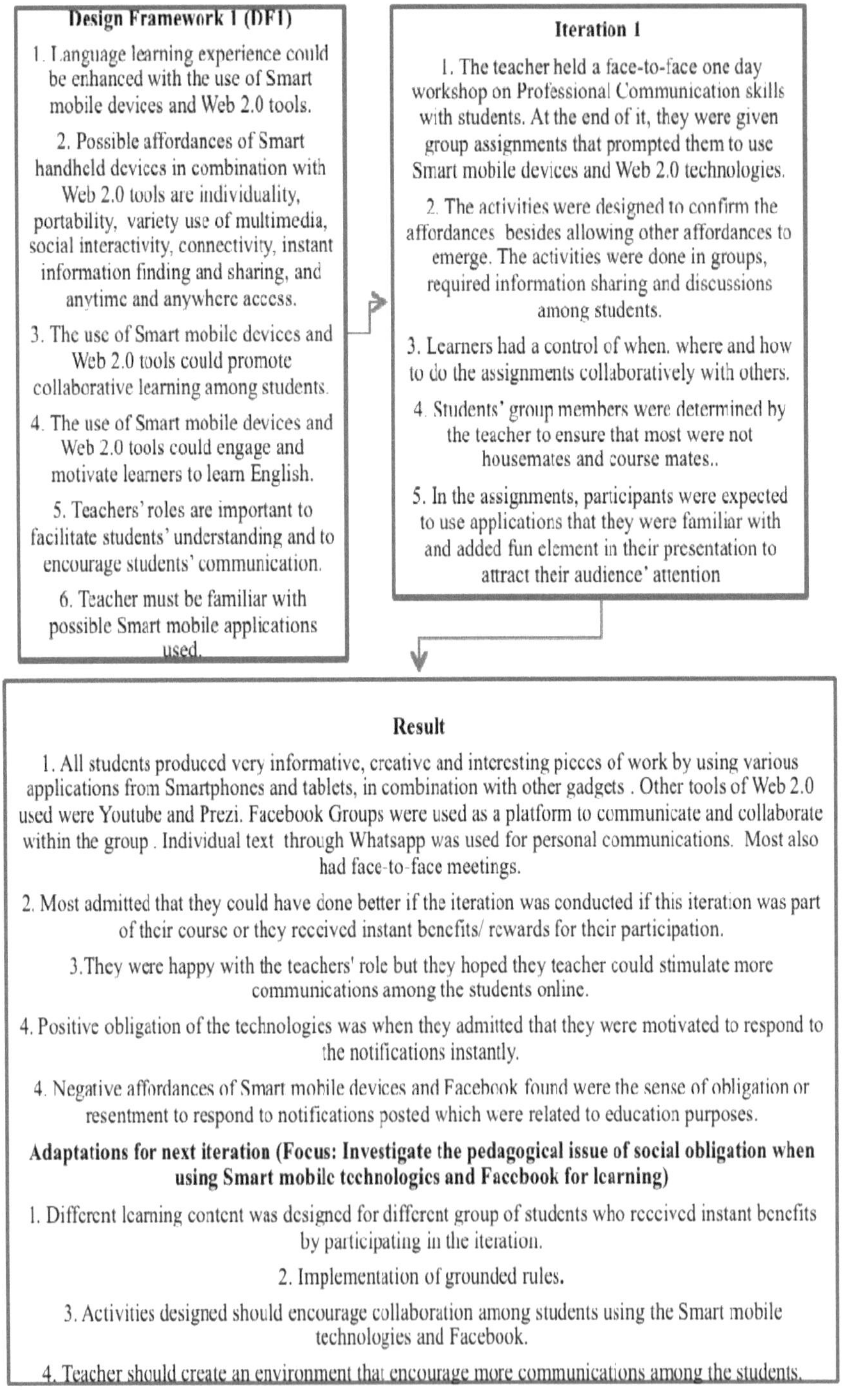

Fig. 2. The first iteration, main findings and the adaptation for next iteration

4 Findings and Discussion

The findings from the iteration confirmed the positive affordances of mobile learning and social networking in enhancing teaching and learning. Generally, all the respondents of this study used their Smartphones and tablets and social networking site, Facebook to do the tasks due to the affordances of the technologies as found in the literature. Besides the online communication via Facebook groups, Facebook messafe and personal texts via Whatsapp, most relied on face-to-face meetings to discuss about the tasks. To share the work that they did collaboratively with each other, they used Facebook and other tools of Web 2.0 such as YouTube and Prezi. All used their laptop to prepare their presentations and to find information. Smartphones were used to take photos and videos and also to find information while they were on the go. With the fact that the Smartphones and tablets are their personal belonging, they had a total control of when and where to do the research and discussed about it with their group mates as the devices provided anytime anywhere access through portable/wireless delivery mechanisms. Using various applications found in the gadgets, most produced a very informative, creative and interesting piece of work. All students were excited to be collaborative in their learning as they could use various kinds of multimedia in their presentations. But all admitted that they would commit to the iteration more if it their participation was part of their course, where it rewarded them credits. They were satisfied with the teachers' role but they hoped they teacher could stimulate more communications among the students online.

The use of Facebook Groups was found to be a suitable platform for students to share their work and to supply interpersonal and social communication network between students to communicate and also promote their work. Through investigating this combination of m-learning with social networking, the main finding of the iteration was a powerful social obligation effect in the combination. Positive affordance of the technology was that it motivated the students to respond to the notifications received as soon as possible as they got the notifications on their phones. They admitted that they would not do it later if they procrastinated. Nevertheless, as well as motivating, the social obligation effect can also have some negative consequences for learning. Smartphones and tablets may facilitate learner self-initiation and control to do the activities assigned but when they were expected to share their work to a bigger audience, some students were not happy. Postings that were made to Facebook wall can be seen by anybody on the web if the settings to limit the viewing were not changed. They felt embarrassed to show their work to the public as it might reveal the mistakes that they might have done in their work. They felt pressured as they knew they were competing with other participants whom they felt might be more advanced learners. Some participants also admitted that they felt embarrassed to ask questions in their Facebook Group page. They felt that their questions were to too simple to be asked in public. So, rather than sharing their doubts to their groups and also their teacher, most participants chose to ask questions personally to their friends by sending personal messages through their mobile devices using Whatsapp application. Some also had face-to-face meetings. Other negative aspect of the technologies was when students felt obliged to respond to notifications that were automatically sent to their smart mobile devices. About half of the participants were uncomfortable to receive notifications about the research on their mobile phones and they experienced this as an intrusion into their social space. Some reported that they changed their phone setting so that they did not receive any notification about their work on their phone. Some also ignored the notifications because they were in the midst of doing their revision for their exams. Another issue that was related to the negative affordance of the technologies was that since the work can be done anytime and anywhere, nearly all participants also admitted that they completed the work assigned at the last minute. The participants were given two weeks to do the tasks but based on their postings and discussions in their Facebook Groups, most only started to do the work at the end of the second week. This finding suggested that some changes need to be done in the next design to avoid participants to produce work at last minute.

It is interesting to find in the next the iteration of how these negative aspects of m-learning and social networking can be addressed. Focussing more on pedagogical issues, next iterations will investigate how mobile learning teaching that incorporates the idea of social obligation should be conducted. Firstly, the participants of next iteration will be carefully selected to ensure that they gain benefits by taking part in the iteration. Secondly, grounded rules will be imposed to guide the participants on how to participate in the iteration. Thirdly, the activities designed should encourage collaboration among students using the Smart mobile technologies and Facebook. Fourthly, the teacher should create an environment where students were welcome to ask questions and discuss anything to clear any misunderstanding about the task.

5 Significance of the Study

The findings of the first iteration so far suggested the importance of social obligation effect for the design of m-learning with social-networking. It was not just mobile learning but it was the integration of mobile learning and web 2.0 tools (Facebook) that lead to the social obligation effect because it involved social network and learning. As this study researched on the pedagogical affordances of mobile learning integrated with Facebook, it hopes to explain the situations where it is not good to use the technologies, when to use it, when not to use it and also how to use it for teachers. The next stage of this study will focus on exploring the motivating power of social obligation in combination with m-learning issue further. Particularly, the design for the next iteration will focus on how teachers should create a motivating and supportive online learning environment, how much notifications are just right and how much is too much and how the activities should be designed to explicitly demonstrate collaborative work among students.

6 Conclusion

Overall, the affordances of Smart mobile devices and Web 2.0 tools which were tested in this study confirmed the conjecture on the abilities of these technologies to enhance collaborative learning of English among ESL learners. However, respondents' uses of the technologies were shaped by the learning activities that they were engaged in and this study found that teachers' roles were important to facilitate students' understanding. The findings from the iterations in this study also revealed that the integration of mobile learning and Web 2.0 tools has an effect of giving obligation to its users to respond to notifications of message that wre sent to their mobile devices which can be both positive and negative for engagement and learning. The next iteration of this ongoing design-based research explored how to make the social obligation effect of combining m-learning with social networking positive for learning and how to avoid its potential negative consequences.

References

1. Bryman, A.: Social Research Methods, 4th edn., p. 808. Oxford University Press, Oxford (2012)
2. Cobb, P., Confrey, J., DiSessa, A., Lehrer, R., Schauble, L.: Design Experiments in Educational Research. Educational Researcher 32(1), 9–13 (2003), doi:10.2307/3699928
3. Kukulska-Hulme, A., Shield, L.: An overview of mobile assisted language learning: From content delivery to supported collaboration and interaction. ReCALL 20(3), 271–289 (2008), doi:ISSN: 0958-3440
4. Purcell, K., Buchanan, J.: How Teachers Are Using Technology at Home and in Their Classrooms Summary of Findings. In: Pew Research Center's Internet & American Life Project (2013), http://www.pewinternet.org/~/media/Files/Reports/2013/PIP_TeachersandTechnologywithmethodology_PDF.pdf (retrieved November 07, 2013)
5. Wang, F., Hannafin, M.: Design-based research and technology-enhanced learning environments. Educational Technology Research and Development 53(4), 5–23 (2005)

Multimodality in Mobile-Assisted Language Learning

Volker Eisenlauer

University of Salzburg, English and Applied Linguistics
Salzburg, Austria
volker.eisenlauer@sbg.ac.at

Abstract. Current mobile language learning applications are the latest link in a chain of learning materials that are designed to trigger self-directed and holistic learning experiences. The interactive and visually appealing learning materials provide contextualized input and offer various options for enhancing a learner's productive and receptive language skills. Turning to practice, the multimodal potential of mobile assisted language learning applications appears far from exhausted: Various mobile Apps use multimodality only in a limited way as they either employ images only for illustrative purposes or exhibit large discrepancies between the ideational meanings of different representative modes.

To assess the potential of multimodal meaning creation for mobile learning, my paper reflects on specific semiotic characteristics of verbal and pictorial signs and investigates the semantic relations holding between these modes. The integration of insights from educational and multimodal theory with findings from mobile learning helps to identify a set of intermodal relations that are particularly suited for analyzing text-image links in mobile assisted language learning environments. Applying an empirical lens, the paper investigates patterns of multimodal meaning creation in the vocabulary tasks of one of the most popular language learning Apps, i.e. Duolingo. Finally, my descriptive and empirical findings are summarized and integrated into a set of guidelines on how to exploit text-image links for vocabulary/language learning purposes in mobile environments.

Keywords: Multimodality, Mobile Assisted Language Learning, Digital Literacy, Semiotics, Text-Image Relations, Multimodal Vocabulary Instruction.

1 Multimodality in Mobile Learning Environments

Though there is extensive research on mobile-assisted language learning [5], [11] as well as on intersemiotic relations in educational contexts [16], [24], multimodality in mobile language learning applications represents a somewhat understudied area. The limited studies in this line of research situate the concept of multimodality either in a broader understanding relating to user-activity, see 'content-related MALL studies' in [11], address technical issues relating to the integration of non-verbal data [6], or consider general design-related questions that interrelate with the shift from stationary to mobile devices. In relation to the latter, Clarke's study [4] explores how

M. Kalz et al. (Eds.): mLearn 2014, CCIS 479, pp. 328–338, 2014.

multimodal augmented-reality contexts[1] may trigger engagement among pupils, whereas Magal-Royo et al. [12] investigate different opportunities for interfacing with mobile language learning devices, i.e. voice, stylus or keyboard.

Researchers in media and educational studies describe various key affordances of mobile learning technologies that orbit around several functional and interactional features, i.e. portability, multi touch interfacing, social and instructional interactivity, context sensitivity, connectivity and individuality [7], [14]. These mobile-specific conditions interrelate in various ways with emergent multimodal literacy practices: The high-resolution and limited-sized touch-screens as well as the in-built cameras prompt users to produce and peruse rich verbal and non-verbal texts. The mobility, the innovative designs and appealing surface feels as well as the intuitive and easy to use interfaces contribute to the omnipresence of mobile devices in people's daily lives. Therefore, mobile technologies afford the formation of intimate connections between user and machine and provide rich resources for integrated and holistic learning experiences. Multimodal teaching materials can be connected to various real world contexts that are, again, multimodal by nature. The focus of this study is therefore twofold: Firstly, on the teaching material itself, i.e. on how meaning making through combinations of language and images can be exploited for vocabulary learning purposes in mobile environments. Studies in second-language acquisition have repeatedly shown that the combination of language with non-verbal resources can enhance significantly the comprehension and retention of L2 forms [19], while supporting self-directed and self-motivated language learning [16]. The second (somewhat broader) focus of this study is on how communicative practices afforded by mobile devices, such as taking pictures or going online, may enhance established imagery-based vocabulary instruction methods. Multimodal theory provides a range of analytical tools that help to describe and understand meaning making and learning in mobile, touchable data environments.

2 Multimodal Meaning Making and the Linking of Text and Image

Meaning making through the combination of language with images and other modes has been approached from various disciplines and theoretical standpoints. An approach that has been widely adopted across linguistics, education and media studies is 'multimodality theory' [8,9,10] which reflects on the "use of several semiotic modes [...] together with the particular way in which these modes are combined" [9, p. 177]. Comparable to the linguistic turn that explained the recognition and processing of reality through language structures [17], multimodality theory seeks to expose meaning making experiences by grounding its paradigms on the interweaving of disparate representative modes. Though researchers agree that meaning making spreads by default across different modes, it is less clear what modes are made of. Approaches in social semiotics view mode in a more general sense as a "resource for making signs in

[1] A physical, real-world setting supplemented with digital data.

a social-cultural group" [10, p. 346], whereas the categorical approach to multimodal discourse analysis [3], [22,23] draws on a more sign-system oriented comprehension: it connects the concept of mode to specific sign-related properties in cognitive information processing. The focus of this paper is laid on multiple relationships between pictorial (iconic) and verbal (symbolic) modes and its potential for mobile assisted language learning. In my discussion of the text-image links and their specific semantic and semiotic properties, I draw therefore largely on sign-related approaches to mode and multimodality.

Relations between text and image are manifold and can be studied from formal-syntactic, semantic or pragmatic perspectives. Nöth [13] proposes an accurate and applicable typology of semantic links holding between text and images and distinguishes in five different linking types:

1. *Redundancy* relies on a logocentric epistemology and applies for text-image relations, where the image echoes information (or parts thereof) provided in the text.
2. *Dominance* applies for text-image links where either the text or the image are prevalent in the presentation and processing of the data. In multimodal texts with a prevalence of textual information, images may serve for illustrating, decorative or didactic purposes.
3. *Complementarity* designates integrated text-image relations. Both representative modes complement each other and need to be integrated in order to understand the overall meaning of the multimodal text.
4. *Discrepancy* may occur in cases where pictorial and textual signs lack a semantic connection or where the text-image link is not recognized as such by the recipients of the multimodal text.
5. *Contradiction* relates to text-image links, where textual and pictorial data are incompatible and in conflict with each other.

With regards to language learning, in particular redundant text-image relations serve as effective means for acquiring new concepts and expressions. The dual coding of the same data using different channels stimulates the human mind and has positive effects on the acquisition, processing and retention of new information [15]. On the other hand, imagery data that serve, first and foremost, decorative purposes may distract from textual input and thus decrease the retention of information [20]. The logocentric bias of redundant text-image links chimes in with the dominance of the verbal mode in language teaching materials. However, the idea of images echoing, first and foremost, verbal information has been widely criticized. Due to their iconic ontology the semantic interpretation of pictorial signs is less fixed. Pictorial signs are characterized by similarity to what they represent. They facilitate a fast, simultaneous and holistic processing, but tend to be imprecise and vague, as pictures typically display a surplus of meaning, when compared with language. Although the meanings arising from picture and text may be semantically close, pictures do not simply duplicate the verbal information (and vice versa). Barthes [1] points out two fundamental semantic relations, in which verbal text and image(s) can be organized: In anchored sense relations "the text directs the reader through the signifieds of the image, causing him to avoid some and receive others" whereas in relay links "text [...] and image stand in a complementary relationship [...] and the unity of the message is realized at a higher level" [1, p.40]. In other words, language and image may elaborate on each other, when the verbal input anchors meanings emerging from imagery data or when the image illustrates meanings arising from verbal data. Imagery and verbal meanings may extend each other, when both text and image exist as relatively independent parts of a syntagma, so that the overall meaning results from a complementary meaning interplay [9]. A more specific model on semantically close text-image relations is proposed by Unsworth [24] who approaches the ideational equivalence between text and image in terms of four subcategories: Clarification accounts for cases where the imagery data helps to understand the verbal input and "can be glossed as 'viz', 'to be precise'" [24, p.1175). Exposition comes closest to information redundancy and occurs when the image serves as an alternative way of expressing the same idea. Exemplification applies for cases where the image provides a specific example of a more general concept provided by verbal signs. Homospatiality is achieved when text and image are merged into one cohesive and coherent whole, that is the lettering of the verbal mode enters the pictorial mode and becomes part of the image. According to Reinfried [16], such word-image blends are particular useful in language learning contexts, namely when acquiring and memorizing new lexical components.

When relating these concepts to the focus of this study, we can generally assume a strong dominance of the verbal mode in language teaching materials. Image(s) serve primarily for didactic purposes, cf. when compensating for deficiencies in the target language – their main objective is to support the learning and retention of L2 forms and concepts. Insights from semiotics and visual literacy have identified various sense-relations that may hold between imagery and verbal data. The most important findings presented above research can be combined and integrated in the following way:

(a) *Elaboration* describes text-image relations, where the meanings of the verbal input elucidate the meanings of the imagery meanings, or vice versa. In language teaching materials, we can embark on/assume a default prevalence of the verbal mode in the learning and processing of new information. Elaborated sense relations in text-image combinations can be further distinguished in four linking types

- o *Representation* relates to intersemiotic links, where the image displays and reiterates the literal and/or associated meanings established by the verbal input.
- o *Specification* applies to text-image links, where the image illustrates a particular aspect or instance of the verbally established meanings.
- o *Clarification* operates in text-image links, where the image serves to explain more abstract concepts arising from the verbal input.
- o *Integration* describes intersemiotic relations, where text and images are merged into each other. In such blends the letters/words are displayed in such a way that they establish some kind of iconic representation e.g. the word 'smoke' rising in a smoke-like fashion above a fire.

(b) Extension relates to intersemiotic sense relations, where textual and imagery meanings expand each other. Each mode exists as a relatively independent part of a larger multimodal syntagma. In language teaching materials text-image links that reciprocally extend each other may be useful in tasks that aim at associative learning, i.e. when a particular stimulus is associated with a new concept or situation. Extended text-image combinations can be specified along three linking types.

- *Addition* relates to text-image links, where the verbal and imagery meanings complement each other by giving extra information.
- *Incongruence* applies to cases of text-image combinations, where there is a mismatch between verbal and imagery meanings.
- *Contradiction* relates to intersemiotic sense relations, where verbal and imagery meanings are incompatible or in conflict with each other.

Note that I do not claim these categories to be exhaustive. Even though there might be gaps or overlaps, the model warrants heuristic access to a critical discussion of multimodal language teaching materials in mobile environments.

3 Dual Coding in Vocabulary Teaching Materials

Multimodal texts enjoy a growing popularity in language teaching materials and prove to be an effective method for instructing vocabulary. In line with the dual coding theory [15], research in vocabulary learning has shown that word meanings that are presented both verbally and pictorially are better memorized by learners than word meanings that are presented only verbally. However, the linking of text and image can also have negative effects on the acquisition of new words. Following the focal attention hypothesis, learners always focus on the sign system that encodes the meaning in the most immediate fashion: "[W]hen pictures and printed words are presented together, readers attend to whichever one helps them produce the response with the least effort" [19, p. 225]. It is therefore recommended to incorporate pictures in such a way that they do not distract from building a mental model of the precise word form. As Shen [21] has shown for the teaching of Chinese vocabulary, word instructions based on verbal plus pictorial coding had positive effects on memorizing the form and meaning of abstract words (i.e. ideas, qualities, and conditions), while they did not significantly enhance the retention of concrete words. Sadoski [19] reviewed a range of studies on imagery-based methods in vocabulary instruction. He highlights in particular self-generated drawings of verbal definitions as an effective method in vocabulary learning. Another valuable way of vocabulary instruction producing enhanced retention is the so called keyword method: Learners are asked to associate a familiar, concrete word with the foreign language word to be learned. Thereby the familiar word must share some acoustic elements with the unfamiliar word: For example, when learning the Australian capital 'Canberra', learners may benefit from generating the mental image of a 'can' that was put on the top of a map of Australia (see memory-key.com).

4 Analysis

While a range of studies deliver empirical evidence for the effectiveness of dual coding in second language acquisition [21], most authors neglect to question how this correlates with the ideational (in-)equivalence between text and image. The gap between multimodal and educational theory appears to be reflected/reproduced in various language learning application design practices. A broad analysis of some of the major language learning Apps (i.e. Duolingo, Fun Easy Learn, Test Your English) gave the impression that several learning environments did not employ image-text links at all or only in limited and inefficient ways. To get more specific information on meaning making through text-image links in MALL contexts, I conducted a more detailed multimodal analysis of one of the most popular language learning Apps, i.e. Duolingo.

Duolingo has more than 25 million users subscribed to the App[2] and was selected as the 'free iPhone App of the Year' in 2013 by Apple. The application allows a game-like approach to language learning, where learners can gain points and can connect to other users. The language learning platform comprises more than 55 units with each unit comprising several lessons. Each lessons consists of different types of tasks that let users practice receptive and productive language skills, such as listening, writing, speaking, translating. Scanning through the different tasks, I found Duolingo's vocabulary tasks utilizing to some extent text-image combinations. The vocabulary tasks are divided into different thematic areas, such as sport, family, travel or places, and present different questions to stimulate the learning and retention of new words. From a quantitative point of view the App makes only moderate use of text-image links. Out of the first 11 vocabulary units comprising a total of 762 questions, only 154 questions incorporate images. Among the image-supported vocabulary tasks two different types can be distinguished:

1. The 'Word Matching' task asks learners to match an L1 word to its corresponding L2 word and photograph. The learner can choose among four text-image sets, with only one being the correct answer.
2. The 'Word Translation' task stimulates learners to translate a word, from L1 to L2, and to type it in a template form. Up to three photographs illustrate the word meaning(s), or parts thereof, and thus facilitate the activation of corresponding mental imageries.

A closer look at the overall layout of the word matching task reveals a significant difference regarding the presentation of the assembled input data. By default, the L2 expression and the corresponding photograph is presented together on a 'playing card like' background, thus signifying that text and image relate in some way to each other. On a formal level we find a strong dominance of the photograph in relation to the relative size of the linked word and image. It is therefore likely that learners make their choice primarily on the basis of the imagery input information, while neglecting

[2] As of January 2014.

the processing (and learning) of the written L2 word form.[3] Greater attention to the written word form is given in the word translation task, where users are required to type in the L2 equivalent. In both task types the combination of verbal/word stimuli with semantically related photographs is likely to activate all kinds of mental imageries that may enhance or be in conflict with the learning and retention of the L2 word.

In order to discover multimodal meaning creation in mobile vocabulary learning in greater detail, the word-image stimuli in Duolingo's vocabulary tasks were subjected to a closer analysis. The analysis was based on the intersemiotic linking types introduced in section 2. Following the quantification and categorization of text-image links in Duolingo's vocabulary tasks, the disclosed sense relations at work were discussed in terms of the findings from educational research put forward in section 3.

The sample comprises a total of 79 tasks. The text-image stimuli differed in relation to the number of attached photos. The majority of text-image links (73) involved a combination of three photos with a single verbal prompt, in five cases the verbal stimulus was combined with two photos, and in only one task the verbal prompt was linked to a single photo. In cases of tasks with multiple images each photo was classified separately in its sense relation to the verbal stimulus. In a second step, the intersemiotic relations were examined together/evaluated in sum and assessed in relation to the predominant semantic connection of the verbal-pictorial prompt.

When focusing on the tasks as a whole and from a quantitative perspective, the multimodal stimuli employed slightly more extended text-image relations (40) than elaborated text-image relations (39). A more narrow focus on the individual word-photo links, disclosed photos with a representational function as the most prominent pattern (29,1 %), followed by photos with an additional function (27,4 %), that is photos where the imagery meanings complement the meanings established by the verbal input. The chart below provides more specific information on the percentage distribution of the linking patterns that were deduced in the multimodal analysis of the word-image stimuli in the word translation tasks.

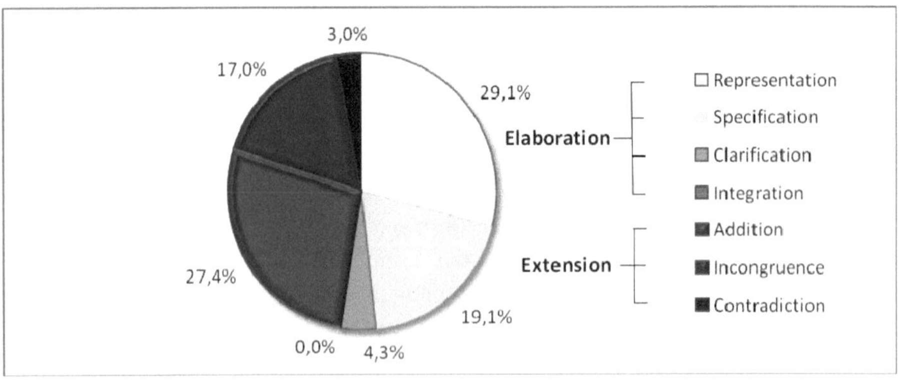

Fig. 1. Patterns of intersemiotic sense relations in Duolingo's translation tasks

[3] On the other hand, the learning of the corresponding pronunciation is supported through an audio recording of the word that is being played when tapping on the text-image link.

5 Discussion

In view of the fact that previous research advocates that a high ideational equivalence between verbal and pictorial prompts would promote the learning and retention of vocabulary, the high quantity of intersemiotic meaning extension came somewhat as a surprise. In most cases, the extra information did not establish any systemic interconnections, that is associative links to meanings established in preceding or subsequent tasks. There is no evidence that intersemiotic meaning extension within Duolingo would prompt associative learning among users by actively establishing conceptual relationships between (unrelated) tasks and units. In photos assigned to the meaning Addition type, the photos displayed, first and foremost, specific contextual conditions, i.e. participants, actions or settings, that connected somehow to the lexical meanings of the verbal input. However, almost one fifth of the images established Incongruent sense relations to the verbal stimuli, as they illustrated very general or far-fetched lexical meanings. A total of seven cases corresponded to meaning Contradiction, as the predominant lexical and imagery meanings were either incompatible or in conflict with each other. An example of incongruent and contradictory text-image links in Duolingo is given below.

Fig. 2. A case of Incongruence and Contradiction in the text-image stimulus 'Erde'

The figure illustrates the presented multimodal input for finding the English translation equivalent of the German word 'Erde' (English: earth) denoting either 'the planet on which we live' or 'the material in which plants grow'[4]. At closer examination, we quickly observe an inconsistency in the ideational equivalence between verbal and pictorial input data: The salient fence like structure in a landscape is in conflict with the literal meaning of the verbal expressions. Likewise, the semantic

[4] "Earth." *Merriam-Webster.com*. Merriam-Webster, 2014. 8 May 2014.

connection between the image depicting a river scenery and the dictionary definition of Erde/earth is less apparent, if at all. It follows that the ideational meanings proposed in the images are prone to distract learners from building a mental model equivalent to the literal meaning of earth.

In Duolingo's pictorial prompts that elucidated the meanings of the verbal stimuli (Elaboration), the lexical meanings (or parts thereof) were displayed in unambiguous and generic ways. Photographic Representations of the verbal stimuli portrayed the lexical meaning in a way that contextual conditions, i.e. associated actions, settings or participants, faded either into the background[5] or were omitted completely. In Specifications the image portrayed specific details or aspects of the verbal meanings. However, in some cases, the selection of the image details appeared to be largely pre-set by the layout properties of the learning environment and less by the motivated choice of the learning material designers/developers: The imagery translation prompt of the German word 'Rock' (English: skirt) displayed an enlarged detail view of a skirt's waistband, so that the garment was hardly recognizable. Images that corresponded to the Clarification type, were used to illuminate the meanings of more abstract qualities, ideas and concepts with fewer semantic features. Here, the intersemiotic sense relations emerged from more figurative and culturally biased connections. Although verbal-pictorial blends (Integrations) are considered to be particularly useful for acquiring new lexical components, text and images were by design presented more or less separately from one another.

6 Conclusion: Designing Text-Image Links in MALL Contexts

My descriptive and empirical findings can be summarized along a set of guidelines on how to exploit text-image links for vocabulary/language learning in mobile data environments. By not just transferring previous findings in multimodal learning to mobile environments, but allowing for mobile-specific literacy practices these principles promise to trigger effective and sustainable learning processes. The scientific contribution and aim of these principles is twofold: On the one hand, they aim at guiding developers when designing and implementing mobile vocabulary applications. On the other hand, they provide a conceptual framework towards more empirical research.

1. The dual coding of the same ideational meaning can have positive effects on the acquisition and retention of vocabulary. However, according to the focal attention hypothesis, images should be incorporated into MALL environments in such a way that they give room for the acquisition and retention of the precise word forms. This can be reached by a consecutive alignment of text and image combined with a repeated presentation of the word form. Alternatively, a discreet design of the image may contribute in drawing the attention to the textual input.
2. In language teaching materials language is the dominant mode – images serve for didactic purposes and aim at the learning and retention of new language forms. Pictorial prompts may elucidate the verbal input, when they display lexical mean-

[5] So that their display did not disturb the activation of the lexical meanings.

ings in unambiguous and generic ways. When organized independently from each other in larger multimodal syntagma, textual and imagery meanings may expand each other. Such extended text-image combinations may be useful in tasks that aim at associative learning.

3. Semiotic practices afforded by mobile devices tie in with and enhance established imagery-based vocabulary instruction methods: In-built cameras enable learners to generate photos that relate either in a more direct way to the verbal input data, or visualize memory aids that go in accordance with the keyword method. The connectivity of mobile technology allows users to upload and connect their photos to the respective entries of the mobile vocabulary learning environment.

The next steps in this research field are to assess the (in-)efficiency of the indentified text-image linking patterns for the acquisition and entrenchment of different language skills. Future studies could for example focus on the design and development of prototypes and provide empirical understanding for multimodal language learning in mobile contexts from a background in computer science and communication design. Other promising lines of research relate to the question how traditional language learning strategies, such as self-generated drawings or the keyword method, can be integrated and combined with multimodal literacy practices afforded by mobile learning environments.

References

1. Barthes, R.: Image, music, text. The Noonday Press, New York (1977)
2. Beatty, K.: Beyond the classroom: Mobile learning the wider world. The International Research Foundation for English Language Education (2013), http://www.tirfonline.org/wp-content/uploads/2013/12/TIRF_MALL_Papers_Beatty.pdf (last accessed May 05, 2014)
3. Bucher, H.J.: Multimodalität - ein universelles Merkmal der Medienkommunikation: Zum Verhältnis von Medienangebot und Medienrezeption. In: Bucher, H.J., Schumacher, P. (Hrsg.) Interaktionale Rezeptionsforschung. Theorie und Methode der Blickaufzeichnung in der Medienforschung, Wiesbaden, pp. 51–82 (2013)
4. Clarke, J.: Augmented reality, multimodal literacy and mobile technology: An experiment in teacher engagement. In: Proceedings of the 12th World Conference on Mobile and Contextual Learning (mLearn 2013) (2013), http://www.qscience.com/doi/pdfplus/10.5339/qproc.2013.mlearn.28 (last accessed May 05, 2014)
5. Godwin-Jones, R.: Emerging Technologies. Mobile Apps for Language Learning. Language Learning & Technology 15(2) (2011), http://llt.msu.edu/issues/june2011/emerging.pdf (last accessed May 05, 2014)
6. Joseph, S., Uther, M.: Mobile Devices for Language Learning: Multimedia Approaches. Research and Practice in Technology Enhanced Learning 4(1), 7–32 (2009)
7. Klopfer, E., Squire, K., Jenkins, H.: Environmental Detectives: PDAs as a window into a virtual simulated world. In: Kerres, M., et al. (Hrsg.) Didaktik der Notebook-Universität, Münster, pp. 259–274 (2004)
8. Kress, G.: What is a mode? In: Jewitt, C. (ed.) The Routledge Handbook of Multimodal Analysis, London, pp. 54–67 (2009)

9. Kress, G., van Leeuwen, T.: Reading Images: The Grammar of Visual Design. Routledge, London (2006)
10. Kress, G., van Leeuwen, T.: Colour as a semiotic mode: notes for a grammar of colour. Visual Communication 1(3), 343–368 (2002)
11. Kukulska-Hulme, A., Shield, L.: An overview of mobile assisted language learning: From content delivery to supported collaboration and interaction. ReCALL 20(3), 271–289 (2008)
12. Magal-Royo, T., et al.: Multimodal application for foreign language teaching. In: 14th International Conference on Interactive Collaborative Learning, pp. 145–148 (2011)
13. Nöth, W.: Handbuch der Semiotik. Metzler, Stuttgart (2000)
14. Ogata, H., Yano, Y.: Knowledge awareness for a computer-assisted language learning using handhelds. International Journal of Continuous Engineering Education and Lifelong Learning 14(4/5), 435–449 (2004)
15. Paivio, A.: Mental representations: A dual-coding approach. Oxford University Press, New York (1986)
16. Reinfried, M.: Vom 'Stellvertreter' zum 'Türöffner' – Bilder in Fremdsprachen-Lehrwerken. In: Lieber, G. (Hrsg.) Lehren und Lernen mit Bildern. Ein Handbuch zur Bilddidaktik, pp. 198–211. Schneider, Baltmannsweiler (2008)
17. Rorty, R. (ed.): The Linguistic Turn: Essays in Philosophical Method. University of Chicago Press, Chicago (1992)
18. Royce, T.: Intersemiotic Complementarity: A Framework for Multimodal Discourse Analysis. In: Royce, T., Bowcher, W. (eds.) New Directions in the Analysis of Multimodal Discourse, pp. 63–109. Erlbaum, Mahwa (2007)
19. Sadoski, M.: A Dual Coding View of Vocabulary Learning. Reading & Writing Quarterly: Overcoming Learning Difficulties 21(3), 221–238 (2005)
20. Schallert, D.: The role of illustrations in reading comprehension. In: Spiro, R., et al. (eds.) Theoretical Issues in Reading Comprehension, pp. 503–524. Lawrence Erlbaum, Hillsdale (1980)
21. Shen, H.: Imagery and verbal coding approaches in Chinese vocabulary instruction. Language Teaching Research 14(4), 485–499 (2010)
22. Stöckl, H.: In Between Modes. Language and Image in Printed Media. In: Ventola, E., et al. (eds.) Perspectives on Multimodality, Amsterdam, pp. 9–30 (2004)
23. Stöckl, H.: Sprache-Bild-Texte lesen. Bausteine zur Methodik einer Grundkompetenz. In: Diekmannshenke, H., et al. (Hrsg.) Bildlinguistik, Berlin, pp. 43–70 (2010)
24. Unsworth, L.: Image/Text Relations and Intersemiosis: Towards Multimodal Text Desciption for Multiliteracies Education. In: Proceedings of the 33rd International Systemic Functional Congress, pp. 1165–1205 (2006)

The Stages in Mobile-Assisted Language Learning Material Development

Yasemin Bayyurt, Gülcan Erçetin, and Nur Başak Karataş

Bogazici University, Faculty of Education,
Foreign Language Education Department, Istanbul, Turkey
{bayyurty,gulcaner,nurbasak.karatas}@boun.edu.tr

Abstract. Mobile language learning is increasingly becoming an integral part of higher education with a wider availability of lightweight handheld devices, which allow 'anywhere' and 'anytime' learning. This popular medium also implies "complex" stages for the design of suitable and efficient foreign language learning activities [3], [19]. The current paper illustrates this multi-layered process through which MLARG (Mobile Learning in At-Risk Group) project materials were developed. First 81 tourism vocational high school students' perceptions of English language learning needs, lacks and wants were scrutinized under the theoretical foundations laid by Dudley-Evans and St. John [9]. Then an appropriate and stimulating mobile language learning platform was constructed whereby learners' positive attitudes, language proficiency and technological literacy would be boosted in lieu of the traditional time-and place-constrained learning practices.

Keywords: Mobile language learning, Needs-analysis, Vocational language learning.

1 Introduction

Mobile assisted language learning (MALL) via cell phones has, to date, focused on vocabulary instruction through the use of voice and email or SMS [20], [33], games [29] and an intelligent virtual tutor [31]. A few have also created a classroom polling system assessing vocabulary retention [33], while others have attempted to promote listening skills [24], or collaborative activities by means of moblogging where students post words and/or pictures to a website [22].

In broader terms, MALL studies might be classified as content-based versus design-based studies [17]. The former points to the development of learning materials in formal language learning contexts with little focus on human interaction, whereas the latter thrusts into design-based issues to trigger human interaction along with independent language learning. In this regard, content-based approaches to MALL may be exemplified by delivering text or audio/video content through SMS or a website. The design-based approaches to MALL, on the other hand, are principally steered to foster learner collaboration or communication. In a melting pot, Kukulska-Hulme and Shield [17] argue that MALL ought to present multimedia as well as collaborative listening and speaking activities.

M. Kalz et al. (Eds.): mLearn 2014, CCIS 479, pp. 339–350, 2014.

To this end, the paper provides a detailed account of the decision-making process undertaken to construct the materials as well as of the pedagogical challenges encountered during the developmental phase. At this point, it deserves to elucidate how needs-analysis, as the first step of material construction, was implemented. Converging the to-be-developed content of mobile learning platforms with a spectrum of learners' needs was, hereby, the target product of a Leonardo da Vinci (LdV) Transfer of Innovation project entitled "Mobile Learning in At-Risk Group", or MLARG in short. MLARG set out to incorporate Information and Communication Technologies (ICT) and Mobile Learning Technologies (MLT) into language teaching materials and methodology for young people (aged 14-17) with limited exposure to English, which presumably jeopardizes the students' future occupational status. Within the landscape of self-directed, unobtrusive, pervasive, ubiquitous and ambient learning merged into the umbrella term, 'mobile learning' [18], MLARG helped the disadvantaged group of students to be armed with the assistive technology in the sense that they were able to make up for the incomplete process of language education, whilst engaging in 'edutainment'. Accordingly, the breadth of this extremely fast-moving field was expanded and enriched by transparency of competencies for vocational language development.

2 Needs-Analysis

Needs-analysis in second language education was pioneered in the 1960s as English for Specific Purposes (ESP) instruction gained momentum [23], [27]. Being in the limelight for at least three decades in curriculum design or language programme development schemes, learner-oriented diagnosis of language learning needs has so far been applied to diverse groups (e.g. [5], [34], [21], [26]). The current needs-analysis[1] also examines the English language learning needs, wants, and lacks of tourism vocational high school 9th graders, based on the theoretical framework postulated by Dudley-Evans and St. John [9]. Relating the inquiry area to both 'Target Situation Analysis' (TSA) introduced by Chambers [6] and 'Present Situation Analysis' (PSA) posited by Robinson [28], it taps into the idiosyncratic dynamics of the pilot group, in particular.

The study largely relies on the argument that needs-analysis should be the backbone of any ESP course, as strongly argued by Hutchinson & Waters [14] and Hamp-Lyons [13]. Subsequently, a thorough needs analysis was conducted, as perceived from two poles, by adult professionals and by the students themselves. Students' internal subjective needs were determined through questionnaires while hotel managers/academics/teachers' views on students' objective needs were explored through semi-structured interviews. The foremost reason for this two-stage needs-analysis was the fact that, distinct from the needs of regular English language learners, vocational high school students' needs arise from pressing professional needs, particularly the ability to transfer language knowledge to novel situations and to use acquired language skills in real life communication [4], [15].

[1] Here we adhere to the common term 'needs analysis', which can be confused with its interchangeable partner 'needs assessment' by some authors [12], because we believe that 'analysis' assigns value to the data obtained through 'assessment'.

2.1 Participants

The 9th graders of a tourism vocational high school participated in this needs-analysis study as they were assumed to be essentially deprived of the multidirectional access to foreign language learning materials. Indeed those youngsters suffer from the insufficient pedagogical and technological infrastructure though they are surrounded by the wide deployment of intellectual and technological tools in the neighbourhood. Unable to exploit this rich array of opportunities, the school is also left with no option but to shrink the process of English language education within 6 hours a week, due to the mainstream vocational education regulations. It seems that the inefficient language education is hard to be compensated for by the gradual increase in the following grades, hence jeopardizing students' careers in the long run. Closely allied with the above-mentioned situation, MLARG attributed the overarching term "at risk" to those 81 students (40 girls and 41 boys aged between 14–16). Consequently it evoked an awareness of the magnitude and seriousness of the case through the multi-purpose needs analysis, in that it constituted the baseline for the content development of MLARG.

2.2 Instruments

Semi-structured Interviews

In qualitative research, semi-structured interviews are frequently preferred, presuming that they yield reliable data through an 'interview guide', which offers the informants freedom to express their views in their own terms. The researcher was also able to tailor her questions to encourage meaningful two-way communication augmented by empathy, rapport, and trust, as highlighted by Glesne & Peshkin [11]. Interviews with the professionals—two hotel representatives, four English teachers at the pilot school, and two academics in the tourism department of a state university—were conducted in Turkish, their native language. The researcher went to their places of work, where they were comfortable exchanging ideas about the desired profile of vocational students in the tourism sector. In addition to questions intended to fill out their demographic profile, the four English teachers were asked programmatic questions on "time of instruction, opinions about course books, technological resources, and any experience with mobile or e-learning".

Questionnaires

In the second phase of the needs-analysis, the students' perceived needs were pinpointed by questionnaires devised to highlight any consistency among their remarks. It should be stressed that the findings of the interviews with adult professionals helped to develop a comprehensive questionnaire to explore students' English language needs. The questionnaire basically seeks information based on some commonly-known theoretical frameworks such as TSA (Target Situation Analysis), PSA (Present Situation Analysis), and LSA (Learning Situation Analysis), as well as other aspects pertaining to the model proposed by Dudley-Evans and St. John [9].

The questionnaire, which was written in Turkish, the students' native language, consisted of three sections: personal information, a five-point Likert scale, and one open-ended question: “As prospective employees in the tourism sector, please give your suggestions for improving the quality of English language education at your school so that it best fulfills your needs?” Given to 81 students, the questionnaire first elicited personal information such as gender, age, schooling, English language education, estimated proficiency levels. Then the students were asked to rate 20 items related to their language needs, using a scale of 1 (least important) to 5 (extremely important).

2.3 Results

Interviews

The interviews with teachers suggested that six hours of English language education in Grade 9 is not enough for fully-developed competence in all four skills. One teacher said, *“There is no solution to this. There is not enough time to have it more than 6 hours. It’s a pity.”* The teachers found it hard to accept the fact that students could barely speak English once they entered professional life, but *“this is the best they can do with this limited range of facilities; we cannot blame them.”* They were well aware of the employers’ concern about their employee’s need to communicate with people from various linguistic backgrounds: *“The little chance of practising the language results in this inability to speak English fluently, effectively, and appropriately. That’s why they are mostly weak communicators, and that’s what the hotel managers perpetually complain about”*.

All the teachers pointed out that the current trend to leave grammar out of instruction cannot be the remedy for the deficiencies. *“We should give them the chance of using the language in ‘real world’, and grammar is definitely in this. How can you speak without it?”* a teacher commented. Another teacher proposed integrating grammar into skills instruction; *“Things would be better”* if we could relate grammar to their prospective occupations by using field-appropriate texts.

The most commonly used technological device is the cassette player. It was acknowledged, however, that some *“well-off”* schools had started to use *“technology-rooms”*, with data-projectors, television, and so on. *“It would make a lot of difference, if we could use videos or something all the time. But although we try to make up for it with what we have, there is not much we can do,”* a teacher sighed.

Questionnaires

The paramount set of data were gleaned from the five-point scale by boiling down 20 items into three broad categories, linguistic, vocational, and “other” (academic, cultural, etc.). There is likely to be some overlapping among the categories; however, these categories help to give a comprehensive picture of the analysis (see Table 1). In the table, the most important three needs in each category were tabulated along with their individual means (M), standard deviations (SD), and corresponding responses of importance. Table 1 indicates that students’ vocational language needs remarkably outweighed the other types of needs per se (M = 4.25). This accounts for the early awareness of the comparable importance of vocational language demands and hence implies that those demands should be seriously addressed in English lessons.

The parameters of importance formulated in line with the mean scores concomitantly establish the greater influence of vocational language competence. It can be assumed that the limited time for vocational language instruction in the 9th grade does not thwart its perceived importance.

The open-ended question evinced supplementary responses that are similar to the interview findings. Fifty-two per cent of the students suggested that the number of hours of English be increased; 23% proposed that speaking could be taught more effectively; and 12% recommended that Computer-Assisted Language Learning (CALL) materials be utilized.

Table 1. Categorization of language needs by vocational high school students

Categories	Items	M	SD	Students' responses of importance
General language needs	To understand native speakers of English easily.	4.44	.94	Very important
	To use the four basic skills of English (listening/ speaking/reading/ writing) effectively on a daily basis.	4.43	.94	Very important
	To be able to use English grammatical rules and sentence patterns fluently.	4.06	1.11	Very important
Vocational language needs	To use English in the specific contexts such as restaurant, front desk, etc.	4.54	.79	Extremely important
	To have a high position in tourism sector in future.	4.53	.72	Extremely important
	To improve vocational vocabulary knowledge.	4.39	.78	Very important
Other: academic, cultural...	To prepare for English proficiency tests and entrance exams in Turkey or abroad	4.01	1.14	Very important
	To participate in the exchange programs funded by the school or international organizations.	3.93	1.25	Very important
	To be able to express the Turkish way of life and culture in English.	3.74	1.00	Very important

3 Material Development

It is worth noting at the outset that MLARG ventured two main goals: (a) to develop an all-encompassing set of vocational language learning materials rather than small-scale application, (b) to go beyond content delivery with the collaborative and communicative tools. As for the content delivery, a courseware with eight units has been composed, and the content of each unit has been carefully designed so as to achieve coherence across units and continuity within units. With respect to collaboration and communication, blog, discussion, and chat rooms have also been concatenated with this mobile system.

For content delivery, a courseware with eight units were developed and uploaded on a learning management system (LMS), namely Moodle. The units in the courseware included two types of activities: a) recognition-type reading and listening activities (e.g., filling in blanks, matching, multiple-choice, True/False) that provided immediate feedback for learners so that they could access them outside the classroom whenever and wherever they wished, b) collaborative writing tasks based on tools like blogs, wikis, and discussion forums, intended to be incorporated into classroom teaching by the teacher by engaging a group of learners. These activities are discussed in detail in the next section.

Moodle was chosen as the LMS not only for its wide use, openness, and an existing (and modifiable) support for displaying of content on mobile terminals but also for its capability to provide the learning content in the same graphical user interface (GUI) that is used on standard computers. Since the original Moodle system is not suitable for devices with small displays, the MLE (Mobile Learning Engine) plug-in was used to run Moodle on Android systems. The user interface and its functions are based on standard Moodle version with the addition of a specially prepared video server within the system. Video sequences were uploaded to the server and then automatically converted to a format suitable for mobile devices (i.e., mpeg-4).

Initially, the user must enter a username and a password to download the Moodle data. After login, the system displays a screen containing the list of available courses. All of the eight units All of the eight units entitled "*Restaurant, Tourist Information Office, Hotel Check-in, At the Travel Agency, Destinations, On the Tour, Hotel Facilities and Check-out*" are available to the student for access. Each unit is divided into warm-up, listening, reading, vocabulary, and language use sections. Since the materials (except the collaborative writing tasks) are designed to be used by learners individually outside the classroom, the student can access any of the sections in any order, yet s/he is advised to access them in the order presented. In other words, the activity in each section is self-contained for that section but because one section prepares the student for the next section, the student is advised to follow the order of the sections. For collaboration and communication, a "community" tab was created for students to share information, read posts, and send instant messages through Moodle's blog, forum, instant messaging and wiki functions. Moodle on android queries blog entries and brings them to the user's android device. Blogs are shown as a list of blog titles. Students can search existing blogs with a few keywords, and the android application will search and list all relevant blogs in Moodle. They are shown in an easy-to-read text format. Students cannot write new blogs from the android application, mainly because it is difficult and time consuming to send long messages from a mobile device. Hence blogs can only be read. However, students can write comments for the blog posts. The forum feature enables students to post messages and to reply to posts that have been submitted by other people in the community. This feature, like the blogs, allows students to share ideas. With the instant messaging feature, one can see online friends and start instant messaging with them.

Collaborative activities were incorporated into the writing section. Unlike reading and listening activities that could be used outside the classroom individually by the student, collaborative writing activities would be meaningfully used if they were incorporated into classroom teaching the teacher. For instance, one writing activity required the student to provide the best directions from one place to another in Istanbul. A wiki was chosen as the application to implement this activity so that the learners could make corrections on each other's directions. A collaborative activity like this would only be possible to implement if the teacher incorporates the activity as part of classroom teaching.

3.1 Content Description

Syllabus

For MLARG purposes, the miscellany of topical and situational syllabuses, where the content was organized around eight topics/situations appealing to the learners, was deemed to be the most felicitous. These eight topics/situations were as follows: at the restaurant, at the tourist information office, hotel check-in, at the travel agency, destinations, on the tour, hotel facilities, and hotel checkout. In accordance with the students' needs, all of the topics were selected, which was accordingly surmised to arouse their interest in the materials. Under each topic/situation, functions, structures, vocabularies, and tasks were listed and aligned to the 9th and 10th grade curriculum in terms of their linear progression so as to ensure the congruence between MLARG syllabus and school curriculum [8].

Unit Structure

Identified with a different topic and congruent lessons, each unit had a similar structure composed of listening, reading, vocabulary, and language use sections (see Figure 1). The consistency of the structure was of key importance to embark on a clear reading path, that is, accessibility [10].

Concerning listening and reading skills, the gist and specific information of texts were inquired. Assumedly these two sub-skills for listening and reading would render revisiting the same text for different purposes in a more meaningful way. Lastly, drills with domain specific vocabulary and language functions in selected topics/situations and structures were inserted into each unit.

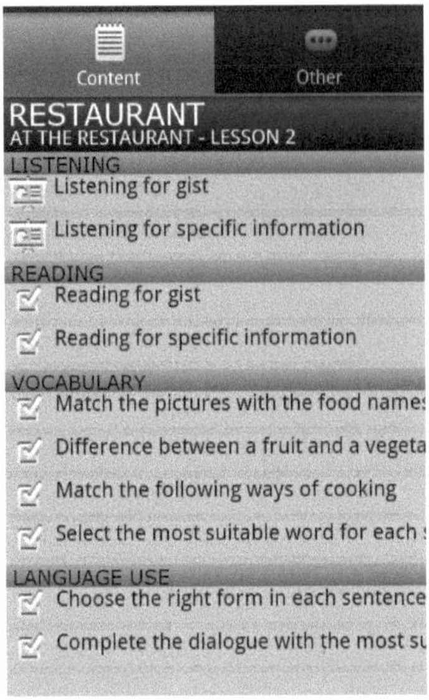

Fig. 1. The layout of 'restaurant' unit

Input

For English for Specific Purposes (ESP), input is referred to as the primary facet of the materials design model offered by Hutchinson and Waters [14]. It may involve text, dialogue, video/audio recording or diagram which supplies new language items, proper models of language, and opportunities for learners to employ their information processing skills and background knowledge of both target language and subject matter [14]. An exceptionally influential characteristic of input in MLARG materials was the inclusion of target learners' culture, in other words, the cultural context for the given input comprised the local elements so that learners were able to deploy their resources to be engrossed on language instead of unfamiliar cultural components [1].

The types of input in MLARG materials ranged from reading texts to dialogues or monologues coalesced into audio or video recordings, and polished by visuals such as diagrams and pictures. For example, with a domain-specific language in their dialogues and monologues, the listening texts entailed the roles the target students were likely to take on as service providers. Likewise, the reading texts were attuned with the sorts the students were likely to come across with in real life such that they were empowered with brochures, advertisements, leaflets, signs, and notices.

Activity Types

As stated above, listening/reading for gist and listening/reading for specific information were two broad categories for listening and reading activities. The former asked for the grasp of the overall meaning of the text, yet the latter required non-or-linguistic responses to the questions.

In general, the materials included the following exercises that exploited top-down skills:

- Listen to a part of a conversation and infer its topic.
- Listen to a set of conversations and match them with given pictures.
- Listen to conversations and identify the speakers, location, and the topic.
- Listen/read and order a sequence of pictures.
- Listen/read and complete a document or diagram.
- Listen/read and answer direct or inferential questions in multiple-choice or true/false format.
- Listen/read and select pictures, akin to the topic.
- Listen/read and follow the directions on a map.
- Read and determine the topic, source and audience of the text.
- Read and select a title for the text.
- Read and match the given descriptions with pictures.

Recognition type activities (e.g. filling in blanks, matching pictures with words, etc.) together with vivid and stirring visuals underpinned the vocabulary section. In a similar vein, language use section was prevailed by multiple-choice questions, matching, and filling in sentences or dialogues. Also, communicative and collaborative activities like blogs with text and multimedia messages, or discussion forums with text-based comments were built into the system (See Figure 2).

Fig. 2. Course community section within 'restaurant' unit

3.2 Design Features

A striking aspect of the MALL materials in the present project is that they enable different types of media to be reconciled into a single application. The collage of verbally presented information and visuals in contextualized learning [16] and users' control over their interaction with multimedia input [7], [25] are the prominent virtues of multimedia environments. Referring to Scheiter and Gerjets's [30] classification of learner control, it was asserted that learners' control on the pace of the material (pace control), the content and the time to be spent on it (content control), the order of the available topics to study on (sequence control), and on the layout of the content (in a verbal or pictorial format, labeled as representation control), increases interest and motivation in learners [2], [7]. Therefore, the current materials were accredited with pace and content control; for instance, the audio and video control menu became active during listening for specific information so that learners could pause, play, rewind, or forward the files. Sequence control was also allowed, to a great extent, except for re-ordering listening and reading sub-skills. Listening/reading for gist was essentially required to be completed prior to listening/reading for specific information, since understanding the gist was anticipated to ease detailed comprehension. Finally, representation control was merged in reading texts through multimedia glosses, in which words were defined with pictures. Since the MLARG materials were intended to be used by individual students outside the classroom as supplementary to the class materials, providing such learner control was deemed necessary. Therefore, access to all the units within the courseware and activities within a unit was allowed upon login into the system.

3.3 Challenges

Small screen size posed certain challenges in terms of manipulating the cognitive load imposed on working memory resources [32], for poorly designed instructional materials could increase the load on working memory. For example, in listening activities, a scheme where students could see and answer the questions while listening was projected but small screen size did not allow it, which made both users' and designers' job painstaking. Unfortunately, there is a dearth of empirical research to guide the design of MALL materials by optimizing this load with an efficient user interface.

Small screen size could also have forced to curb necessary paragraphs and so distort the nature of reading as a discourse level activity. However, ranging from 76 to 310 words, the content of the reading texts was not overshadowed by the screen size. Still it warrants further research to enlighten the appropriateness of the length of reading texts in MLARG.

4 Conclusion

This paper has shown that needs-analysis ought to be the starting point for curriculum and materials development as it provides an accurate profile of a target group. Based on the needs analysis model developed by Dudley-Evans & St John [9], it administered two primary data collection instruments, namely questionnaires and interviews, which offered some insight into the students' profile and language needs. In this respect, the present study took the first step in exploring a disadvantaged group's internal needs and some external demands that informed the development of a suitable mobile language learning platform supplementary to the students' in-class vocational English education.

Alongside this evolving issue, MLARG envisages to foster English teaching/learning practices cognizant with students' real needs by embedding and institutionalizing mobile learning at a tourism vocational high school in Turkey. All in all, these results can guide some worthy modifications to the curriculum of vocational high schools in Turkey, where the use of mobile phones is claimed to become a facilitating tool in support of English language learning. Despite its impacts on attitudes towards English, technological literacy, motivation to study 'anywhere' and 'anytime', and increased self-esteem that comes from feeling valued, MALL has a long way to go with empirical studies vetting the factors for 'learnability' and 'usability'.

Acknowledgement. This study was supported by European Commission Education and Training Lifelong Learning Leonardo da Vinci Transfer of Innovation Programme, "MLARG: M-Learning for young people in at-risk groups" Project (project number: 2009-TR1-LEO05-08674). We would like to thank the administrative staff, the English language teachers, and 9th and 10th grade students at Istanbul Etiler Tourism Vocational High School for their valuable contribution to the project.

References

1. Alptekin, C.: Cultural familiarity in inferential and literal comprehension in L2 reading. System 34, 494–508 (2006)
2. Alexander, P.A., Jetton, T.L.: Learning from traditional and alternative texts: New conceptualization for an information age. In: Graesser, A.C., Gernsbacher, M.A., Goldman, S.R. (eds.) Handbook of Discourse Processes, pp. 199–241. Erlbaum, Mahwah (2003)
3. Atwell, J.: A technology update and m-learning project summary. Learning and Skill Development Agency, London (2005)
4. Bayyurt, Y., Karataş, N.B.: Needs analysis as the first step when developing content for mobile language learning. Boğaziçi University Journal of Education 28, 13–21 (2011)
5. Brown, J.D.: Foreign and second language needs analysis. In: Long, M.H., Doughty, C.J. (eds.) The Handbook of Language Teaching, pp. 269–293. Blackwell, Oxford (2009)
6. Chambers, F.: A re-evaluation of needs analysis. ESP Journal 1(1), 25–33 (1980)
7. Deimann, M., Keller, J.M.: Volitional aspects of multimedia learning. Journal of Educational Multimedia and Hypermedia 15, 137–158 (2006)
8. Dubin, F., Olshtain, E.: Course design: Developing programs and materials for language learning. Cambridge University Press, Cambridge (1991)
9. Dudley-Evans, T., St John, M.J.: Developments in English for specific purposes. Cambridge University Press, Cambridge (1998)
10. Ellis, J., Ellis, P.: Learning by design: Some criteria in EFL coursebooks. In: Sheldon, L.E. (ed.) ELT Textbooks and Materials: Problems in Evaluation and Development, pp. 90–98. The British Council, Oxford (1987)
11. Glesne, C., Peshkin, A.: Becoming qualitative researchers: An introduction. Longman, White Plains (1992)
12. Graves, K.: Teachers as course developers. Cambridge University Press, England (1996)
13. Hamp-Lyons, L.: English for academic purposes. In: Carter, R., Nunan, D. (eds.) The Cambridge Guide to Teaching English to Speakers of Other Languages, pp. 126–130. Cambridge University Press, Cambridge (2001)
14. Hutchinson, T., Waters, A.: English for specific purposes. Cambridge University Press, Cambridge (1987)

15. Kavaliauskiene, G., Uzpaliene, D.: Ongoing needs analysis as a factor to successful language learning. Journal of Language and Learning 1(1), 4–11 (2003)
16. Kramsch, C., Anderson, R.W.: Teaching text and context through multimedia. Language Learning and Technology 2, 31–42 (1999)
17. Kukulska-Hulme, A., Shield, L.: An overview of mobile assisted language learning: From content delivery to supported collaboration and interaction. ReCALL 20, 271–289 (2008)
18. Kukulska-Hulme, A., Traxler, J.: Mobile learning: A handbook for educators and trainers. Routledge, London (2005)
19. Kukulska-Hulme, A., Traxler, J.: Designing for mobile and wireless learning. In: Beetham, H., Sharpe, R. (eds.) Rethinking Pedagogy for a Digital Age: Designing and Delivering E-Learning, pp. 180–192. Routledge, London (2007)
20. Levy, M., Kennedy, C.: Learning Italian via mobile SMS. In: Kukulska-Hulme, A., Traxler, J. (eds.) Mobile Learning: A Handbook for Educators and Trainers, pp. 76–83. Taylor and Francis, London (2005)
21. Long, M.H.: Second language needs analysis. Cambridge University Press, Cambridge (2005)
22. Mielo, G.: The medium is the moblog. ETC: A Review of General Semantics 62, 28–35 (2005)
23. Munby, J.: Communicative syllabus design. Cambridge University Press, Cambridge (1978)
24. Nah, K.C., White, P., Sussex, R.: The potential of using a mobile phone to access the Internet for learning EFL listening skills within a Korean contex. ReCALL 20, 331–347 (2008)
25. Plass, J.L., Jones, L.C.: Multimedia learning in second language acquisition. In: Mayer, R.E. (ed.) The Cambridge Handbook of Multimedia Learning, pp. 467–488. Cambridge University Press, New York (2005)
26. Purpura, J.E., Graziano-King, J.: Investigating the foreign language needs of professional school students in international affairs: A case study. Working Papers in TESOL & Applied Linguistics, Teachers College, Columbia University 4(1), 1–33 (2004), `http://www.hawaii.edu/sls/uhwpesl/25(1)/Watanabe.pdf` (retrieved)
27. Richards, J.C.: Curriculum and development in language teaching. Cambridge University Press, Cambridge (2001)
28. Robinson, P.: ESP today: A practitioner's guide. Prentice Hall, New York (1991)
29. Sandberg, J., Maris, M., de Geus, K.: Mobile English learning: An evidence-based study with fifth graders. Computers and Education 57, 1334–1347 (2011)
30. Scheiter, K., Gerjets, P.: Learner control in hypermedia environments. Educational Psychology Review 19, 285–307 (2007)
31. Stockwell, G.: Vocabulary on the Move: Investigating an intelligent mobile phone-based vocabulary tutor. Computer Assisted Language Learning 20(4), 365–383 (2007)
32. Sweller, J.: Cognitive load during problem solving: Effects on learning. Cognitive Science 12, 257–285 (1988)
33. Thornton, P., Houser, C.: Using mobile web and video phones in English language teaching: Projects with Japanese college students. In: Morrison, B., Green, C., Motteram, G. (eds.) Directions in CALL: Experience, Experiments & Evaluation, pp. 207–224. Hong Kong Polytechnic University, Hong Kong (2003)
34. Van Avermaet, P., Gysen, S.: From needs to tasks: Language learning needs in a task-based approach. In: Van den Branden, K. (ed.) Task-based Language Teaching in Practice, pp. 17–46. Cambridge University Press, Cambridge (2006)

Challenges in Context-Aware Mobile Language Learning: The MASELTOV Approach

Mark Gaved[1,*], Patrick Luley[2], Sofoklis Efremidis[3], Iakovos Georgiou[3], Agnes Kukulska-Hulme[1], Ann Jones[1], and Eileen Scanlon[1]

[1] The Open University, Milton Keynes, United Kingdom
(mark.gaved,agnes.kukulska-hulme,ann.jones, eileen.scanlon)@open.ac.uk
[2] DIGITAL, Joanneum Research Forschungsgesellschaft mbH, Graz, Austria
patrick.luley@joanneum.at
[3] Athens Information Technology, Athens, Greece
(sefr,iage)@ait.gr

Abstract. Smartphones, as highly portable networked computing devices with embedded sensors including GPS receivers, are ideal platforms to support context-aware language learning. They can enable learning when the user is en-gaged in everyday activities while out and about, complementing formal language classes. A significant challenge, however, has been the practical implementation of services that can accurately identify and make use of context, particularly location, to offer meaningful language learning recommendations to users. In this paper we review a range of approaches to identifying context to support mobile language learning. We consider how dynamically changing aspects of context may influence the quality of recommendations presented to a user. We introduce the MASELTOV project's use of context awareness combined with a rules-based recommendation engine to present suitable learning content to recent immigrants in urban areas; a group that may benefit from contextual support and can use the city as a learning environment.

Keywords: context-aware learning, language learning, location-based learning, recommender systems, immigration, urban informatics.

1 Introduction

Smartphones enable context-aware learning – real-world support that takes into account the learner's present and historical preferences, learning activities and places visited – on devices that many people use daily. Smartphones have sensors that can capture location and activity, they have network connectivity to enable interaction, and they have the computing power to manage learning activities. Context plays an important role in language learning, and context-aware technologies have been identified as potentially enhancing language learning [1]. Context refers to a number of

* Corresponding author.

M. Kalz et al. (Eds.): mLearn 2014, CCIS 479, pp. 351–364, 2014.

parameters that may affect learning, and inform the selection of relevant resources to support a learner "including location, social activity and learning goals" [2, p. 4]. Lave and Wenger [3] argue knowledge is best gained in the situation in which it is to be employed, and since language use is typically social, enabling communication with fellow learners and native speakers is important.

This paper reviews approaches to context-aware mobile language learning, and introduces the MASELTOV approach: exploring how contextually aware language learning could be used to support recent immigrants to Europe to help them with social inclusion. MASELTOV, an EU FP7 funded project (http://www.maseltov.eu), is considering how a system could support learners in a city-wide landscape, operating at scale (across a large area and potentially thousands of users), and applied to any urban area in Europe. Our target audience has a limited educational background and likely to have work and family commitments that will make attending formal educational classes difficult; therefore a smartphone based service which can be accessed anywhere and make use of the lived environment as a contextual resource is particularly suited to this group. In the MASELTOV project, the concept of the learner's location triggering activity is important, in particular how the services can support incidental learning, "unintentional or unplanned learning that results from other activities" [4, p.1]. An Incidental Learning Framework was produced to analyse mobile incidental learning in detail and to facilitate the communication of learning designs appropriate to the situation in which an immigrant wishes to use the service [5]. The Framework considers the place an incident occurs, task(s) the learner is carrying out, the tools the learner uses, the social support that the learner makes use of, the learning outcomes to be achieved and the (relative) time the incidental learning occurs.

To achieve the goals of language learning and social inclusion, MASELTOV is building a number of integrated mobile apps, under a single dashboard (the MApp) that our target audience can use in their daily activities to both resolve immediate needs and also to enable reflection and further planning of learning goals. The MApp includes the following services:

- Language learning activities: focused around key inclusion challenges such as employment, healthcare, and negotiating bureaucracy
- TextLens: converts images taken with the phone's camera to text, and translates them using a third party translation tool or onboard dictionary
- Navigation tools: public transport planning and pedestrian sat-nav
- Help radar: finds nearby volunteers to help the user solve immediate problems
- Places of interest service: listing places of interest in the city
- Serious game: providing playful learning about cultural differences
- Information service: information about services in the new city e.g. health, employment
- Social tools: to enable contact with other learners, and sharing knowledge
- Context awareness service to make sense of users' activities
- User Profile to store user preferences, records of activities, show usage statistics and display progress indicators

- A Recommender system to provide contextually relevant targeted and personalized recommendations, including learning resources

The MASELTOV system uses mobile phone sensors and user inputs to establish context, and interrogates data collected by these services with a recommendation engine to provide relevant prompts for immediate or future learning. By drawing on user activity across a number of widely differing services, and multiple attributes of context, the MASELTOV system aims to provide highly relevant learning resources. By offering users a number of ways of improving their knowledge and understanding of the target language and culture, the system offers flexible and personalized learning that also takes account of the local environment and people as learning resources.

In this paper we consider five key contextual aspects (location; mode of activity; history of activities; social interactions; and learning goals and personal interests) and report on how the MASELTOV project addresses each, and how, through the use of a contextually aware service, recommendation engine and a user profile, they support contextually aware language learning. We conclude by discussing outstanding challenges and future work.

2 Location

Early work in context-aware computing identified location as a key aspect to providing relevant resources to users, e.g. Abowd et al.'s conceptualisation of a tour guide [6].The development of smartphones with affordable and compact embedded sensors in the last decade has made such theoretical tools a reality. Approaches can be divided into those that 'automatically' identify location by the use of sensors built into the phone, and those that require manual input from the smartphone user.

2.1 Location Identified through Automated Input

Most smartphones can calculate geographical position using onboard sensors. Location can be calculated in a number of ways:

- GNSS receivers: calculating location from satellite transmissions, often using the US GPS (Global Positioning System). They can often identify location when outside to approximately 5-10 metres, but can be affected by large buildings and offer poorer quality accuracy where there is less satellite coverage.
- WiFi access points: location can be calculated from the comparative signal strength of nearby WiFi access points with known geographical coordinates, "WiFi fingerprinting". This can give accuracy at around 30m when used in urban areas where there is a sufficient density of WiFi networks, and requires internet connectivity.
- Cell phone tower signals: triangulating location based on signal strength of nearby mobile phone towers. However, this method only identifies location to approximately 300m accuracy in urban areas (less in rural areas due to lower coverage).
- Inertial Measurement Units: smartphone build-in sensors capable of measuring motion as well as orientation. Given a starting position, the subsequent locations

can be iterated e.g. by step length and step frequency detection or double integration of the determined acceleration. This approach is also known as Dead Reckoning (DR) and stays stable within up to 1 metre for a 100 metre distance travelled.
- Transponder Beacons: dedicated transmitters which are sending a periodic beacon signal based on the Bluetooth Low Energy profile. The transponders have to be installed on site prior to location measurements. Position can be determined by either trilateration or fingerprinting (identifying the transponders by unique information they provide). Depending on deployment granularity provides accuracy between 70 to 0.1 metres.

In some cases, more than one of these methods is used in combination to increase the accuracy of position sensing.

Other services are also required to enable meaningful interpretation of the location identified, often provided as a series of geographical coordinates, such as a map service to show where the user is, or a database of Places Of Interest (POIs), to show what significant places are near to where the user is currently located. These services might be stored on the smartphone, or on remote servers (e.g. Google Maps) and in the latter case require a network connection. With some services users can download resources (like local maps) while connected, for later offline use.

The LOCH system [7] enabled Japanese learners to carry out practical tasks in a typical everyday environment such as asking for information at a train station. GPS receivers were used to identify location, and based on this information, remote tutors would offer location-specific feedback. Similarly, the PALLAS system, designed for language learners' interest-based informal learning around a city uses GPS to identify position and recommend relevant places of potential interest [8]. Zhou and Rechert's prototype personalized e-learning system for use in a botanical garden draws on both WiFi and GPS positioning to establish the learner's location, and position of nearby plants of probable interest [9]. In the MASELTOV app, we identify location through the "Android Location API" which uses a combination of GPS, WiFi fingerprinting and cell tower positioning.

2.2 Location Identified through Manual Input

Location may be identified by the learner manually engaging with the environment, or with their phone. Markers can be placed on objects or locations in the physical environment, and taking a photo or otherwise scanning these with the phone can capture information, such as coordinates or a URL that can then access information from a remote server. For example, the KLIV project [10] enabled nurses to scan barcodes on equipment in an intensive care unit to be successfully instructed on their use. Ogata and Yano's JALEPAS system for learning the correct forms of politely addressing other people in Japanese [1] used RFID tags fixed to doorways when indoors to understand when a learner entered a particular room and scanned the tags, which could then trigger corresponding appropriate content objects to indicate position.

These approaches require the prior 'marking up' of locations, to ensure that the necessary tags are in place in all the likely locations learners might visit. While this can work for small and controlled study areas, there is a challenge of scalability if a larger area such as a whole city is to be the environment where context aware learning may be triggered.

A different approach is taken in Edge et al.'s 'MicroMandarin' language learning system [11] which makes use of an existing social network service, the popular Foursquare app (http://www.foursquare.com), to confirm the user's current physical location and likely activity. The MicroMandarin service takes the information about the Foursquare location chosen by a learner to offer context-relevant small learning resources. If a user checks into a location that has been previously recorded as a café by Foursquare, the MicroMandarin service will offer vocabulary resources based on dining, such as the name of common café choices, paying for orders, etc. Linking through to social networks also enables the learner to access an existing community and draw upon their resources.

As MASELTOV seeks to provide a service that can support users across whole cities, the prior marking of locations is not an approach that we consider achievable, and similarly, reliance on third party locational services (e.g. Foursquare) may only give limited coverage. Furthermore, we want to make the service as easy to use as possible, and hence not require users to manually identify their location before being able to provide locational services. Even places like home and work are identified automatically by analysing daily routines. For this reason our approach is to identify location 'automatically' rather than via user input, deriving data from sensors onboard the smartphone. While this currently limits accuracy, advances in technology are likely to continue to improve accuracy rapidly over the next few years, for example with systems achieving greater accuracy by triangulating data from different GNSS systems (GPS, the Russian GLONASS, and the soon to appear European GALILEO).

3 Mode of Activity

The user's current mode of activity can provide contextual information which indicates which resources may be appropriate to offer to a learner, or the best times to provide resources. Like locational data, this can either be automatically derived from a smartphone's sensors (e.g. accelerometers), or by the user entering their activity manually (e.g. from a menu). Bristow et al. [12] identified that body position (e.g. sitting, standing, or walking) was key in defining the user's context when considering what resources to recommend to them, and used accelerometers to calculate this information.

The MASELTOV system has an activity recognition module which collects data from sensors built into the smartphone (movement and tilting) and interprets these to understand when the user is walking, is in a vehicle, or idle. This information is passed to the user profile, and adds contextual information to enable a better recommendation of learning resources: for example, a user who has been stationary for a period of time might be interested to receive a recommendation to try a language learning activity, while a user who is moving rapidly is not so likely to be receptive to an immediate recommendation, but might be interested to check later. Furthermore, the mode of activity also includes the recognition of different kinds of transportation like riding a bike, driving a car or using public transport. This knowledge is also shared with the user profile to learn about users' daily behaviours in order to recommend learning content on public transport or different kinds of vehicles.

4 History of Users' Activity

Return visits to locations, or other repeated activities can provide contextual information that can trigger learning resources: if a learner frequently visits a place we might assume it has some significance to them. Like the MOBIlearn project, we see "context as a *dynamic process with historical dependencies*" [13, p. 116]. A user's previous activities and choices can provide contextual information as well as their current actions.

For example, frequent visits to a train station might result in a recommendation to learn language about the public transport system and repeated visits might lead to different material being offered. The MOBIlearn project recognised that context included historical interactions as well as an interpretation of the current, dynamically changing information: resources provided to someone visiting a museum for the fourth time might not be as suitable as for someone's initial visit [14].

Another example is the SCROLL system [15] that stores a learner's images and text notes to remind them of what language lessons they have learned in a real life situation (such as seeing a doctor in the hospital), and associates these notes with locational data from the GPS receiver. On revisiting the location, the previous visits' notes are presented to the user, to remind them of what they learnt before, encourage consolidation and further learning.

Determining which elements of the environmental data are relevant and most important for informing the learner's goals is a significant challenge, as some data passed to a user profile may be of little importance to the learner. This is highly problematic in busy urban environments where there may be many places of interest nearby, and there are large amounts of contextual information that can potentially be gathered. Cui and Bull's TenseITS system [16] considered this challenge and included historical, cumulative preferences so that the user could register their context manually, to enable the system to infer what materials would be most appropriate for the learner. Learners could enter information about where they were, their current concentration level (e.g. high or low), how much time they had available, and how often they were likely to be interrupted.

The MASELTOV user profile (described in more detail below) records both user preferences, and also historical activity to enable more accurate contextual recommendations.

5 Social Interactions

Social interaction is critical for language learning [17], and smartphones which can connect learners to peers may encourage interaction with other learners and natives in authentic environments, enabling learners to "co-construct knowledge to solve problems and fill information gaps" [18, p. 283].

Recommender systems have used records of social interactions to help inform further recommendations, such as the 3a system [19]. Specific services may encourage social interactions, for example the PERKAM prototype that enables learners to find relevant peer learner-helpers [20].

The MASELTOV app has two explicit social tools: a forum, that allows users to share learning experiences, contact other learners, and socialise; and the geo-social radar, which enables them to identify local volunteers who may be able to help resolve a specific problem (such as translating at a local government office to resolve a bureaucratic need). The language learning activities also include social interactions: tasks are set that encourage the learner to interact with native speakers, and also to post small texts on the MASELTOV forum and get feedback from other learners. Furthermore, the TextLens tool allows users to upload images they have captured to social spaces, to get help for understanding their meaning. MASELTOV captures social activity through two methods: capturing usage of the relevant tools (forum, language learning activities, and geo-social radar), and a social interaction detection module, which, with the users' explicit permission, can generate anonymous statistics on communication behaviour including phone calls and typed text messages. This contextual information might identify appropriate times to prompt users to try a learning activity, or encourage them to further participate in socially-focused learning activities: for example, if little use of the forum is identified.

6 Learning Goals and Personal Interests

Brown et al. [2] identify learning goals as providing important data for context-aware learning systems. Recommendations for learning resources that aim to support daily living, rather than the completion of a structured curriculum need to take into account personal goals and interests as well as providing resources appropriate to the learner's current level of learning competency and progress through materials. The PERKAM system, for example, identifies personal interests through a learner profile: on registration, learners are asked to enter their personal information and topics of interest, and as they engage with the system, their actions are also recorded to the profile [20].

The MASELTOV service, which aims to support immigrants learning through their daily activities, and motivate an audience which might have limited or poor prior experiences of formal education, takes into account their personal interests and activities to provide relevant learning content. The MASELTOV system asks users to set up a user profile when initially registering, indicating their preferences and interests. While all fields are optional (to allow a user to take advantage of the MASELTOV services yet remain as anonymous as they wish), the user is informed that providing preferences will help improve recommendations for learning resources. This information is combined with the history of activities described previously and similarly reported to the user profile, in order to improve the quality of recommendations. On the production of recommendations, users are asked to rate their quality to enable better defined future recommendations. Future versions of the MASELTOV system will also enable learners to indicate their learning goals, and match recommendations against these ambitions.

7 The MASELTOV Approach to the Challenges of Context-Aware Learning

In MASELTOV we provide contextually aware language learning services through: (1) the use of sensors (e.g. GPS, accelerometers), (2) a number of services that generate context aware information, (3) the user's profile to store users' identities, preferences and records of activity, and (4) a recommendation service that analyses contextual information and produces meaningful recommendations. We have described the use of sensors above, and will now turn to describe our context aware service, and the user profile and recommendation services.

7.1 Context Aware Service

The context recognition service is designed to run locally on the user's smartphone as a ubiquitous background service implemented for the mobile multisensory interpretation of user behavior, as a foundation to support immigrants in host urban environments. This module within the MASELTOV app enables filtering of relevant context information, and provides the background data for language learning recommendations based on the situation-dependent context of the user's environment.

A geo-contextual event analysis sub-module incorporates geographic information of the user's surrounding environment. The geographic information facilitates recommendations connected to places visited and places that are of special interest to a user. If the MASELTOV system detects that the user is near a specific place, common phrases for predicted communication scenarios (i.e. at the doctor, at the supermarket, etc.) or instructions for proper behaviour/communication specific for this kind of place can be provided. The module provides a collection of identified interests as well, which is deduced from geo-contextual analyses and can be used to deliver highly relevant language learning lessons only, which are strongly connected to real situations faced during the day. The knowledge about the current state of movement of users can be used to improve the acceptance of recommendations for a user, e.g. to determine the proper moment to send information to the user by detecting idle or high activity periods.

It should be noted that the user's context is a wider notion and encompasses information in addition to the user's geolocation. For example, information that is searched, or a topic that is discussed in a forum defines additional contextual information pertaining to the current interests of the user. Such contextual information can also form the basis for generating targeted recommendations for learning.

7.2 User Profile

In MASELTOV, a User Profile is used to store the learner's personal preferences and a history of their actions and activities, in order to inform recommendations for learning resources. On registering, the learner enters their basic profile including their personal information, interests and competencies. As they engage with MASELTOV services, e.g. search for a place of interest, or complete a language lesson, their usage and progress is recorded. If they have agreed to have their location tracked the context aware services will report their journeys and interpret their mode of activity to the user profile. This data forms the basis of the information used to present the user with recommendations for learning resources.

7.3 Recommendation Service

A context aware learning system requires a service that will provide relevant con-tent to the user based on the contextual cues it has been given. Recommender systems for technology enhanced learning offer "some specific characteristics that are not met by today's general purpose recommendation approaches" [21, p. 319] particularly that each learner has their own learning path, and may be using their own preferred combination of tools in different environments. Traditional recommendation systems draw from two types of entities: users, and items. The majority of TEL based systems rely on contextual information drawn from personal profiles and learning progress, and do not draw on locational or activity based contextual information (ibid.). Wanaskar et al. [22] distinguish recommender systems as falling into one of two major categories: collaborative recommendation systems, and content based recommendation systems. In collaborative filtering systems a user is recommended items based on the past ratings of all users collectively; whereas in content-based recommending systems, a user is recommended items that are similar in content to items the user has liked in the past, or matched to attributes of the user. Melville and Sindhwani [23] identify that the two approaches can be combined to form a hybrid approach. The approach taken by the MASELTOV system is to employ a rule based recommender system.

Underpinning such a system is needed to interpret contextual data; this is achieved in MASELTOV by employing a rule based system as noted above. Such systems take as input a set of rules of the form precondition $\rightarrow$ action, as well as a set of data and produce their recommendations in three phases [24]:

1. Data collection – data on the user's interests and activities must be collected and combined to provide information for the recommendation system
2. Pattern discovery – rules (a number of conditions) are applied to the collected information. If a match occurs, then a recommendation might be triggered. Alternatively, the information might activate a different rule or bring a rule closer to being triggered
3. Recommendation – when the conditions of a rule are met a recommendation (action) is generated and sent to the user, suggesting an action they can take or a link to some resources they may use.

Zaldivar et al. [25] explain how rules can offer specific recommendations with even no usage information but caution that that large rule sets are hard to maintain, reengineer, and adapt to user preferences.

All the services in the MASELTOV app send information ('events') periodically when they are used to the learner's user profile on a remote server. These events are temporarily stored on the smartphone if a network connection is not available. As soon as the smartphone goes online again, the events are sent to the backend server where the recommender system is running and checked to see if they match a rule and trigger a recommendation, or contribute towards moving a rule towards being triggered. If a recommendation is produced then the recommender system will send this to the user's smartphone as a notification (similar to a text message notification) for the user to read at their convenience, and follow to the recommender learning resources.

A range of recommendations are triggered by the MASELTOV system, including:

- Contextually suitable resources: identifying a user's location may result in a recommendation to visit a nearby place, or try a language learning activity associated with categories of places they are near to (e.g. healthcare, transport).
- Progress in a task: achieving a higher level of language competency (identified by completing language lessons and associated assessment) may trigger a competency-related activity
- Identifying complementary services: encouraging learners to try services that may support their activities. For example, if a learner is using the TextLens service to photograph notices in a doctor's surgery to understand medical services, they may be encouraged to share the images with fellow learners on the MASELTOV forum, or in a dedicated Facebook learners' group.
- Recognising inactivity: if a learner has not used a particular service, or starts to use a service then stops using it, they may be encouraged to try using it again (a common strategy employed by online learning environments like Khan Academy, or busuu).

Some recommendations are time-critical: suggesting that a user may like to visit a nearby place of a type they are interested in is only valid as long as they are within a certain distance of the place, so rules can include the concept of expiration. If somebody has moved elsewhere, or their phone has been switched off and a recommendation can't be sent, then it may be deleted and not sent forward to the user. However this information (being near a place) is still stored in the User Profile as it may be useful for another rule: noting that a person regularly visits train stations may trigger a recommendation to suggest some language activities around travelling on public transport. Even though the learner is not at the train station at the time, they may wish to view their recommendations when at home, when they have time to follow up the learning suggestions.

Learners are asked to provide feedback on the value of the recommendations they are sent (selecting or rejecting them), and this information is stored in the user profile to help further improve the quality of future recommendations. However, this creates the challenge of requiring the system to understand why a recommendation may have been rejected. For example, the user might no longer be interested in a domain they have selected (e.g. restaurants), and a follow-up question needs be sent to the user e.g. "Do you still want to get recommendations for restaurants?" Alternatively, the user may not really be interested in this specific recommendation (a particular restaurant), but is still interested in the general category, and can be asked to provide further detail (they may not like a particular restaurant, or prefer Italian restaurants to French restaurants).

8 Challenges

Contexually aware language learning is becoming a reality with the rapid development of smartphones, and their increasing availability to the public. However, there are still outstanding challenges to be addressed:

- Identifying location accurately: the majority of sensor based approaches (e.g. deriving location from GPS data) do not offer enough accuracy to distinguish the users' focus when there are multiple points of potential interest in close proximity and could lead to very different recommendations, some of which will be seen as irrelevant. Additionally, geographical data systems such as OpenStreetMap that are employed by recommender systems to indicate points of interest based on locational data may not provide enough detail or be comprehensive enough in their coverage to provide accurate recommendations. Identification of location via prepopulated markers may give greater confidence in establishing the user's preferred focus, but faces the challenge of scalability across large areas.
- Timely recommendations: it is important that recommendations to users are given whilst they are still relevant (e.g. current location, mode of activity, or progression though learning materials). Recommendations that are no longer appropriate to the user's context may be seen as irrelevant and a distraction. If the system relies on passing contextual information to a remote server to generate the recommendations, environments where network connectivity cannot be assured may mean recommendations are delivered too late to be of use. The recommendations that are produced by the MASELTOV recommender have an optional expiration tag. An expired recommendation is no longer presented to the user. Recommendations may expire at an absolute time (e.g., Jan 25, 20:30), at a relative time (e.g., 20 minutes from now), or upon the reception of an event.
- The overhead of defining and keeping rules up-to-date for a rules-based recommendation system. Defining rules is a complex task, and needs domain experts (e.g. language teachers) rather than programmers to create and update rules. This implies the requirement for an authoring interface that is usable by domain experts, and not just the programmers who are building the underlying software systems.
- Ethical issues of tracking users' activities and locations: users may be concerned about how their contextual information is stored or who has access to it, and how it may be used. Contextually aware language learning systems must inform users clearly how their data will be used and stored, and offer the opportunity to opt out of some or all data gathering (though users will need to understand this may reduce the quality of recommendations that can be made to them). The User Profile component of the MASELTOV platform gives the user to option to switch on and off the collection of a number of contextual parameters, thus placing the user at control for what information he/she allows to be collected.
- Enabling smartphone users to shift from using their smartphone as purely a communication and entertainment device to exploiting it for learning. This conceptual shift will be facilitated by teachers or mentors encouraging learners to use services such as the MApp for specific learning activities.

- Extending a context-based learning experience into a more prolonged or reflective learning experience over time. This may be achieved through application designs by prompting learners to continue learning or revisit past learning.
- Overcoming various barriers to learners practicing just-in-time oral language skills on the phone in public. Some speaking practice could be designed so that it mimics natural phone conversations.

9 Conclusions

We have reviewed approaches to context aware mobile language learning services, and presented the MASELTOV approach. A number of prototype contextual learning systems have been developed to date, however these have been mostly limited in scope and deployment. MASELTOV has developed a demonstrable prototype that aims to provide services to immigrant users within the challenging context of cities, and moreover to do this at scale, so that potentially the services could be used by thousands of users across major European cities. To support immigrants in a number of different areas of their everyday lives such as finding their way around, supporting social contact and language learning, all whilst they are going about their everyday lives, ten different services have been developed within the overall app. The challenge here, then, is to provide the user with what they are likely to want, when it is most useful, without overwhelming them or irritating them with constant reminders.

To provide relevant 'context-aware learning' multiple dimensions of context must be supported, underpinned by a recommendation service that can make sense of the contextual information, and respond with appropriate and timely recommendations to suitable learning resources. The MASELTOV project considers key aspects of context to be location, mode of activity, history of activity and progress, social interactions, and learners' interests. In the MASELTOV smartphone app, MApp, we derive contextual data from a number of sources. We incorporate a user profile in the registration process to gather initial user preferences, and capture dynamically generated locational data through sensors built into the user's smartphone. We have developed a context awareness service to understand what activity sensors are reporting, and record the user's activities, usage and progression in the MApp services. These data sources are passed to a user profile which is interrogated by a rule-based recommendation system and generates relevant recommendations for resources, information, and services that will support the users' learning activities.

The MASELTOV project is testing the MApp at scale in urban environments, with field trials in London, Madrid and Vienna in 2014. The first field trials are testing services in the wild for one week as a proof of technical services, with users guided through a range of semi-structured tasks. The second field trials will last over two months to test the services adoption and appropriation and understand what incidental learning and unexpected usage might take place.

By using sensor-based detection of location linked to open geographical data systems we believe our system overcomes the scalability challenge of marker-based location systems. By utilising a wide range of contextual data derived both from a user's activities and their environment, we believe we offer highly relevant recommendations for learning and can take advantage of the city as a resource for supporting contextually-aware learning.

Acknowledgements. The research leading to these results has received funding from the European Community's Seventh Framework Programme (FP7/2007-2013) under grant agreement n° 288587 MASELTOV.

References

1. Ogata, H., Yano, Y.: Context-Aware Support for Computer-Supported Ubiquitous Learning. In: Proceedings of the 2nd IEEE International Workshop on Wireless and Mobile Technologies in Education, pp. 27–34. IEEE Computer Society, Washington, DC (2004)
2. Brown, E., Brner, D., Sharples, M., Glahn, C., de Jong, T., Specht, M.: Location-based and Contextual Mobile Learning. A STELLAR Small-Scale Study. STELLAR European Network of Excellence in TEL (EU) (2010)
3. Lave, J., Wenger, E.: Situated Learning: Legitimate Peripheral Participation. Cambridge University Press, Cambridge (1991)
4. Kerka, S.: Incidental Learning: Trends and Issues Alert No. 18 (2000), http://www.eric.ed.gov/PDFS/ED446234.pdf
5. Gaved, M., Jones, A., Kukulska-Hulme, A., Scanlon, E.: A Citizen-Centred Approach to Education in the Smart City: Incidental Language Learning for Supporting the Inclusion of Recent Migrants. International Journal of Digital Literacy and Digital Competence 3(4), 50–64 (2012)
6. Abowd, G., Atkeson, C., Hong, J., Long, S., Kooper, R., Pinkerton, M.: Cyberguide: a Mobile Context-Aware Tour Guide. Wireless Networks 3, 421–433 (1997)
7. Paredes, R.G.J., Ogata, H., Saito, N.A., Yin, C., Yno, Y., Oishi, Y., Ueda, T.: LOCH: Supporting Informal Language Learning Outside the Classroom with Handhelds. In: Proceedings of the Third IEEE International Workshop on Wireless and Mobile Technologies in Education, pp. 182–186. IEEE Computer Society, Los Alamitos (2005)
8. Petersen, S., Miarkiewicz, J.-K.: PALLAS: Personalised Language Learning on Mobile Devices. In: Proceedings of the Fifth IEEE International Conference on Wireless, Mobile, and Ubiquitous Technology in Education, pp. 52–59. IEEE Computer Society, Los Alamitos (2008)
9. Zhou, R., Rechert, K.: Personalization for Location-Based E-learning. In: Proceedings of the Second International Conference on Next Generation Mobile Applications, Services and Technologies, pp. 247–253. IEEE Computer Society, Los Alamitos (2008)
10. Brandt, E., Björgvinsson, E., Hillgren, P., Bergqvist, V., Emilson, M.: PDA's, Barcodes and Video-Films for Continuous Learning at an Intensive Care Unit. In: Proceedings of the Second Nordic Conference on Human-Computer Interaction, pp. 293–294. ACM, New York (2002)
11. Edge, D., Searle, E., Chiu, K., Zhao, J., Landay, J.: MicroMandarin: Mobile Language Learning in Context. In: CHI 2011 Proceedings of the SIGCHI Conference on Human Factors in Computing Systems, pp. 3169–3178. ACM, New York (2011)
12. Bristow, H., Baber, C., Cross, J., Knight, J., Woolley, S.I.: Defining and Evaluating Context for Wearable Computers. International Journal of Human - Computer Studies 60, 798–819 (2004)

13. Lonsdale, P., Baber, C., Sharples, M., Byrne, W., Brundell, P., Beale, R.: Context Awareness for MOBIlearn: Creating an Engaging Learning Experience in an Art Museum. In: Proceedings of the World Conference on Mobile and Contextual Learning (mLearn 2004), pp. 115–118. LSDA, Rome (2004)
14. Syvänen, A., Beale, R., Sharples, M., Ahonen, M., Lonsdale, P.: Supporting Pervasive Learning Environments: Adaptability and Context Awareness in Mobile Learning. In: Proceedings of the Third IEEE International Workshop on Wireless and Mobile Technologies in Education, pp. 251–253. IEEE Computer Society Press, Los Alamitos (2005)
15. Li, M., Ogata, H., Hou, B., Uosaki, N., Yano, Y.: Personalization in Context-aware Ubiquitous Learning-Log System. In: Proceedings of the 2012 Seventh IEEE International Conference on Wireless, Mobile and Ubiquitous Technology in Education, pp. 41–48. IEEE Computer Society, Los Alamitos (2012)
16. Cui, Y., Bull, S.: Context and Learner Modeling for the Mobile Foreign Language Learner. System 33(2), 353–367 (2005)
17. De Jong, T., Specht, M., Koper, R.: A Study of Contextualised Mobile Information Delivery for Language Learning. Educational Technology & Society 13(3), 110–125 (2010)
18. Kukulska-Hulme, A., Shield, L.: An Overview of Mobile Assisted Language Learning: from Content Delivery to Supported Collaboration and Interaction. ReCALL 20(3), 271–289 (2008)
19. El Helou, S., Salzmann, C., Gillet, D.: The 3a Personalised, Contextual, and Relation-Based Recommender System. Journal of Universal Computer Science 16(16), 2179–2195 (2010)
20. El-Bishouty, M.M., Ogata, H., Yano, Y.: PERKAM: Personalized Knowledge Awareness Map for Computer Supported Ubiquitous Learning. Educational Technology & Society 10(3), 122–134 (2007)
21. Verbert, K., Manouselis, N., Ochoa, X., Wolpers, M., Drachsler, H., Bosnic, I., Duval, E.: Context-Aware Recommender Systems for Learning: a Survey and Future Challenges. IEEE Transactions on Learning Technologies 5(4), 318–335 (2012)
22. Wanaskar, U., Vij, S., Mukhopadhyay, D.: A Hybrid Web Recommendation System Based on the Improved Association Rule Mining Algorithm. Journal of Software Engineering and Applications 6, 396–404 (2013)
23. Melville, P., Sindhwani, V.: Recommender Systems. In: Sammut, C., Webb, G.I. (eds.) Encyclopedia of Machine Learning, pp. 829–838. Springer, New York (2010)
24. Mobasher, B., Cooley, R., Srivastava, J.: Automatic Personalization Based on Web Usage Mining. Communications of the ACM 43(8), 142–151 (2000)
25. Zaldivar, V.A.R., Burgos, D., Pardo, A.: Meta-Rules: Improving Adaption in Recommendation Systems. In: Santos, O.S., Boticario, J.G. (eds.) Educational Recommender Systems and Technologies: Practices and Challenges, pp. 211–231. IGI Global, Hershey (2012)

Using Mobile Devices to Help High School Students Improve Their Oral Presentation Skills

Susan Gwee[1] and Hwee Leng Toh-Heng[2]

[1] English Language Institute of Singapore, Singapore
gwee@mac.com
[2] James Cook University, Singapore
daffidowndilly@yahoo.co.uk

Abstract. Video review is a useful tool for formative assessment in higher education settings. However, not much is known about the impact of video review as a tool for bringing about better student outcomes in oral presentation in a school setting. This paper examines the effectiveness of using video review in student learning of English oral presentations in both in-class and out-of-class settings for Grade 11 Research Education students in a Singapore high school. Three conditions were: (a) in-class video review (n = 21), (b) out-of-class video review using mobile devices (n= 18), and (c) no video review (n = 19). We used both parametric and non-parametric analyses to determine whether in-class video review and out-of-class video review had an impact on student outcomes in terms of (a) fluency and clarity of speech, (b) audience awareness, (c) response to questions, and (d) effectiveness of group presentation. The findings indicate that having in-class video review of students' presentations in English results in significantly higher scores for effectiveness of group presentation.

Keywords: video review, assessment, oral presentation, Mobile.

1 Introduction

The purpose of this study is to explore how in-class and out-of-class video review can impact student outcomes in oral presentations in a Grade 11 formal education setting. At a sharing session between two high schools in 2011, the teachers in one school who advised students to use mobile devices for video recording student presentations claimed that students found them useful for improving their oral presentations skills. However, there was no empirical evidence that using such devices contributed to better student outcomes in oral presentations. Prior research [1] examining the use of technology in improving students' use of English has found that infrequent in-class participants were more active and deliberative contributors online and that the removal of an audience in a formal classroom context encouraged participation. Housee [2] found that students were more confident in speaking in an informal context than in a formal context and participated at a higher level. While the investigation of students' level of participation would have added a deeper dimension to the study, we

M. Kalz et al. (Eds.): mLearn 2014, CCIS 479, pp. 365–376, 2014.

did not investigate this aspect as the focus of our study was on the improvement of oral presentation skills arising from video review with the aid of mobile devices. In this paper, we will examine studies that looked into how video review, and self and peer assessment have been used in in-class and out-of-class contexts. We will then describe the methodology of the present study before presenting and discussing its findings.

1.1 In-class Video Review

Video review has become increasingly common for learning purposes. In this paper, we define video review as the use of video recordings to review the performances that have been video recorded using equipment such as mobile devices or video cameras. Video review has been increasingly used to assess skill development [3,4,5] in both in-class and out-of-class contexts in higher education settings. In the in-class context, video review has been found to be effective (effect size of .40) [6]. Participants found video review to be better when they are given feedback in small groups [3] and when they know the desired target behavior [6].

1.2 Out-of-class Video Review

In the out-of-class context, one proposed advantage of the sharing of videos is that it can produce reflective practice and peer-to-peer learning [7]. Students have also found video review to be effective in the out-of class context [8]. One of the main problems identified for video review, however, is that students tended to give higher scores to their peers than teachers [8,9]. Students also had problems giving accurate scores to themselves [4]. Low performing students overestimated their performance while high performing students underestimated theirs [4]. Consequently, some students may not devote enough time to their areas of weakness as they have overestimated their skills [4]. In contrast, students found that teachers tended to provide objective, concrete, and focused feedback compared to peers [9].

1.3 Theoretical Underpinnings

Informing our inquiry is the notion that formative assessment encourages incremental learning [10,11]. It has been shown that video review can lead to learning gains [6]. Having video recordings help students to self-assess, leading to improvement of skills [12]. Having a structured observation form, along with video review, is more effective (effect size = .55) than other instructional components such as explaining, modelling, and practicing the target skills (effect size = .21) [6]. In addition to self-assessment, learners have also found using videos for peer assessment in micro-teaching to be very useful for their learning process [9].

The studies reviewed above have been restricted to the impact of using video review for formative assessment in higher education settings. Moreover, they did not evaluate the impact of using video review in the classroom context compared to using it in the out-of-classroom context. To address these two issues, this study sought to investigate the relationship between using video review as formative assessment on student presentations both in in-class and out-of-class contexts, and its impact on student outcomes in oral presentation in Grade 11 classrooms. In this quasi-experimental study, there were three conditions: (1) in-class video review, (2) out-of-class video review, and (3) no video review. The research question is:

1. Do Grade 11 students from the three conditions (in-class video review, out-of-class video review, and no video review) differ in oral presentation scores (fluency and clarity of speech, audience awareness, response to questions, effectiveness of group presentation) after controlling for teacher effect?

2 Methodology

2.1 Participants

The participants from three Grade 11 classes in Singapore were chosen because they shared the same Research Education teacher, thus controlling the teacher variable. The age range of the participants was between 16 to17 years. There were 21 (16 males, 5 females), 18 (9 males, 9 females), and 19 (5 males, 14 females) students in Conditions 1, 2, and 3, respectively. There were no significant differences in overall academic ability across the three classes. In each class, groups of four or five were formed for the oral presentations as they presented in groups.

2.2 Materials

There was an observation checklist with space for open-ended comments (see Appendix A). Based on the summative assessment instrument (see Appendix B), this observation checklist assessed (a) fluency and clarity of speech, (b) audience awareness, (c) effectiveness of group presentation, and (d) response to questions, that is, the individual student's ability to answer questions.

A consent form was given to parents of all the students, who consented to let their children participate in the study. Parents were told that their children could withdraw from the study at any point.

2.3 Procedure

Table 1 summarizes the sequence of activities for all the students in the three conditions:

Table 1. Sequence of activities

Session	Condition 1 In-class video review	Condition 2 Out-of-class video review	Condition 3 No video review
1 (Combined session for all classes)	Lesson on individual presentation skills		
2 (Combined session for all classes)	Lesson on group presentation skills		
3 (First Dry Run)	Student presentations and assessment of presentations with observation checklist		
		Teacher encouraged students to video record their performance outside the classroom and give one another feedback to improve their performance.	
4 (Second Dry Run)	Student presentations were video recorded.	Teacher encouraged students to video record their dry runs in the classroom. Teacher gave students written feedback using observation checklist.	Teacher gave students written feedback using observation checklist.
5	Students and teacher assessed each student presentation. Teacher gave students written feedback using observation checklist.	N.A.	N.A.
6 (Third Dry Run)	Teacher gave students written feedback using observation checklist.		
		Teacher encouraged students to video record their dry runs in the classroom.	
7 (Fourth Dry Run)	Teacher gave students written feedback using observation checklist.		
		Teacher encouraged students to video record their dry runs in the classroom.	
8 (National Oral Presentation Examination)	Two assessors assessed each student presentation using four examination standards: fluency and clarity of speech, audience awareness, effectiveness of group presentation, and the individual student's ability to answer questions.		

In October 2011, the chief oral presentation assessor, who was a teacher in the school in which this study was conducted, conducted a three-hour training session with all oral presentation assessors, all teachers of the school, to align their assessment of oral presentations. He showed them three video clips of the previous year's student presentations of high, medium, and low quality. He told them to grade them based on the three assessment standards for the four assessment criteria set by the examining board. He then showed them the actual grading and comments as determined by the examining board, after which he discussed the oral presentation of each student and the accompanying assessment grade and comments so that the assessors could align themselves to the standards set by the examining board. The assessors discussed openly any disagreements and settled them during the training session.

For the oral presentation examination conducted in November 2011, two assessors, who were not the teachers of the students, assessed the oral presentations of all the students using the rubrics shown in Appendix B. For the first two student oral presentation examinations assessed by the assessors, a moderator assessed the students to ascertain whether the assessors adhered to the assessment criteria and standards. The moderator discussed discrepancies in scores and the assessors would have to make the necessary changes to the oral presentation scores. The assessors agreed on a final score for each component of the assessment criteria for oral presentations. If any difference in the scores arose, the assessors would discuss and would call the chief oral presentation assessor to adjudicate if they could not resolve their difference in the scores awarded to the students. An email interview with three students, one from each class, was conducted after the oral presentation examinations.

2.4 Data Analysis

With the help of *SPSS*, we used two types of analyses, parametric (*ANOVA*) and non-parametric (Kruskal-Wallis), to determine whether there were significant differences across the three conditions for fluency and clarity of speech, awareness of audience, response to questions, and effectiveness of group presentation. We used descriptive statistics to examine the use of mobile devices to video record oral presentations in and outside the classroom. The interview data will be used below to discuss the quantitative findings.

3 Findings

The Levene's statistic showed that the assumption that variances were equal was supported. Table 2 summarizes the descriptive statistics regarding the scores obtained in the three conditions.

The results for the parametric and non-parametric analyses suggest that there were differences across the three conditions in terms of the impact on effectiveness of group presentation, respectively ($F(2, 55) = 4.84$, $p = .012$ (see Table 3); H (2) = 8.93 , $p = .012$). There was a significant pairwise difference in the effectiveness of group presentation scores between students in Conditions 1 ($M = 5.19$, 95% CI [4.85, 5.53]) and 3 ($M = 4.26$, 95% CI [3.79, 4.74]), $p = .010$. Figure 1 shows that students who were in the in-class video review condition had the highest score followed by students in the out-of-class video condition, and in the no video review condition. The effect size of the conditions on the effectiveness of group presentation was relatively large at .150.

Table 2. Descriptive statistics of oral presentation sub-scores

	Condition	*n*	*M*	*SD*	*SE*	95% Confidence Interval	
						Lower Bound	Upper Bound
Fluency and clarity of speech	1	21	4.52	.873	.190	4.13	4.92
	2	18	4.72	.826	.195	4.31	5.13
	3	19	4.63	.895	.205	4.20	5.06
	All	58	4.62	.855	.112	4.40	4.85
Audience awareness	1	21	4.76	.831	.181	4.38	5.14
	2	18	4.61	.608	.143	4.31	4.91
	3	19	4.42	.902	.207	3.99	4.86
	All	58	4.60	.793	.104	4.39	4.81
Response to questions	1	21	3.95	1.117	.244	3.44	4.46
	2	18	4.17	1.043	.246	3.65	4.69
	3	19	4.00	1.000	.229	3.52	4.48
	All	58	4.03	1.042	.137	3.76	4.31
Effectiveness of group presentation	1	21	5.19	.750	.164	4.85	5.53
	2	18	4.56	1.149	.271	3.98	5.13
	3	19	4.26	.991	.227	3.79	4.74
	All	58	4.69	1.030	.135	4.42	4.96
Overall oral presentation score	1	21	18.43	2.420	.528	17.33	19.53
	2	18	18.06	2.689	.634	16.72	19.39
	3	19	17.32	2.888	.662	15.92	18.71
	All	58	17.95	2.658	.349	17.25	18.65

4 Discussion

Consistent with previous studies [6] [12], using video review as formative assessment in the present study resulted in slightly higher overall student scores and had significantly better effectiveness of group presentation scores. From the interview data, it appears that some students could independently adjust their oral presentations to the desired outcomes after viewing their own performance while others could not do so independently. Further analyses performed to determine whether there was a significant correlation between students' general academic ability (the score was determined by the sum of the points awarded for each grade given for each subject) and the areas assessed in the oral presentation examination revealed that there were significant correlations between students' general academic ability, and fluency and clarity of speech, audience awareness, and effectiveness of group presentation as well as the overall oral presentation score (see Table 4).

Table 3. Summary tables analyzing students' oral presentation sub-scores

		Sum of Squares	*df*	*Mean Square*	*F*	*Sig.*
Fluency and clarity of speech	Between Groups	.385	2	.192	.256	.775
	Within Groups	41.270	55	.750		
	Total	41.655	57			
Audience awareness	Between Groups	1.160	2	.580	.919	.405
	Within Groups	34.719	55	.631		
	Total	35.879	57			
Response to questions	Between Groups	.479	2	.239	.214	.808
	Within Groups	61.452	55	1.117		
	Total	61.931	57			
Effectiveness of group presentation	Between Groups	9.047	2	4.524	4.843*	.012
	Within Groups	51.367	55	.934		
	Total	60.414	57			
Overall oral presentation score	Between Groups	12.652	2	6.326	.892	.416
	Within Groups	390.193	55	7.094		
	Total	402.845	57			

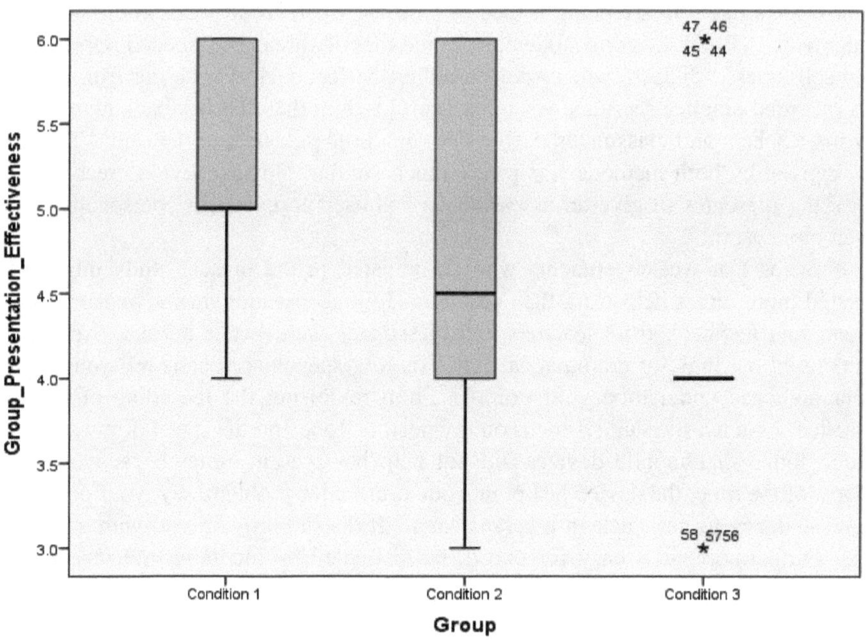

Fig. 1. Mean scores of effectiveness of group presentation in the three conditions

Table 4. Correlation between overall academic ability and fluency and clarity of speech, audience awareness, response to questions, effectiveness of group presentation, and overall oral presentation scores

	Overall academic ability	Fluency and clarity of speech	Audience awareness	Response to questions	Effectiveness of group presentation	Overall oral presentation score
Pearson Correlation	1	.318*	.425**	.228	.334*	.436**
Sig. (2-tailed)		.020	.001	.101	.015	.001
Sum of Squares and Cross-products	3238.208	110.731	126.203	97.665	143.703	478.302
Covariance	62.273	2.129	2.427	1.878	2.764	9.198
n	53	53		53	53	53

Indeed, Joseph (a pseudonym) from Condition 1, an A student, used mobile devices to "better review" his group's performances and to "isolate areas which require further improvements, for e.g., sentence structures, hand movements and vocals." He said, "I basically look out for my mannerisms, gestures and also listen to how I elaborate certain points when viewing my own recordings."

He found that it was more engaging using a mobile device to learn to present better because "there is that additional pressure on presenters when being recorded which almost feels like you are being filmed in a movie where your peers would be able to judge you." Being a strong student, he could identify areas that needed improvement and self-assess. Indeed, when asked whether the feedback that he got from viewing his recorded practice sessions was more useful to him than the feedback given to him by his teachers and classmates during dry runs in the classroom, he said, "The feedback given by both methods was pretty much similar. However, video recording allows the presenter to give an assessment of himself too, which I personally feel is even more useful."

It seems that weaker students who participated in the present study might have needed more direct help from their teacher. Julie (a pseudonym), a weaker student, found that feedback from teachers and classmates were better because "recordings can't point out logic or grammatical flaws in your speech so it can't tell you whether your audience understood your point." From reviewing the recording of her own practice sessions, she said, "I focus on my posture, tone and the speed of my speech."

Julie found that mobile devices did not help her present better because she said, "Most of the time, the device just points out your major problem, say, your posture, or your tendency to get stuck in a certain area. It doesn't provide you with advices or tips. Furthermore, it is easy for one to be distracted by incoming messages." She further elaborated, "Most of the time I know what went wrong so watching the recording only confirms it. It doesn't help me much if it doesn't point out to me the subtle mistakes I made, or the subtle things I can add in to improve my presentation."

Julie found practice during lesson time most useful, saying "I get instant feedback and I am able to practice almost instantly after getting this feedback. I also get to watch others practice, which additionally allows me to learn from them."

Students with Julie's profile may find it difficult to self-assess, and may need teacher or peer assessment. As noted by Lane and Gottlieb [12], unskilled individuals consistently overestimate their ability because of metacognitive deficits (i.e., the skills that engender competence in a particular domain are often the very same skills necessary to evaluate competence in that domain). They cited a previous landmark study in the psychological literature by Kruger and Dunning [13], which suggested that poor performers show little insight into the depth of their deficiencies relative to their peers. This is also congruent with a later study [14] where the authors found that poor performers grossly overestimated their performances regardless of the performance metric examined, even after accounting for error due to measurement unreliability.

The oral presentation dry runs using the checklist for assessing self and peer performance and having teacher assessment might have also accounted for the non-significant results among the students in the three conditions for most of the sub-scores. This is because having a checklist might have helped students improve their performance during dry runs in their regular lessons. This is consistent with the results of the meta-analysis by Fukkink et al. [6] who found that having an observation form while reviewing videos was much more effective than viewing videos alone.

To ensure that students review their performances using the video review method outside the classroom, teachers might have to ask students to submit video recordings and the corresponding observation checklists. Otherwise, the less motivated students, who probably would need more practice, might not be motivated enough to conduct the video review by themselves. For weaker students, teachers might have to ask them to write their speeches and proofread them before practice sessions. For those shown to have pronunciation errors, teachers could highlight these errors either using an online system or in face-to-face consultation sessions with individuals.

5 Conclusion

Video review for formative assessment purposes can be helpful to students for the development of effective group presentation. However, for individual oral presentation skills such as fluency and clarity of speech, audience awareness, and the ability to respond to questions, video review, especially for weak students using mobile devices for informal learning in out-of-class settings, needs follow-up actions by the teachers to help students, especially weaker students, identify problem areas and work out how the improvements could be made for better oral presentations.

References

1. Uysal, E., Bayyurt, Y.: Analysis of Student Participation in Classroom and Bulletin Board Discussions in an EFL Context. Boğaziçi University J. Educ. 22, 1–13 (2005)
2. Housee, S.: When Silences are Broken: An Out of Class Discussion with Asian Female Students. Educ. Rev. 62, 421–434 (2010), doi:10.1080/00131911.2010.486475
3. Brimble, M.: Skills Assessment Using Video Analysis in a Simulated Environment: An Evaluation. Pediatr. Nurs. 20, 26–31 (2008)
4. Mort, J.R., Hansen, D.J.: First-Year Pharmacy Students' Self-Assessment of Communication Skills and the Impact of Video Review. Am. J. Pharm. Educ. 74, 1–7 (2010)
5. Zick, A., Granieri, M., Makoul, G.: First-Year Medical Students' Assessment of Their Own Communication Skills: A Video-Based Open-Ended Approach. Patient Educ. Couns. 68, 161–166 (2007)
6. Fukkink, R.G., Tienekens, N., Kramer, L.J.C.: Video Feedback in Education and Training: Putting Learning in the Picture. Educ. Psychol. Rev. 23, 45–63 (2011)
7. Pimmer, C., Pachler, N.: Mobile Learning in the Workplace: Unlocking the Value of Mobile Technology for Work-Based Education. In: Ally, M., Tsinakos, A. (eds.) Mobile Learning Development for Flexible Learning, Athabasca University Press, Edmonton (2013), http://www.christoph.pimmer.info/wp-content/uploads/2008/08/Pimmer-Pachler-Mobile-Learning-in-the-Workplace1.pdf
8. Thanasingam, S., Soong, A.S.K.: Interaction Patterns and Knowledge Construction Using Synchronous Discussion Forums and Video to Develop Oral Skills. In: Proceedings of the Ascilite Singapore 2007 (2007), http://www.ascilite.org.au/conferences/singapore07/procs/thanasingam.pdf
9. Wu, C.-C., Kao, H.-C.: Streaming Videos in Peer Assessment to Support Training Pre-Service Teachers. Educ. Technol. Soc. 11, 45–55 (2008)
10. Shepard, L.A.: The Role of Assessment in a Learning Culture. Educ. Res. 29, 4–14 (2000)
11. Wiliam, D., Black, P.: Meanings and Consequences: A Basis for Distinguishing Formative and Summative Functions of Assessment. Brit. Educ. Res. J. 22, 537–548 (1996)
12. Lane, J.L., Gottlieb, R.P.: Improving the Interviewing and Self-Assessment Skills of Medical Students: Is It Time to Readopt Videotaping as an Educational Tool? Ambul. Pediatr. 4, 244–248 (2004)
13. Kruger, J., Dunning, D.: Unskilled and Unaware of It: How Difficulties in Recognizing One's Own Incompetence Lead to Inflated Self-Assessments. J. Pers. Soc. Psychol. 77, 1121–1134 (1999)
14. Ehrlinger, J., Johnson, K., Banner, M., Dunning, D., Kruger, J.: Why the Unskilled are Unaware: Further Exploration of (Absent) Self-Insight among the Incompetent. Organ. Behav. Hum. Decis. Process. 105, 98–121 (2008)

Appendix A

Observation checklist

Assessment Criteria	Indicators to take note of	Check item	Remarks/Comments
INDIVIDUAL			
Fluency and Clarity of Speech	Am I speaking clearly and loudly enough?		
	Am I using too many fillers while presenting (Er … Um … Ok, ok, ok …Yah …, etc.)?		
	Am I varying my tone?		
	Am I speaking at an appropriate pace and not too fast?		
	Am I presenting with appropriate emphasis placed on key words/phrases/points?		
Awareness of Audience	Am I reading word-for-word from cue cards/notes/script/slides/screen?		
	Am I using appropriate gestures and directing the audience to the presentation aids when necessary?		
	Am I maintaining eye contact with the audience?		
	Am I maintaining a friendly and relaxed atmosphere, e.g. smiling, using humor?		
	Am I engaging the audience by using rhetorical/leading questions?		
Response to Questions	Am I answering the question?		
	Am I including my personal thoughts/insights/experiences in my answers?		
	Am I providing sufficient details in my answers?		
Assessment Criteria	**Indicators to take note of**	**Check item**	**Remarks/Comments**
GROUP			
Effectiveness of Group Presentation	Is there smooth transition between speakers in our presentation?		
	Is our presentation cohesive as a whole and does not appear as several disjointed sections?		
	Is the format of our presentation consistent (same background, font, style)?		
	Are the words on the slides appropriately sized? Are they too small?		
	Are there too many points/words in one slide?		
	Are we over-using bullet points?		
	Is the selected font clear and the easiest for an audience to read quickly?		
	Are we incorporating photographs/tables/graphics in our presentation slides?		

Other Comments

Appendix B

Rubrics for the scoring of oral presentations during the examination

Criterion	No mark	Approaching Expectation	Meeting Expectation	Exceeding Expectation
Individual				
Fluency and clarity of speech	Criterion has not been met	Speaks haltingly and/or mumbles, and is difficult to understand at times	Speaks clearly and intelligibly most of the time	Speaks clearly and fluently throughout, at an appropriate pace
Awareness of audience		Shows little awareness of audience	Shows some awareness of audience	Shows personal engagement with audience
Response to questions		Answers are limited with little or no elaboration	Answers are relevant and contain some elaboration of ideas	Answers are relevant, well thought out and elaborated on
Group				
Effectiveness of group presentation	Criterion has not been met	Presentation has limited effect due to lack of cohesion and organisation. Presentation aids do not enhance the presentation	Presentation is generally effective with some degree of cohesion and organisation. Presentation aids used appropriately to enhance presentation	Presentation is highly effective, cohesive and well-organised. Presentation aids used to effectively enhance overall effect

Mobile Literacies

Navigating Multimodality, Informal Learning Contexts and Personal ICT Ecologies

Jessica Katherine. Frawley and Laurel Evelyn. Dyson

School of Software
University of Technology, Sydney
Sydney, Australia
{Jessica.Frawley,Laurel.E.Dyson}@uts.edu.au

Abstract. The affordances and ubiquitous uptake of technically convergent devices has led to widespread change in communicative practice. Mobile devices and their wider ICT ecology have afforded people with the means through which to consume and produce multimedia content. In such an environment, literacy can no longer be limited to the reading and writing of word and character-based texts. The emergent field of New Literacies research has contributed early understandings of these new practices. This paper contributes to understanding how these new literacies occur within a mobile and informal learning space. This study details *mStories* a creative, participatory, digital mobile storytelling project comprised of nine adult participants who created "stories" with their mobile phone device. These stories were shared on the *mStories* project website, which became a repository for: fiction, non-fiction, poetry and diary-style content. Stories used a range of written text, visual images, sound, music and video. Using content analysis, interview and survey methods this paper describes mobile literacy as characteristically situated and experiential in nature. The mobile device was catalytic to furthering digital writing on other devices within the individual's wider ICT ecology. This research contributes understandings of multimodal mobile literacies as part of a foundation for framing and understanding mobile learning in informal settings.

Keywords: Mobile learning, informal learning, new literacies, ICT ecology.

1 Introduction

Social media, web 2.0 applications and mobile devices have come to characterise a digital landscape that affords people with new ways to interact, communicate and learn. The communication culture and the artefacts that comprise it are increasingly described as participatory, visual, and multimodal in nature [1,2,3,4]. Mobile devices are just one tool that enables people to navigate these new semiotic surrounds. The technical convergence that typifies later generations of mobile devices has privileged digital media (e.g. video) and multimodal content (e.g. image, video, sound) over traditional written text. People now have the tools with which to produce and share

M. Kalz et al. (Eds.): mLearn 2014, CCIS 479, pp. 377–390, 2014.

their own multimedia culture and meanings [5]. User-generated content platforms form an environment that supports widespread participatory culture: non-experts are able to create and share new content online [3]. Mobile devices are one gateway to participating in such a culture.

1.1 'New' Literacies

Shifts within the socio-technical landscape, challenge traditional concepts of literacy and what it means to be literate. In this context traditional definitions of literacy that privilege the written word to the exclusion of all else, fail to account for the many ways in which people go about interacting, communicating, and making meaning with the tools they have. Recent literacy research has begun to redefine and explore these phenomena. Discussions have shifted from the singular 'literacy' to the plural 'multiliteracies' [6]. Terms such as 'visual literacy', 'multimodal literacy', 'digital visual literacy' and 'design literacy', attempt to account for both the mode and skill associated with such behaviors [4], [7,8,9,10]. New literacies present educators and technologists with more than just a change in terminology. The acquisition and learning of these skills differs to that of traditional literacy. Where reading and writing were acquired within the educational institution and/or home, new literacies may be informally learnt, self-taught and formed outside the classroom. As such, the concept of new literacies is part of a wider paradigmatic shift in learning. Educators may find themselves to be less literate in these new skills than their students [18]. The rapid innovation of technologies brings rapid change to the modes and medias that comprise these new literacies. In these conditions, achieving a complete understanding of these new communication skills may be impossible for any one individual – teacher or student, adult or child [19]. These changes disrupt traditional models of learning.

The importance of such research is not only in its theoretical contribution to understanding existing practice but also in its recognition of the importance of such skills. Like reading and writing, understanding of digital technologies and multimodal literacy practices is essential for participating in a society that utilizes both [11]. The relationship between digital technologies and new literacies has been recognised as a significant area for research [12]. To date, however empirical studies of new literacies have predominantly focused on either formal learning environments, child or youth practice or digital desktop technologies [7], [9], [11], [13]. New research into the literacies afforded by mobile device challenge some of the assumptions made in prior studies and indicates that there is a real need for further research [14,15].

1.2 Digital Storytelling

Digital storytelling is especially relevant to discussions on new literacies. Pioneering work has demonstrated the extent to which digital storytelling can empower and give voice to its participants [16,17]. This work, however, is largely under-theorized and not framed in terms of literacy. Notable exceptions to this include recent studies of mobile digital storytelling that are framed by theoretical perspectives, such as social semiotics, that are derived from new literacies research [18,19]. However, these studies derive their empirical data from work undertaken in a formal educational setting [18] or with adolescents [19]. The informal adult learner, though a major adopter and user of mobile technology, is not accounted for. Though still an emerging field, mobile digital storytelling presents practical strategies that are well suited to exploring and researching new literacies.

1.3 Mobile Digital Literacies

The question for mobile learning is how these literacies are enacted within the mobile space. Mobile technologies support learning in informal places, and can situate a learning conversation within the individual user's work and personal time and space [20,21,22]. To understand what form this learning conversation might take, we need to understand new literacies from a mobile perspective. This means not only practically grounding the research, for example in studies involving mobile devices, but orientating such studies theoretically within existing contributions of mobile learning. From such a vantage point studies may better account for human and computer mobility, informal practice and how literacy practice form part of the wider ecological mobile complex [23].

1.4 Contributions

This paper contributes an understanding of new literacies that emphasises the informal adult learner within the mobile complex. Despite the widespread adoption of smartphone technology, research on mobile literacies has confined itself to young persons or those enrolled in formal education. This paper presents *mStories*, a participatory mobile digital storytelling project comprised nine adults from across Australia and the UK. Through qualitative methods, this paper describes the mobile meaning making practices of adults in informal settings. This paper draws attention to the inherently multimodal (e.g. image and text, sound and video) nature of mobile device use, and how this usage relates ecologically to other devices owned by the individual. This paper describes how participants engaged in creating multimodal content within the ever-changing context of the mobile space and the learning behaviours that participants presented. From these contributions future research may be better placed to explore how such skills are acquired or best applied to facilitate learning.

2 The *mStories* Project

mStories was established as a participatory creative project. Creative perspectives have been shown to be a valuable approach to researching multimodal semiotics and practice [24]. The project was originally structured around the concept of a writers' group: a community group where people could share, engage and learn about creative writing within a community group setting. The project recruited participants through writers' groups and community networks. The group was originally established to work face to face, but the digital aspects of the project changed the dynamic in which people wanted to participate [25]. Secondly, though established on a writers' group model it was largely non-writers who expressed interest in joining. The final *mStories* participant group was composed of nine people. Participants were from Australia (n=5) and the UK (n=4), the age range spanned from 21-25 to 46-55 and there were both male (n=4) and female (n=5).

Whilst pre-existing user generated platforms (e.g. Facebook, YouTube, Flickr) offer places for people to create multimodal content, such forums already have firmly established genres. In contrast, *mStories*, as an independent project, provided a creative space from which people could engage informally in *new* meaning making and practice.

2.1 Methodology

The project-based approach was able to accommodate a diversity of mobile device types and individual participant needs. We address the research question through qualitative data on both 'what is made' and 'how it is made'.

Research Question

How can we best describe adults' informally acquired mobile literacy practices?

Data Collection and Analysis

The research question was addressed through three iterative phases of data collection and analysis. The stages are depicted in Table 1.

Table 1. Iterative staged data collection and analysis

Stage	*Method*	*n*	*Aims and data collected*
1	Preliminary survey	9	Who participated and why? • Participants' demographics • Type of smartphone and existing usage • Occupation and hobbies • Motivations for participating
2	Content analysis of mStories	9	What was created? • Modes and media used • Genres and content created
3	Post-story interview and survey	9	How did people create the *mStory* analysed in Stage 2?

3 Findings

The findings from the *mStories* project are presented sequentially by stage. Qualitative data is summarized and supported by example quotations. In instances where it is both clearer and more informative to present findings numerically (e.g. common themes) we use numeric data, though as a qualitative study this data has no statistical significance.

3.1 Stage 1: Preliminary Survey

The preliminary survey was designed to find out what mobile phone devices people owned and how people used such devices. There are several different ways of distinguishing between different Smartphone devices, but not all of these are useful to understanding how the device shapes users' communication practices. Of most use to understanding questions of device affordance is the mobile's user interface (UI). Differences in device UI may indicate the affordances and constraints of the device from the user perspective. The preliminary survey categorized devices through two distinguishing UI features: screen-size and primary mode of user input. Three categories were identified:

- *Type A*: devices with large screens and touch-screen input *(e.g. an iPhone)*
- *Type B*: devices with medium sized screens and QWERTY keyboard input *(e.g. a BlackBerry)*
- *Type C*: devices with smaller screens and numeric keyboard input

Participants were asked to select one of three simplified images (depicted in Table 2.) and identify what they used their device for. The results from this demonstrate that whilst all participants used their phone for calls and text messaging, there were differences in phone usage across the three device ownership groups. As seen in Table 2, those *mStories* participants who owned a Type A mobile device participated in more multimodal producing behaviors than those who owned other device types. None of the participants had ever used their device to create a mobile story before.

Hobbies, Interests and Motivation to Participate

Although writers' groups were initially approached, members from writers groups accounted for only one third (n= 3) of the final group. Whilst there were two professional writers in the group, other participants had a range of occupations (Table 3). Participants' hobbies were similarly diverse and included things such as photography, drama and acting, sports and outdoor activities. Interest in the project was gauged through the open question: "What interested you in: participating in the mStories project?" Participants' responses were coded thematically. From this coding three dominant themes emerged:

- *To be creative (n= 5):* e.g. "A chance to be creative and conduct my own e-show"
- *It sounds interesting or fun (n=4):* e.g. "sounds fun, interesting and a tiny bit silly"

- *It is different or new (n=5):* e.g. "Writing a short story with a mobile phone isn't something many people would consider and all the better a way to inject some stimulus into the writing world"

Table 2. Phone ownership and existing usage

Literacy practice as categorized by the researcher		***Activity***	***Participants' Device Type***		
			Type A n= 5	**Type B n=1**	**Type C n=3**
			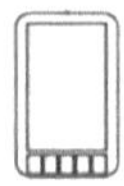		
Multimodal and multimedia *dominant* literacy	*Consumption*	Surfing the net	5	1	1
		Watching online video	2	1	1
		Downloading music or video	1	-	-
		GPS and maps	5	1	1
		Playing games	3	-	-
	Production	Taking photos	5	1	1
		Making videos	3	-	-
		Recording sound	2	-	-
	Both	Social networking	4	-	-
Text *dominant* literacy	*Reading*	Reading eBooks	2	-	-
	Reading and Writing	Text	5	1	3
		Email	5	1	-

3.2 Stage 2: Content Analysis

The completed mStories and their inclusion within the *mStories* website provided data about the media, genre and context that create the employed by users (Table 3).

Media and Mode

Of the nine creative stories generated by participants all stories employed the visual mode through either still or moving image. Text and image combinations accounted for the majority of stories uploaded. One example of this is "What am I wearing today?" a sequence of daily self-portraits that are accompanied by text offering the author's personal reflections and comments (Figure 1). Participants demonstrated the ability to not only use different semiotics (e.g. image and text) but to combine these to create a single comprehensive semantic unit in which different semiotics "spoke" to each other. An example of this can be seen in Figure 1., where the pronoun "this" (in the written text) can only be understood in relation to the visual image to which it refers.

Table 3. Participants, their mStories, media and device usage

ID No.	Participant/Author			Genre	Media	Context	Phone type
	Occupation	**Sex**	**Age (Years)**				
1	Fire fighter Sydney (Australia)	M	21 – 25	Documentary	Video	Journey from fire station to fire.	A
2	Writer Cambridge (UK)	M	26 – 30	Poem "iambic textameter"	Text Photo	Poem and photo inspired by view of a churchyard.	C
3	Fundraiser London (UK)	F	26 – 30	Themed Photos	Photo	Mornings in London taken from flat and commute to work.	C
4	Management Consultant London (UK)	M	26 – 30	Photo diary	Photo	Commute to work in London.	B
5	Accountant Sydney (Australia)	M	31 – 35	Mixed genre (Composite Diary)	Photo Video Music	Idealised Saturday compiled of many Saturdays at markets, beaches, art galleries, gardens, and a fireworks display.	A
6	Interaction Designer Sydney (Australia)	F	31 – 35	Diary	Text Photo	A daily portrait and diary entry about the clothes a person wears.	A
7	Medical doctor neonatal intensive care Cambridge (UK)	F	31 – 35	Photo diary	Photo	Moments from a "special day" in London taken at markets, on the tube, in the shops, at the theatre and at home.	A
8	Writer and mother Sydney (Australia)	F	36 - 45	Short story fiction	Text Photo	'Spooky story' set in many different locations: playground, the street, a preschool classroom etc.	A
9	Researcher Sydney (Australia)	F	46 -55	Poems	Text Image	Poems set to photos of flowers.	C

Today it's raining when I woke up so I decided to wear this outfits:same jeans as yesterday; I only do laundry on the weekend so it's not unusual for me to keep wearing the same jeans for 3 days! ;-)

I have this wrap dress or cardigans that can be worn in 3 different styles (as you can see from the photos) sorry the lighting in my bedroom isn't very good. So I wear this today with black t-shirt underneath. Also have my long booths on today. Mainly because it's raining ....

Fig. 1. What am I wearing today? (Excerpt)

Genre

Several genres were represented, including a short speculative fiction story, poetry, documentary and diary forms. Photo or video diaries were the most common genre on the *mStories* website. However, whilst participants adopted various genres, these were appropriated in ways that subtly morphed existing conventions. One poem, written using SMS on a Type C mobile phone, was comprised of four lines of iambic pentameter. However, the author chose to refer to this as the iambic textameter, in recognition of the mobile technology. Similarly, the visible presence of the mobile phone covering the face of the author in the story 'What am I wearing today?' (Figure 1.) diverges from both the traditional conventions of portraiture, and the contemporary convention of a "selfie", which is a photo taken of oneself using a phone held at arm's length..

Though the diary genre was popular, these were sometimes part fiction in nature. One story, entitled "My Saturday" was a composite mix of photo and video from many Saturdays that formed a single narrative structured from morning to evening (Figure 2.). As a pastiche of photos, videos, sound recordings and music, the story of "My Saturday" resists clear categorization. Similarly, the question of what counts as poetry is challenged by the two poetic contributions, both of which use text and image. Given that the poem's meaning is bound to both modes, such content presents a challenge to existing definitions of poetry.

Fig. 2. My Saturday (excerpt)

Context

What is easily observable from Table 3 is the extent to which the stories reflect the context in which they are situated. Though Story 9 was a series of poems that reflected professional photographs of native Australian flowers, the eight other stories all directly reflect a dialogue between the participant and the mobile context in which they are situated. This dialogue can result in different outcomes. In Story 8 photos of local places in Sydney are rewritten by text into a spooky, speculative fiction. In 'What am I wearing today?' (Fig. 1), the participant's choice of clothes is made in the morning before she goes to work; each photo reflects the time of that person's experience, and their place of decision in front of the mirror.

3.3 Stage 3: Post-mStory Survey and Interview

The post-mStory interview and survey explored how participants' created their story and made decisions about mode and process associated with generating that content.

Modal Choice

Participants' choice of mode or media was principally motivated by the perceived needs of either the individual author/creator or the story itself. The dominant themes that emerged from the data were:

- *Practical needs:* "I wanted to illustrate the pace of my journey to work and needed something that was quick to use and easily accessible. The camera works on one click so I could get an image quickly without drawing too much attention to myself
- *Expressive needs:* "I thought about taking a series of photos and adding text but decided that the task could be accomplished more simply if I allowed the pictures and sound to speak for themselves"
- *Interest needs:* "I love photography. I thought it was a good way to register my day like a diary"

Participants were asked whether they supplemented their mStory with any content not created on their mobile phone. Two participants with Type C devices added photographs to their mStory: they chose to use a camera because of the perceived limitations of their device: One person's phone did not have a camera, whilst one chose to use their camera to get a "higher quality of photo" than their phone allowed. Both said that the photo "added a great deal" to the SMS poems that they had written. In addition to this, two participants who created mobile video content with Type A devices resorted to their laptops to format their final submissions. Both participants found it to be too time-consuming and difficult to do on their mobile device.

Participant Process

In describing the process, participants were heavily influenced by the mobile context in which they created their stories.

- "I was inspired by the view out of my window…"
- "I wanted to give the impression that they were doing the commute with me"

However, whilst the mobile space provided creative impetus, the shaping of this into a meaningful story is a result of the individual's own dialogue with that mobile context of use. This dialogue was inseparable from the individual and their personal motivations, interests, likes, and reflections. Thus, decisions on what to create a story about were driven by what the individual perceived to be interesting: e.g. "I wanted to create something fun and exciting". Likewise, reflection on that environment also became an important part of this dialogic interaction. In one example, a fire fighter who used his phone to create a video story about his work describes: "Since making my mStory I have caught myself taking photos of things. I feel this is because capturing these things makes me aware of them".

Did mLearning Occur?

Outside of an educational environment and in an informal setting, it is often difficult for people to recognise or articulate what they are doing in terms of learning. Asking direct questions on learning, especially within a creative project, is potentially problematic. However, from indirect questions such as 'what did you gain from this experience?' and 'what did you like or dislike?' qualities important for learning did emerge:

- *Self-efficacy:* "I gained a lot and I found that I gained the knowledge that I can think on my feet more than I think I can… And yeah that's the thing I learnt […] thinking on your feet you've got nothing backing you up. You've got no permission to write. And I was able to embrace that challenge and I was happy when I produced something that I kind of liked"
- *Adaption:* "On a computer at home I would have a research document, drafts […] you can't do that on a phone, so I had to produce something completely different."
- *Reflection and metacognition:* "Since making my mStory I have caught myself taking photos of things. I feel this is because capturing these things makes me aware of them. The type of thing that became my mStory is fairly common, one that I normally experience and forget. Even though I still haven't looked at my mStory since I created it, I still very clearly remember what happened, whereas I don't remember half of the other similar instances. This is something that has also occurred with other photos since. I think capturing events has made me aware that things I find mundane, may, if looked at closely, actually be worth remembering."

All participants reported to never having created a digital or mobile story prior to joining the project, yet all participants did create something entirely new. From this we can infer that learning, in an informal sense, did take place. However, gauging the extent of this learning is very difficult and constitutes a significant area for future research.

4 Discussion

When combined, the findings from the three stages of data collection allow us to form an elementary understanding of how the wider mobile complex affects new literacy practices. Whilst the design of the device may afford or constrain different multimodal uses, people are not technologically determined. Personal choice exerts itself, whether through the choice of device at purchase or the decision on which features and functions to use. Moreover, mobile devices cannot be considered in isolation from the individual's wider ICT ecology of different technologies and contexts of use [26]. Within the *mStories* project participants turned to other devices to realize design intentions and overcome the limitations of their particular device. It is important to recognise that each individual's personal ICT ecology is, to some extent, shaped by wider socio-economic factors such as cost, availability and access to different devices.

By opening up storytelling to multimodal methods and approaches, the *mStories* "writing group" was able to accommodate a diverse group of people: writers and non-writers. The dominance of visual content is supportive of a society that has shifted from a word-centric to a visual-centric culture [1]. Whilst the affordances of the device do allow for greater multimodal communication, participants' choice of mode was motivated by the perceived needs of what they were creating. The final stories are highly designed, and show a high degree of sophistication in the way images, texts and sound are used to communicate. Participants' stories simultaneously appropriate and challenge existing genre conventions. This aligns with recent research that suggests digital technologies may be changing the structure of stories and narratives

[27,28]. Situated practice was a recurring theme within all participant feedback; the mobile context of use provided creative stimulus to many of the stories. In contrast to other ICTs, the device's mobility provided new opportunities through which participants were able to be creative. Participants adapted their skills to engage in a new practice and engendered a greater sense of self-efficacy and engagement in reflective practice. These qualities are essential for learning.

Throughout this study several themes emerge to address the research question: "How can we best describe adults' informally acquired mobile literacy practices?" In answer to this, adults' informally acquired mobile literacy practices can best be described as:

- Multimodal
- Participant designed
- Situated
- Experiential and reflective
- Motivated.

5 Conclusions and Future Work

Mobile devices afford visual and multimodal communication; by embracing this, the *mStories* project was able to widen participation from that found at usual writers' groups. Mobile devices afforded situated, experiential and reflective practice that engaged directly with the mobile context of use. People came to this project, not as blank slates but as people with motivations, individual interests, attitudes and ideas, and used the mobile context as a place in which to be creative. When encountering a limitation in their mobile phone, participants readily turned to other technologies to fulfil their design intentions. This picture of mobile literacy is place-based, ecological and disruptive to our traditional notions of learning and literacy. Such findings may inform both practice and research in mobile learning. Moreover, from this understanding future research is better placed to address key questions about the acquisition of new literacies by adult users in informal settings, and how and to what extent informal learning occurs.

Acknowledgments. As a participatory project, this research would not have been possible without the nine participants.

References

1. Spencer, S.: Visual Research Methods: in the Social Sciences. Routledge, New York (2011)
2. Kress, G.R.: Literacy in the New Media Age. Routledge, London (2003)
3. Jenkins, H., Purushotma, R., Weigel, M., Clinton, K., Robison, A.J.: Confronting the Challenges of Participatory Culture: Media Education for the 21st Century. The MIT Press, Cambridge (2009)

4. Kress, G.R., van Leeuwen, T.: Reading Images: the Grammar of Visual Design. Routledge, London (2006)
5. Dyson, L.E., Litchfield, A., Raban, R.: Exploring Theories of Learning and Teaching using Mobile Technologies: Comparisons of Traditional Teaching, eLearning and mLearning. In: mLearn 2010, Malta, pp. 354–357 (2010)
6. Cope, B., Kalantzis, M.: Multiliteracies: the Beginnings of an Idea. In: Cope, B., Kalantzis, M. (eds.) Multiliteracies: Literacy Learning and the Design of Social Futures, pp. 3–8. Routledge, London (2000)
7. Jewitt, C.: Multimodality, "Reading", and "Writing" for the 21st Century. Discourse: Studies in the Cultural Politics of Education 26, 316–331 (2005)
8. Kress, G.R.: Multimodality. In: Cope, B., Kalantzis, M. (eds.) Multiliteracies: Literacy Learning and the Design of Social Futures, pp. 179–200. Routledge, London (2000)
9. Spalter, A.M., van Dam, A.: Digital Visual Literacy. Theory into Practice 47, 93–101 (2008)
10. Sheridan, M.P., Roswell, J.: Design Literacies: Learning and Innovation in the Digital Age. Routledge, London (2010)
11. Mills, K.A.: Shrek meets Vygotsky: Rethinking Adolescents' Multimodal Literacy Practices in Schools. Journal of Adolescent & Adult Literacy 54, 35–45 (2010)
12. Mills, K.A.: A Review of the "Digital Turn" in the New Literacy Studies. Review of Educational Research 80, 246–271 (2010)
13. Wolfe, S., Flewitt, R.: New Technologies, New Multimodal Literacy Practices and Young Children's Metacognitive Development. Cambridge Journal of Education 40, 387–399 (2010)
14. Pachler, N., Bachmair, B., Cook, J., Kress, G.: A Social Semiotic Analysis of Mobile Devices: Interrelations of Technology and Social Habitus. In: Mobile Learning: Structures, Agency, Practices, pp. 185–204. Springer, New York (2010)
15. Fails, J.A., Druin, A., Guha, M.L.: Mobile Collaboration: Collaboratively Reading and Creating Children's Stories on Mobile Devices. In: IDC 2010: Proceedings of the 9th International Conference on Interaction Design and Children, pp. 20–29. ACM (2010)
16. Lambert, J.: Digital Storytelling: Capturing Lives, Creating Community. Digital Diner Press, Berkley (2002)
17. Meadows, D.: Digital Storytelling: Research-based Practice in New media. Visual Communication 2, 189–193 (2003)
18. Nordmark, S., Frolunde, L., Milrad, M.: Promoting the Creation of Digital Stories by Teacher Students: A Pilot Project on Curriculum Development in a Swedish Teacher Education Program. In: Proceedings of Society for Information Technology & Teacher Education International Conference 2010, pp. 1153–1160. AACE (2010)
19. Ranieri, M., Bruni, I.: Mobile Storytelling and Informal Education in a Suburban Area: A Qualitative Study on the Potential of Digital Narratives for Young Second-generation Immigrants. Learning, Media and Technology 38, 217–235 (2013)
20. Sharples, M., Taylor, J., Vavoula, G.: A Theory of Learning for the Mobile Age. In: Andrews, R., Haythornthwait, C. (eds.) The SAGE Handbook of e-Learning Research, pp. 221–247. Sage, London (2007)
21. Coulby, C., Hennessey, S., Davies, N., Fuller, R.: The Use of Mobile Technology for Work-based Assessment: the Student Experience. British Journal of Educational Technology 42, 251–265 (2011)
22. Dyson, L.E., Litchfield, A., Lawrence, E., Raban, R., Leijdekkers, P.: Advancing the m-Learning Research Agenda for Active, Experiential Learning: Four Case Studies. Australasian Journal of Educational Technology 25, 250–267 (2009)

23. Pachler, N., Bachmair, B., Cook, J., Kress, G.: Mobile Learning: Structures, Agency, Practices. Springer, New York (2010)
24. van Leeuwen, T., Djonov, E., O'Halloran, K.L.: "David Byrne Really Does Love PowerPoint": Art as Research on Semiotics and Semiotic Technology. Social Semiotics 23, 1–15 (2013)
25. Frawley, J.: mStories: Exploring Modes of Participation in a Creative Storytelling Project. The Journal of the Australian and New Zealand Mobile Learning Group 1, 10–14 (2012)
26. Brady, F., Dyson, L.E.: A Comparative Study of Mobile Technology Adoption in Remote Australia. In: Seventh International Conference on Cultural Attitudes towards Technology and Communication (CATaC), pp. 69–83. Murdoch University, Australia (2010)
27. Alexander, B., Levine, A.: Web 2.0 Storytelling: Emergence of a New Genre. Educause Review, 40–56 (2008)
28. Loveless, A.M.: Report 4 Update: Creativity, Technology and Learning - A Review of Recent Literature. NESTA FutureLab (2007)

Author Index

FSC
www.fsc.org
MIX
Papier | Fördert
gute Waldnutzung
FSC® C083411

Kolibri 360 GmbH
Ferdinand-Jühlke-Straße 7
99095 Erfurt, Deutschland
produktsicherheit@kolibri360.de